Pre-equilibrium Emission in Nuclear Reactions

Fundamentals, measurements and analysis

Pre-equilibrium Emission in Nuclear Reactions

Fundamentals, measurements and analysis

B P Singh

Department of Physics, Aligarh Muslim University, Aligarh-202002, Uttar Pradesh, India

Manoj K Sharma

Department of Physics, University of Lucknow, Lucknow-226007, Uttar Pradesh, India

R Prasad

Department of Physics, Aligarh Muslim University, Aligarh-202002, Uttar Pradesh, India

IOP Publishing, Bristol, UK

ISBN 978-0-7503-5077-8 (ebook)
ISBN 978-0-7503-5075-4 (print)
ISBN 978-0-7503-5078-5 (myPrint)
ISBN 978-0-7503-5076-1 (mobi)

DOI 10.1088/978-0-7503-5077-8

Version: 20220601

IOP ebooks

British Library Cataloguing-in-Publication Data: A catalogue record for this book is available from the British Library.

Published by IOP Publishing, wholly owned by The Institute of Physics, London

IOP Publishing, Temple Circus, Temple Way, Bristol, BS1 6HG, UK

US Office: IOP Publishing, Inc., 190 North Independence Mall West, Suite 601, Philadelphia, PA 19106, USA

Dedicated to our Parents.

Contents

Preface

The book discusses in detail the phenomena of *pre-equilibrium emission* in statistical nuclear reactions. Nuclear reactions at moderate energies are mostly binary in nature and the reaction process is generally explained by Bohr's compound nucleus (CN) reaction mechanism. At energies where a large number of overlapping levels are populated in the compound nucleus, a statistical approach is used to describe the reaction process. Bohr's CN reaction mechanism is based on the assumption that the projectile ion fuses with the target nucleus forming a compound nucleus which remains until a thermodynamic equilibrium is established in the compound system. After the establishment of thermodynamic equilibrium, the CN may decay through energetically possible channels. In an average nucleus, establishment of thermodynamic equilibrium may take times of the order of 10^{-16} s. Bohr's theory does not provide any reason for the stability of the CN during the process of equilibration, for a time which is order of magnitudes larger than the typical nuclear interaction time scale of the order of 10^{-21} s. During the last three decades or so, considerable experimental evidence has been gathered that indicates that nuclear particle or small cluster emission do occur during the process of equilibration. The component of emitted particles during equilibration is termed as pre-equilibrium or pre-compound emission. Many groups of experimentalists all over the world, including our own group, carried out complementary and focussed experiments on pre-equilibrium emission to determine its strength and other characteristics. In the meantime, a few semi-classical and quantum mechanical theories for pre-equilibrium emission have been developed. In the present scenario of the availability of a considerable amount of experimental data and theories of the reaction process, it is required to critically evaluate both the experimental data and theories to draw important conclusions about the process. This book/monograph is an attempt to put together both, experimental measurements and theoretical model-based computer codes with a view to analyse the existing data.

The book is organised to serve as a source of complete information on pre-equilibrium emission in nuclear reactions on one hand and as an experimental manual for beginners in the field of accelerator-based research on the other. The book is divided into six chapters; chapter 1 gives a short review of nuclear reaction studies and spells out the motivation for the work. Important characteristics of binary nuclear reactions, fundamental conservation laws obeyed by them, Q-equation etc, are discussed in chapter 2. Chapter 3 provides formulations that are required to analyse raw experimental data to obtain various important parameters. Sufficient details of several important models and computer codes that are frequently used to evaluate experimental data are provided in chapter 4. Details of measurements employing different experimental set-ups used for obtaining pre-equilibrium component, reaction-wise analysis of experimental data etc, are presented in chapter 5. Analysis of experimental data on pre-equilibrium emission obtained from different types of reactions including heavy ion and radioactive ion beam induced reactions is presented in the chapter 6. An attempt has also been made to obtain systematic trends the in pre-equilibrium component with different parameters of the interacting systems. With these details, the book will help as a resource to the students and researchers working in the field of nuclear reaction dynamics at low energies and may open up new avenues to extend these studies at relatively higher energies.

B P Singh
Manoj K Sharma
R Prasad

Acknowledgments

To have an idea and get it converted into the form of a book is as hard as it sounds. We especially want to thank the individuals who worked and helped to make this happen. With immense pleasure, the authors wish to put on record their sincere thanks, to all the members of our research group, including (Late) Dr H D Bhardwaj, Dr (Mrs) Sunita Gupta, Professor M M Musthafa, Dr (Mrs) Unnati Gupta, Dr Devendra Pal Singh, Dr Pushpendra P Singh, Dr Abhishek Yadav, Dr Vijay Raj Sharma, Dr Mohammad Shuaib, Mr Ishfaq Majeed and Mr Mohd. Shariq Asnain.

We also thank the Director and the staff of Variable Energy Cyclotron Centre (VECC), Kolkata, India, for their support during the experimental work carried out using proton and alpha particle beams. Our thanks are also due to the Director and staff of the Inter University Accelerator Centre (IUAC), New Delhi, India for providing necessary facilities for carrying out experiments with heavy-ion beams and for sponsoring the thesis projects of some students. We thank Professors G K Mehta, Amit Roy, and Dinakar Kanjilal, former directors of the IUAC, New Delhi, India for their support during the experiments using heavy-ion beams. The authors wish to put on record their special appreciation for the support extended by Dr Ranjan K Bhowmik, Senior Scientist, Inter University Accelerator Centre, New Delhi, and Dr Ajit K Sinha, Senior Scientist and Ex-Centre Director, UGC-DAE-CSR, Kolkata Centre and Ex-Director, UGC-DAE-Consortium for Scientific Research, Indore, India. Thanks, are also due to the scientists at IUAC, New Delhi, especially Dr S Muralithar, Dr Rajesh P Singh, Dr Rakesh Kumar, Dr (Ms) K S Golda, and Dr (Mrs) Indu Yadav for all their support during the experiments at the Pelletron accelerator facility.

Our special thanks are due to Aligarh Muslim University, Aligarh, Uttar Pradesh, India for supporting our research projects through the Department of Physics of the University. Manoj K Sharma extends his thanks to Professor N K Pandey, Head, Department of Physics, University of Lucknow, Lucknow for all his help and moral support.

The authors are also thankful to various Indian Government Agencies, like the UGC-DAE Consortium for Scientific Research, Kolkata Centre, Kolkata; University Grants Commission (UGC), New Delhi; and Department of Science and Technology for providing financial support for carrying out the research projects for these studies. Thanks are due to Mr Mohd Shuaib and Mr M Shariq Asnain for all the help in typesetting and formatting of the manuscript.

On behalf of all the research group members, two authors BPS and MKS would like to extend their appreciation and sincere thanks to Professor Rajeshwari Prasad, for his commitment and dedication for initiating the projects in the fascinating field of experimental studies of nuclear reaction dynamics at the Physics Department, Aligarh Muslim University, Aligarh, India. His leadership qualities have always been a source of inspiration to many. In fact, we want to say a big thank you for being the inspiration for the work presented here.

Last but not the least, the authors wish to put on record sincere thanks to their wives in particular, and to all the other family members for their unending encouragement and moral support during the course of the work. Further, thanks are due to all the members of the publishing team.

Author biographies

Professor B P Singh

B P Singh is a Professor and Chairperson of the Physics Department, Aligarh Muslim University, India. He has more than three decades of experience in teaching nuclear physics courses and has more than hundred research publications in international journals of repute to his credit. Apart from this, he has taught courses on thermal and statistical physics, electricity and magnetism, electronics, heavy-ion physics etc, at graduate and postgraduate level. Professor Singh's area of research is experimental nuclear physics with special interest and contributions in the particle accelerator-based studies on pre-equilibrium emission in nuclear reactions and incomplete fusion in heavy-ion interactions at low energies. Professor Singh has been the Chairman of the Accelerator User's Committee of the Pelletron Accelerator facility, Inter University Accelerator Centre, New Delhi, India, and a member of the National level Cyclotron User's Committee of the Variable Energy Cyclotron Centre (VECC), Kolkata, India. He has worked at various prestigious national/international institutes and visited IAEA (Austria), RIKEN (Japan), JINR Dubna (Russia), ICTP (Italy), Venice International University (Italy), NTU (Singapore), Addis Ababa University (Ethiopia), Ruder Boskovic Institute (Croatia), Yangon University (Myanmar), Paris and Saint Malo (France) in connection with academic programs. Professor Singh has successfully completed several major research projects including international collaboration. Professor Singh is also one of the authors of the book *Fundamentals and Applications of Heavy Ion Collisions: Below 10 MeV/ Nucleon Energies*, published by Cambridge University Press.

Professor Manoj K Sharma

Manoj K Sharma is a Professor at the Department of Physics, University of Lucknow, Lucknow, India. Previously he served at the Department of Physics, Shri Varshney College, Aligarh, India. He obtained his PhD degree in experimental nuclear physics from the Aligarh Muslim University, India. After completing his doctorate, he worked as a Young Scientist under the DST fast-track scheme at the Physics Department, AMU, Aligarh. His thrust area of research is investigations of pre-compound emission and incomplete fusion processes in low energy light and heavy-ion induced reactions. Dr Sharma has successfully completed several major research projects. He has published more than eighty research articles in international journals of repute and has presented more than a hundred papers in international/national conferences within India and abroad. Dr Sharma was selected as a participant to the Nobel Laureate Meeting, Lindau, Germany.

Professor R Prasad

Professor R Prasad is an Emeritus Professor of Physics, Ex-Dean, Faculty of Science and Ex-Chairman, Department of Physics, Aligarh Muslim University (AMU), India. He has more than 40 years experience of teaching nuclear physics, thermal physics and electronics to postgraduate and graduate students. He has published more than 100 research papers in international journals of high impact factor. Professor Prasad has authored several books including *Classical and Quantum Thermal Physics*, and *Fundamentals and Applications of Heavy Ion Collisions: Below 10 MeV/Nucleon Energies* published by Cambridge University Press and *Nuclear Physics* published by Pearson Education. The latest publication of Professor Prasad is the book entitled *Analog and Digital Electronic Circuits: Fundamentals, Applications and Analysis* published by Springer Nature.

Symbols

e^-	Electron
n	Neutron
β^+	Positive beta particle (Positron)
β^-	Negative beta particle (Negatron)
ν	Neutrino
$\bar{\nu}$	Antineutrino
Q	Reaction Q-value
m_0	Rest mass of electron
ω	Cyclic frequency
τ	Mean life time
Λ	Wavelength (m)
h	Plank constant
γ	Gamma rays
α	Alpha particle
σ	Cross section
A	Mass number of nucleus
D	Deuteron
P	Proton
J	Angular momentum
π	Parity
θ	Scattering angle of ejectile in laboratory frame
Z	Atomic number
ϕ	Scattering angle of residual nucleus in laboratory frame
Ψ	Wave function
K	Wave number
T	Transmission coefficient
Λ	Wavelength (m)
U^*	Excitation energy
$\vec{j}$	Total angular momentum
I_0	Moment of inertia of nucleus (as a rigid rotator)
ρ	Nuclear level density
E^*	Excitation energy
B_p	Binding energy of particle
ε_h	Binding energy of hole
μ	Reduced mass
Δ	Back shifted energy term or Pairing energy
$\Theta_{effective}$	Effective momentum of inertia
W	Statistical probability of population of a state
Θ_{rigid}	Defined as Rigid body moment of inertia as a free parameter in code ACT
E_F	Fermi energy
K	nuclear level density parameter
n_0	Initial exciton number
f_R	Pre-equilibrium fraction
V_b	The Coulomb barrier
E_{CM}	Energy in centre of mass (c.m.) frame
B/A	Binding energy per nucleon
ΔV	Potential gradient

ω_{rf}	RF frequency
W_λ	Eigen values
Γ_λ	level width
X_λ	Compound nucleus state wave function
$S_{\beta\alpha}$	S-matrix element
Ψ_α^+	Total wave function
$\langle\sigma_{\beta\alpha}\rangle$	Average cross section for inverse reaction
$<D>$	Average level separation
R_β	Branching ratio
CN	Compound nucleus

Symbols

E_x	Kinetic energy of nucleus x
MV	Linear momentum
T_3	Third component of isospin
q	Electric quadrupole moment
E_{rel}	Collision energy
Φ_B	Flux of incident beam
N_T	Number of target nuclei
G_ϵ	Efficiency of the detector
$\vec{l}$	Orbital angular momentum
$\vec{s}$	Spin angular momentum
S	Scattering matrix element
r_0	Nuclear radius parameter
β_s	Surface component of the single particle level density
T	Nuclear temperature
I	Moment of inertia of the heated nucleus.
U_C^*/A	Excitation energy per nucleon of compound nucleus
T_α	Transmission coefficient
U_0	Energy backshift
g	Density of single particle states
W	Energy shift
p_F	Fermi momentum
$\overline{\mid M^2 \mid}$	Square of average matrix element
$\overline{\rho_0(\ell)}$	Average nuclear density
OMP	Optical model potential

IOP Publishing

Pre-equilibrium Emission in Nuclear Reactions
Fundamentals, measurements and analysis
B P Singh, Manoj K Sharma and R Prasad

Chapter 1

Introduction to the history of nuclear reactions and motivation for this book

This chapter gives a short review of the history of nuclear physics in general and of nuclear reactions in particular, along with the motivation for the present work.

1.1 Background

A nuclear reaction may be defined as an interaction between two bodies, at least one of which has a nuclear field associated with it and that the field plays a role in the interaction. Thus, photo disintegration, Coulomb excitation, and other similar interactions may form a part of nuclear reactions. More than 90% of reactions in general and nuclear reactions in particular in the Universe are of binary nature; two entities in the initial state and two in the final, because the probability of finding more than two particles at the same point in space and at the same time is very small. Although nuclear reactions played a crucial role in synthesis of matter right from the very beginning of the Universe, the history of nuclear reaction studies is only about a century old. The scientific quest, though not intended, began after the discovery of x-rays by Wilhelm Conard Roentgen in 1895. Henri Becquerel on 26–27 February, 1986 demonstrated that radiations emitted by potassium uranyl sulphate are not x-rays, but very different from them. He used an apparatus similar to the one shown in figure 1.1 and showed that radiation emitted by the salt of potassium are of three distinct types: positively charged; negatively charged; and electrically neutral. Becquerel used the observed deflection of radiations in a magnetic field to decipher their electrical nature and carried out these experiments in an environment of low air pressure.

The name 'radioactivity' was actually coined by Marie Curie in 1898, who together with her husband Pierre Curie, was investigating the phenomenon recently discovered by Becquerel. It is interesting to note that the maiden name of Marie Curie, who was originally from Poland, was Maria Salomea Sklodowski. Maria

doi:10.1088/978-0-7503-5077-8ch1

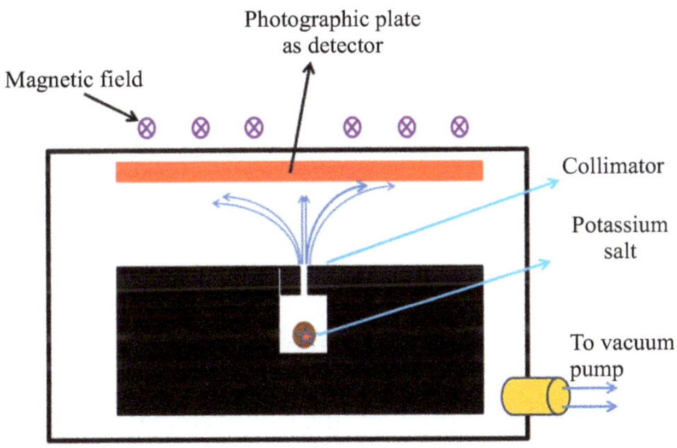

Figure 1.1. A typical layout of the apparatus used by Becquerel.

graduated from High School at the age of 15, standing first. Maria and her elder sister, Bronia, both wanted to go for University education but in those days the University of Warsaw did not accept women. At the age of 17 Maria took the job as a governess to help her older sister Bronia to study medicine in France. Ultimately, Maria registered at the Sorbonne, in Paris and signed her name as Marie to seem more French. She got her first Master's degree in physics in 1893 and a second Master's degree in Mathematics in 1894. She married Pierre Curie, an experimentalist who devised several sensitive instruments for recording magnetic and electric fields in 1895. The couple discovered the two elements polonium and radium. Together they were awarded the Nobel Prize for Physics in 1903 and for Chemistry in 1911. During her research work she exposed herself to extreme doses of radiations from radioactive materials and she expired in 1934 probably from blood cancer.

1.2 Nuclear atom

In a parallel development, Ernest Rutherford, a New Zealand born British physicist who migrated to Britain in 1895 on a Royal Society for the Exhibition of 1851 Science Scholarship to join Trinity College, Cambridge, as a research student at the Cavendish Laboratory under J J Thomson, got interested in the work on radioactivity. Rutherford carried out experiments on radiation emitted by several different radioactive materials that had been isolated by that time, using apparatus similar to the one used by Becquerel, with the change that the photographic plate was replaced by a fluorescent screen. He named the positively charged radiation as alpha particles, negatively charged radiation as beta particles and electrically neutral radiation as gamma rays. He also showed that the charged components of radiation are easily absorbed or scattered by a few cm pressure of air, as the bright spots produced by their hitting the fluorescent screen disappeared when the air pressure was increased in the instrument. Rutherford further established that both alpha and beta radiation are particles in nature and that gamma radiation is of wave nature. From his experimental studies Rutherford determined the charge of alpha particles as of $+2e(=2 \times 1.6 \times 10^{-19}$ Coulomb) units, while the Curies had already indicated that alpha particles are doubly ionised helium atoms. However, to be doubly sure, Rutherford with Thomas Royds collected alpha particles in a sealed class tube and passed a spark to study the emission spectra. Analysis of the observed spectra proved that alpha particles were ionised helium atoms. Within few months after this, Rutherford was awarded the Nobel Prize in Chemistry in 1908. Some of his work on radioactivity was carried out at McGill University, Montreal, Canada which he joined as Macdonald Chair in 1898. With R B Owens he studied the 'emanation' of thorium and discovered a new noble gas, an isotope of radon, later to be called 'thoron'. In 1900 Frederick Soddy, arrived at McMaster from Oxford and he collaborated with Rutherford to outline the disintegration theory of radioactivity, indicating that radioactivity is an atomic phenomena and not a molecular process, which was the belief till then. Otto Hahn, the discoverer of the process of fission also worked under Rutherford at Montreal Laboratory during 1905–1906. Rutherford returned to England in 1907 to join as Langworthy Professor of Physics at the University of Manchester. Here, in conjunction with Hans Geiger, who was already known for inventing the charge particle detector with Walther Muller called Geiger–Muller counter, or in short the G M Counter, Rutherford developed the technique to count a single alpha particle which he used to study the elastic scattering of alpha particles (emitted from radioactive sources) by other materials.

In his unique technique, Rutherford counted the scintillations produced by charged (alpha) particles when they hit a fluorescent screen. Counting these tiny flashes of light in a dark room required training and acclimatization and was quite eye tiring. However, Rutherford continued with it and in 1910–1911 carried out the famous large angle alpha scattering experiments [1], the basic layout of which is shown in figure 1.2. Large angle alpha scattering experiments were carried out by Geiger and Marsden in which they bombarded thin foils of gold and platinum by a collimated beam of alpha particles obtained from a moderately strong radioactive source of radium. As shown in the figure, the complete experiment was carried out in an evacuated chamber under low pressure so as to avoid the scattering of the alpha particle beam by air molecules. The design of the experimental chamber was such that its top cover which was fitted with a microscope with a ZnS screen could be

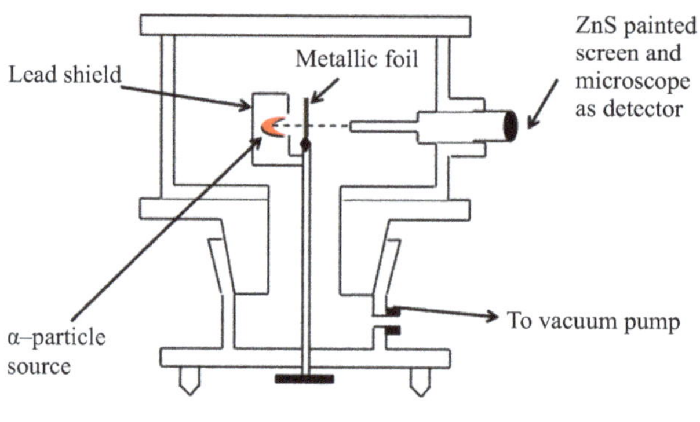

Figure 1.2. Sketch of the apparatus used by Rutherford and co-workers to study scattering of alpha particles with thin metal foils.

rotated around the target foil in the horizontal plane without disturbing the vacuum of the chamber. It was also possible to change the inclination of the microscope with respect to the centre of the target foil. With this arrangement it became possible to record the number of alpha particles scattered at different angles with respect to the direction of the incident alpha beam. The experiment was perhaps designed to test the so called 'plum pudding' or 'watermelon' model of the atom proposed by J J Thomson, a British scientist in 1904, seven years after the discovery of electrons. According to the watermelon model, it was assumed that atoms are spherical and filled with a positively charged jelly-like matter of charge density ρ^+ just like the red jelly of a watermelon, and negatively charged electrons were assumed to be embedded in this positive jelly like the seeds in a watermelon. It was further assumed, to maintain electric neutrality of atom, that the total positive charge of the atom $Q^+=\int \rho^+ dv$ is counterbalanced by the total negative charge $Q^-=N \quad (-e^-)$, where N is the number of electrons each with a negative charge e (= 1.6×10^{-19} C).

Figure 1.3 shows a sketch of the watermelon model of the atom and the trajectories of positively charged alpha particles on scattering by the positive jelly of the atom. The first point to remember in this analysis is that electrons are not likely to affect the trajectory of the scattered particle because of their negligible mass and negligibly small interaction time, the swift alpha particle passes from the vicinity of the electron before any effect of the electron's presence is felt by the alpha particle. The scattering of alpha particles, therefore, will be essentially due to the Coulomb repulsion from the positive jelly. Those alpha particles that pass through the centre of the atom, bisecting the positive jelly in two equal semi-circular halves will pass undeviated along their original direction, as shown in the figure. Those alphas which divide the jelly into two unequal parts will suffer scattering at some angle θ and if

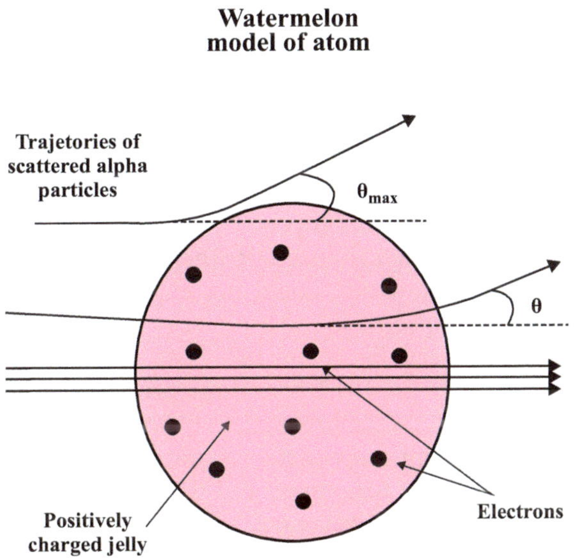

Figure 1.3. Watermelon model of atom and trajectories of scattered alpha particles.

this argument is further extended then maximum angle of scattering θ_m will occur for those alpha particles that graze the periphery of the atom. There might be a counter argument that alpha particles that pass grazing the atom will not suffer any deviation as for such particles total positive and total negative charges distributed over the volume of the atom will appear as if they are concentrated at a point at the centre. However, it really does not matter whether particles with grazing incidence will suffer maximum or no deflection, the important prediction of the watermelon model was that no incident alpha particle should be scattered at angles larger than 90°. In experiments carried out by Geiger and Marsden, most of the incident particles suffered scattering at angles smaller than 90°, however, a few, around one to five alpha particles for almost every 20 000 incident particles, were found to be scattered at angles larger than 90°, or were back scattered. In normal course such small abrasions of five particles in over 20 000 particles, may be attributed to experimental uncertainties and are neglected. However, a scientist of Rutherford's genius who was a hundred percent confident of his experiment, concentrated on these few alpha particles and looked for a plausible reason for them being scattered at backward angles. Later while discussing the results of the experiment Rutherford made the following statement about the back scattered alpha particles:

> '...quite the most incredible event that has ever happened to me in my life. It was almost as incredible as if you fired a fifteen-inch shell at a piece of tissue paper and it came back and hit you. On consideration I realized that this scattering backward must be the result of a single collision, and when I made calculations, I saw that it was impossible to get anything of that order of magnitude unless you took a system in which the greater part of the mass of the atom was concentrated in a minute nucleus. It was then that I had the idea of an atom with a minute massive centre, carrying a charge...'

Thus, a new branch of science, nuclear physics, came into being. Rutherford's large angle alpha scattering experiments conclusively proved that most of the atom is empty, which is why most of the incident alpha particles passed straight through; that total positive charge and almost the whole of the mass of the atom resides in a small volume at the centre called the nucleus. In order to explain the presence of electrons in the atom, Rutherford proposed that electrons move in fixed orbits around the nucleus at different distances. Rutherford's model of the atom, also called the planetary model, however, could not explain the stability of the atom. Accelerated electrons were expected to radiate energy in the form of electromagnetic waves, lose energy and were likely to fall back in the nucleus. Rutherford collaborated with Niels Bohr in 1912 and Bohr proposed the idea of stationary orbits for electrons and put forward his model of the atom.

Rutherford's alpha scattering experiment was important in more than one way; it not only established that almost total mass and total positive charge of the atom is concentrated in a small volume at its centre, which Rutherford named the nucleus, but it also suggested a new way of exploring microscopic systems: bombarding them with energetic particles and looking for the changes that have occurred in the direction of motion or in some other properties of the incident probe. Rutherford and his laboratory steward, William Kay, in 1919 started experiments in which alpha particles were made to strike molecules and atoms of different gases like hydrogen, nitrogen etc. These experiments were carried out in a brass tube which had provisions for a moveable holder to hold the radioactive source of alpha particles (Ra, C or ^{214}Bi) at any desired distance, an inlet for introducing desired gas, and a connection to a vacuum pump. At one end of the brass tube there was a slit which could be covered with fine silver foil. A ZnS scintillation screen could be put in front of the slit to record tiny light flashes. Before introducing any gas, the brass tube was evacuated and scintillations produced by alpha particles coming from a radioactive source were recorded. Next, hydrogen gas was introduced in the brass tube up to a pressure such that alpha particles emitted from the source are absorbed in the gas. However, it was unexpectedly observed that scintillations were still produced on the screen. Rutherford explained it by saying that as alpha particles traverse hydrogen gas they occasionally collide with the nucleus of hydrogen atom and produce 'swift hydrogen atoms' which travel forward in the direction of original alpha particles. This by far was the first report of a nuclear reaction produced in a laboratory by a human being.

1.3 The first transmutation reaction

Having established the presence of the nucleus at the centre of each atom, Rutherford concentrated on the constitution of the nucleus. He asked his colleague Darwin to theoretically investigate the details of the collision between a point nucleus of helium (the alpha particle) with two units of positive charge and the nucleus of hydrogen with one unit of positive charge under Coulomb's inverse square law. Darwin found that all alpha particles that may reach within 2.4×10^{-13} cm of the hydrogen nucleus will produce a 'swift hydrogen atom'. However, the estimated number of 'swift hydrogen atoms' expected to be produced in the experiment carried out by William Kay was much smaller than the actual number observed in the experiment. Several suggestions, including that inverse square law does not hold in such collisions at such short distances and that the charge on different helium nuclei is different, were put forward to explain the discrepancy between the observed and the predicted number of 'swift hydrogen atoms'. Rutherford, however, rejected all of these arguments and assigned the observed discrepancy to the deformation of the complex nucleus of helium which, according to Rutherford, has four nuclei of hydrogen and two electrons, and to the fact that at such short distances it is not valid to treat the alpha particle (nucleus of helium) as a point particle. Kay and Rutherford jointly broke nuclei of several light atoms including that of nitrogen to produce, 'swift hydrogen atoms' which were later identified as protons. In this regard Rutherford wrote [2]:

> 'From the results so far obtained, it is difficult to avoid conclusion that the long range atoms arising from the collision of α particles with nitrogen are not nitrogen atoms but probably atoms of hydrogen…If this be the case, we must

conclude that nitrogen atom is disintegrated under the intense forces developed in a close collision with a swift α particle, and the hydrogen atom which is liberated formed a constituent part of nitrogen nucleus.'

In his lecture at Royal Society of London in 1920 Rutherford speculated about the possibility of having two nuclei that: (i) may have zero charge with a mass as much as the mass of the hydrogen nucleus (P+e−), which he called the combination of a proton and an electron and named the neutron; and (ii) a nucleus with charge +1e and mass almost twice the mass of hydrogen atom (2 P+e−). These speculative nuclei were required to explain the results of experiments carried out by Rutherford's group on the interaction of alpha particles with light elements.

During the intervening period from 1920 to 1932, three major scientific events took place. In early 1920, the German group of Max Born, Werner Heisenberg and Wolfgang Pauli, working at the University of Gottingen coined the word (Quantomechanic) or Quantum mechanics to describe the new theoretical framework being developed for understanding the behaviour of microscopic systems. The special characteristic of the new framework was the dual nature of matter and discrete energy states for microscopic systems. A coherent formulation of quantum mechanics was eventually developed by the years 1925–1926, principally by the work of Schrödinger, Heisenberg and Dirac. On the basis of the newly developed theory, Dirac predicted an electrically neutral particle of nearly the same mass as that of proton. If discovered, this particle was going to replace the (P+e−) particle speculated by Rutherford and named the neutron.

1.4 Discovery of neutron

In 1928, a German physicist, Walter Bothe and his student Herbert Becker bombarded a thin sheet of metal beryllium with alpha particles emitted from a radioactive source of polonium. They found that electrically neutral radiation of high penetrating power was emitted from the beryllium sheet when it was hit by alpha particles. Since it was already known that electrically neutral gamma radiation is emitted in natural radioactive decay, Bothe and Becker thought that the chargeless radiation emitted from beryllium foil was also gamma rays. In 1932, one of Marie Curie's daughters, Irene Joliot Curie and her husband Frederic Joliot Curie repeated the experiment carried out by Bothe and Becker using their stronger and more intense polonium source. To further investigate the properties of this radiation of high penetrating power, Irene and Frederic allowed the radiations emanating from beryllium to fall on a sheet of paraffin wax, which contained a large percentage of hydrogen atoms. The result of the experiment surprised the Joliot Curies as they found that penetrating uncharged radiation knocked out protons from the paraffin wax sheet. Irene and Frederic tried to explain the process by saying that penetrating radiation is high energy gamma ray photons that ejected protons from the atom of hydrogen, as electrons are ejected by photons in the photo-electric effect. However, it was not possible because of the considerably larger mass of the proton as compared to the mass of the electron.

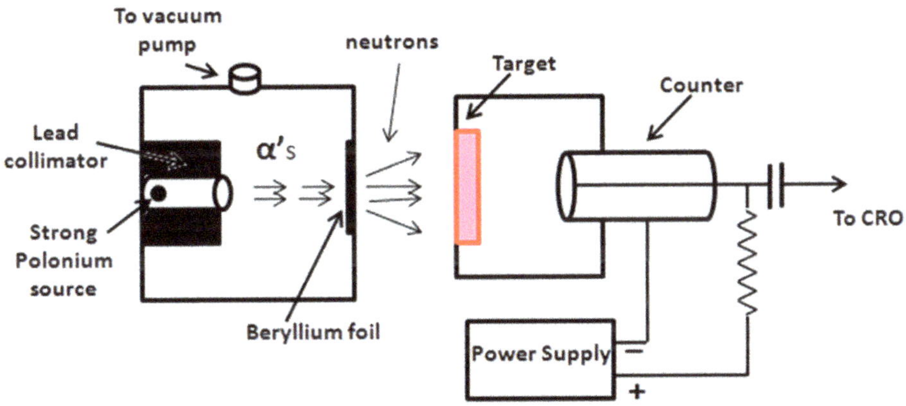

Figure 1.4. Sketch of the experimental setup used by Chadwick.

James Chadwick, a student of Rutherford, reported the result of Irene and Frederic Curie's experiment to Rutherford who immediately said 'I do not believe it'. Chadwick immediately repeated the experiment of Irene and Frederic Curie using the apparatus, a sketch of which is shown in figure 1.4. He not only let the beryllium radiation hit paraffin wax but also other materials like helium, nitrogen and other elements. Chadwick recorded the energy of emitted charged particles in each case and using kinematical formulations calculated back the mass of radiations emitted from beryllium. In each case he found their mass almost equal to the mass of proton. Chadwick published his findings in a paper in *Nature* on 17 February, 1932 announcing [3] the discovery of a neutral particle of mass nearly equal to the mass of the proton, which he called the neutron.

The nuclear reaction responsible for emission of neutron from beryllium foil may be written as,

$$^{4}_{2}\mathrm{He} + {}^{9}_{4}\mathrm{Be} \rightarrow {}^{12}_{6}\mathrm{C} + {}^{1}_{0}n$$

Chadwick was awarded Nobel Prize of Physics for his work in 1935. It is sad to note that Irene and Frederic Joliot Curie missed a Nobel Prize for wrong interpretation of the result. However, it is heartening that they received the Nobel Prize of Chemistry in 1935 for their discovery of artificial radioactivity, again demonstrating the potential of nuclear reaction studies.

1.5 Artificial radioactivity

In a separate experiment Irene and Frederic Joliot Curie in 1934 bombarded aluminum foil with 5.3 MeV alpha particles emitted in natural radioactive decay of polonium. They observed that on irradiation of the Al-foil positively charged particles were emitted from the foil and that the emission of positively charged particles continued for some time even after the polonium source of alpha particles was removed. On removing the alpha particle source, it was found that the intensity of positively charged particles decreased exponentially, just like the decay of the intensity in the case of natural radioactivity. From the experimental data they determined the half-life of the observed activity as 2.34 min. They attributed this activity to the formation of a new unknown isotope ^{30}P according to the following nuclear reaction,

$$^{4}_{2}\text{He} + ^{27}_{13}\text{Al} \rightarrow ^{30}_{15}\text{P} + ^{1}_{0}n$$

Later, they confirmed it by chemically separating phosphorous from the irradiated aluminium foil within a few minutes of irradiation and observing that the activity comes from the separated phosphorous. The isotope ^{30}P was found to undergo the following radioactive decay;

$$^{30}_{15}\text{P} \rightarrow ^{30}_{14}\text{S} + ^{0}_{1}\text{B} + v$$

This was the first production of a radioactive isotope by artificial means and also the first observance of positron decay in radioactivity. Development of the technique for fast chemical separation of phosphorous from the irradiated aluminium foil, and experimental confirmation that the observed β^+ activity belongs to phosphorous gained the Nobel Prize for chemistry in 1935 for Mrs and Mr Joliot Curie.

1.6 Strong nuclear force

Discovery of the neutron immediately led to the neutron–proton model of the nucleus by Heisenberg and Iwanenko in 1932. The stability of this nucleus was ascribed to a new force, the strong nuclear force, which was postulated in addition to the classical gravitational and electromagnetic forces already known at that time. Gravitational force does not play any role in the stability of the nucleus; however, electromagnetic force of repulsion between protons counters the strong attractive nuclear force and could thus split a heavy nucleus into two parts. Already in 1930, taking a clue from the almost constant binding energy of nucleons (a term used for both neutrons and protons) a nuclear liquid drop model of the nucleus was developed and described in literature by Gamow (1930) and later also by Weizsacker, Bethe, and Bohr. The liquid drop model played a pivotal role in explaining the process of fission, the most fascinating, useful and dangerous nuclear reaction ever discovered

1.7 New isotopes

Discovery of neutrons gave a big boost to the study of nuclear reactions. Neutrons with no electrical charge can readily enter a nucleus as they do not have to face any Coulomb barrier. As a result, neutrons of energies as low as thermal energies $\approx$0.025 eV may initiate nuclear reactions. Taking advantage of this fact and taking a cue from Frederic and Irene Curie's experiment that new elements produced in a nuclear reaction may be identified through their radioactive properties or by the method of chemical separation, Fermi and his co-workers, F Rasetti and O D'Agostino [4], carried out a large number of experiments in which they bombarded elements of successively higher atomic numbers with neutrons and produced hitherto unknown isotopes. Nuclear reaction initiated by the absorption of an incident neutron by a target nucleus ${}^{A}_{Z}X$ may be written as follows:

$${}^{A}_{Z}X + {}^{1}_{0}n \rightarrow {}^{\mathrm{A}+1}_{Z}X + {}^{0}_{0}\gamma$$

The new isotope ${}^{A+1}_{Z}X$ is, generally, unstable and undergoes β^- decay leading to the formation of the isotope of the next higher element 'Y' of atomic number $(Z+1)$;

$${}^{A+1}_{Z}X \rightarrow {}^{A+1}_{Z+1}Y + \beta^- + \nu$$

Using this method, the Fermi group in Italy produced more than 40 new isotopes in a few months.

1.8 Discovery of fission

Using heavier and heavier targets, Fermi's group bombarded the heaviest nucleus of uranium with neutrons in May 1934. Fermi paid special attention to a 13 min activity produced in the neutron irradiation of uranium. Using radiochemical methods Fermi ruled out that this activity could be due to elements with $Z = 83$, and also from elements with Z between 86 and 93. Following his findings with neutron reactions on light elements, Fermi indicated the possibility of the existence of an element with $Z = 93$ or 94. Thus the transuranium elements were thought to have been found and the transuranium puzzle started. Fermi received the Nobel Prize in 1938 for the 'disclosure of artificial radioactive elements produced by neutron irradiation'. Fermi delivered his Nobel Prize lecture on 10th December, 1938 in Stockholm; that was just a few days before Otto Hahn and Fritz Strassmann, two German scientists working at Kaiser-Wilhelm Institute for Chemistry, Berlin, who were also studying the same reactions were sure that the so called transuranium elements described by Fermi were not transuranium but only barium, lanthanum and cerium. The special 13 min activity that Fermi ascribed to a transuranium element, was due to the radioactive isotope ${}^{101}_{43}$Tc, discovered by C Perrier and E Segre in 1937. Finally, in 1938 Hahn and Strassmann discovered the phenomena of nuclear fission and published their findings in 1939. The discovery of fission not only proved the mass energy equivalence, proposed by Albert Einstein, for the first time but also opened a new method for producing a huge amount of energy both for peaceful use and for nuclear armaments.

Any reference to the fission reaction will remain incomplete without a mention of the names of two physicists; Lise Meitner and her nephew Otto Frisch. Meitner was actually working with Hahn and Strassmann at the Kaiser-Wilhelm Institute, Berlin, on the problem of the transuranium puzzle, when she had to leave Germany, being of Jewish decent, under stressful conditions and much against her wishes, in July, 1938 to reach Sweden. However, Hahn kept her updated on their findings through letters regarding the transuranium puzzle. Her nephew, Otto Frisch a physicist who was working at Bohr's Institute of theoretical physics at Copenhagen, Denmark, visited Meitner in Sweden in December, 1938 and found her excited and restless with a communication from Hahn saying that neutron irradiation of uranium produced lighter nuclei like barium, strontium and technetium etc, and no transuranium heavy elements, as reported by Fermi (figure 1.5). They verified their findings using the radiochemical separation method. Meitner was wondering how a tiny kick by a neutron of negligible energy may rupture the target nucleus uranium in nearly two halves. In his recollections of that December visit to Sweden, Frisch wrote:

> '...And still I don't know how we got to the concept of fission, but I remember Lise Meitner drawing a dotted circle on a piece of paper and saying, "Couldn't it be this sort of thing?" Now she always rather suffered from the inability to visualise things in three dimensions, whereas I had that ability quite well. And I had, in fact, come around to the same idea, and I drew a shape like a circle squashed in at two opposite points. And Lise Meitner then said, "Yes, well, that is what I mean".'

In their further discussion on the topic, both Meitner and Frisch came to the idea that perhaps one must not think of the nucleus being cleaved in half as with a chisel but rather that there was something in Bohr's idea [5] that the nucleus was like a liquid drop, which on a hit with a neutron absorbs it and undergoes shape vibrations which may eventually lead to breaking of the nucleus into two halves, as shown in figure 1.6.

On returning to Copenhagen, Frisch met Bohr who was leaving for America with his colleague Leon Rosenfeld and narrated his discussion with Meitner, on which Bohr exclaimed with excitement 'oh! what idiots we have been that we have not seen that before. Of course this is exactly as it must be, this is very beautiful'.

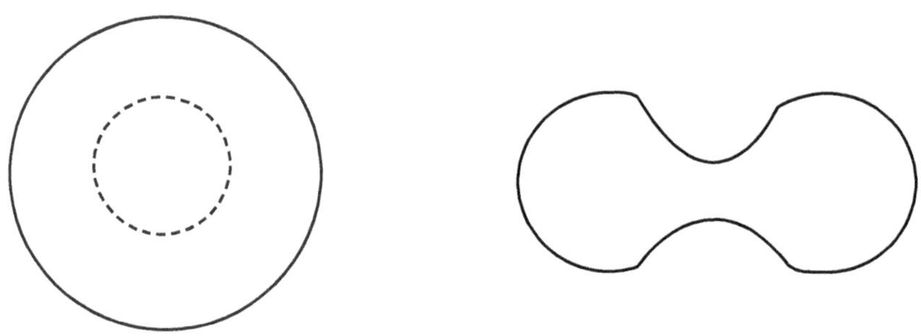

Figure 1.5. Figures drawn by Lise Meitner and Otto Frisch.

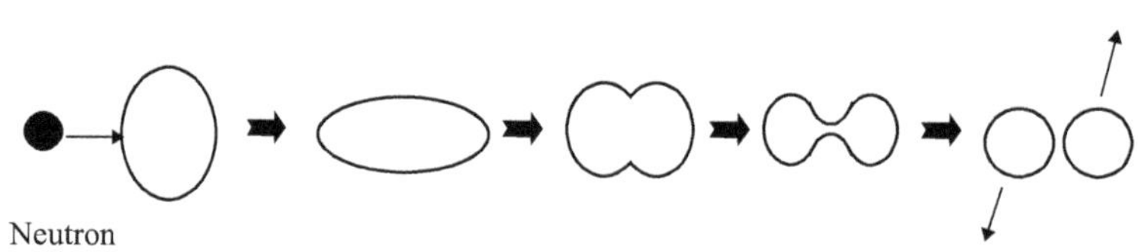

Figure 1.6. Representation of fission on the basis of liquid drop model.

Later, on his sea journey to America which took him six days, Bohr and Rosenfeld developed a more complete theory of nuclear fission based on the liquid drop model.

The evidence that Hahn and Strassmann developed on the nature of fission residues was all based on chemistry. Till then there was no physics-based evidence about fission. Frisch, on returning to Copenhagen used an ionization chamber to record the electrical pulses generated in the chamber when some charged particles passed through it. He was looking for pulses that were considerably larger in size produced by the heavy fission fragments. He successfully recorded such pulses, and in two research papers communicated to the research journal *Nature* for the first time used the name fission for this process. Frisch mentioned that the word fission was suggested to him by an American biologist William A Arnold. In biology, fission is used to represent the process of cell division.

A fission reaction may be represented by the following equation;

$$^{A}_{Z}X \rightarrow\ ^{A_1}_{Z_1}\mathrm{F}_1 +\ ^{A_2}_{Z_2}\mathrm{F}_2 + \text{light particles} + \text{Energy}$$

Here F_1 and F_2 are the two fission fragments, each with their atomic mass numbers A_1 and A_2 of the order of $A/2$ and light particles may include neutrons and/or protons.

The fission reaction immediately grabbed attention of scientists all over the world. It was because of the potential of this reaction as a source of a huge amount of energy. To understand how such a large amount of energy may become available from fission, we have to understand the concept of binding energy (B) of a nucleus. Neutrons and protons in a nucleus are held together by strong nuclear force. Attractive nuclear force does not distinguish between a neutron and a proton and if it is assumed that each nucleon on average contributes, say x units of binding energy, then the total binding energy B of the nucleus with A nucleons will be $B = xA$ units.

However, positively charged protons inside the nucleus repel each other due to Coulomb force and, therefore, reduce the binding energy of the nucleus. The reduction in the binding energy is proportional to the number of proton pairs in the nucleus, that is to $Z\ (Z-1)$, where Z is the number of protons in the nucleus. Therefore, the binding energy B_Y of a nucleus $^{A}_{Z}Y$ may be given as;

$B_Y = xA - k[Z(Z - 1)]$, here k is a constant.

If one divides BY by the total number of nucleons A in the nucleus, one gets a quantity B/A, called the average binding energy per nucleon. B/A for a large number of stable nuclei as a function of the atomic mass number A is plotted in figure 1.7. It may be observed in this figure that that B/A falls both towards small A values and also towards large A values. The reduction in average binding energy per nucleon towards the small values of A is due to small number of nucleons and, therefore, the attractive component of binding energy from strong nuclear force is relatively small. Towards the larger values of A, the B/A falls because the larger number of protons in the nucleus try to disrupt the nucleus and reduce binding energy. The most stable nuclei, with maximum binding energy lie in the range of A values from 60 to 100. In fission, say of ^{235}U, the nucleus breaks into nearly two equal fragments,

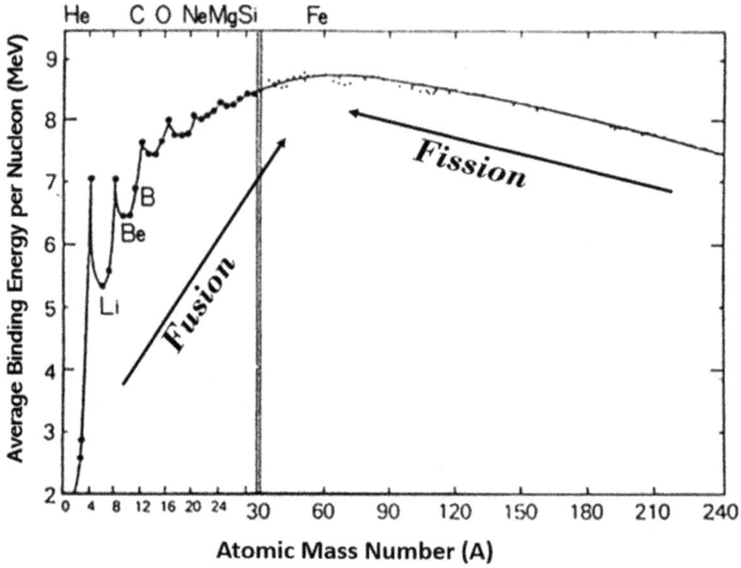

Figure 1.7. Variation of average binding energy per nucleon with atomic mass number A.

each with A value around 118. From the plot in figure 1.7 the B/A value at $A \cong 235$ is around 7.5 MeV, therefore, the total binding energy B_U of ^{235}U will be

$$B_U = 7.5 \times 235 = 1762.5 \text{ MeV}$$

While the total binding energy B_F of the two equal fission fragments of atomic mass numbers $A = 118$ will be, $B_F = 2 \times (118 \times 8.4) = 1982.4$ MeV, here 8.4 MeV is the average value of B/A at $A \cong 118$.

The difference $B_F - B_U$=219.9 MeV of energy will be liberated in the fission of a single nucleus of ^{235}U. Since a large number of nuclei of ^{235}U may be made to undergo fission simultaneously, a huge amount of energy may be released. This energy is released in the form of the kinetic energy of fission fragments and other light particles that are generally emitted with fission. Fission fragments and light particles emitted in fission and moving with large kinetic energy when colliding with the walls of the container vessel produce a large amount of energy in the form of heat which may be trapped and transferred to other locations for converting water into steam which may in turn be used for generating electrical power through turbines.

Soon after the establishment of nuclear fission as a physical process and the realization of its potential as a source of energy, scientists all over the world started looking for the possibility of a self-sustained fission chain reaction. The best example of a self-sustained chain reaction is like a burning matchstick. On striking the head of a matchstick enough heat is produced that sets the head burning. Now the heat produced by the burning of the head is enough to burn the next part of the stick and then the next part and so on. Thus, a self-sustained chain reaction of burning is produced in the matchstick which keeps it lighted. If the heat produced in the burning of the head of the stick is not enough to burn the wooden part of the stick, the fire will die out after the burning of the head and the matchstick will not be of much use. This is what happens if the matchstick becomes moist. Similarly, for fission to be a viable source of energy, it is necessary that a self-sustained chain reaction of fission takes place so that a large number of nuclei undergo fission successively and energy is produced at a constant rate.

1.9 Self-sustained fission chain reaction

Since fission in uranium is initiated by thermal neutrons, the key to a self-sustained fission chain reaction was the production of neutrons in fission in a larger number than are absorbed in it. Experiments were carried out in Germany (Hahn and Strassmann group), in France (Joliet Curie, Halban, Kowarski, called the Paris group) and in New York (Fermi group) to detect fission neutrons. Fermi was already quite sure that fission will be accompanied with additional neutrons, since fission fragments are excited and highly neutron rich, therefore, they are likely to evaporate a few neutrons during their flight. The Paris group was perhaps the first to detect fission neutrons and announced that two to three fast neutrons per fission are emitted. In the meantime, the Fermi group had already started the construction of a fission-based reactor to trap fission energy. Since neutrons emitted in fission were of high energy, it was necessary to reduce their energy to the level of thermal energy (0.025 eV) before they could initiate a fission reaction in other uranium nuclei. A possible way of reducing neutron energy was to let them collide with other nuclei so that at each collision they would lose a part of their energy. It is easy to show that energy loss per collision is a maximum when the masses of two colliding bodies are equal. Therefore, the most efficient energy reducer for neutrons will be a material that contains a large number of protons or hydrogen. Paraffin wax with a high percentage of hydrogen is very efficient in reducing the kinetic energy of fast neutron rapidly in a few collisions. However, it cannot be used in a fission reactor where temperature is likely to go high. Fermi's group used graphite (called a moderator) for neutron energy reduction. The other problem was to stop or regulate the fission reaction rate. The fission reaction rate could be controlled by controlling the supply of thermal neutrons. The element cadmium is found to have a high cross-section for absorption of thermal neutrons. Therefore, Fermi used cadmium rods, called arrester rods to control the fission rate in their reactor. The first fission reactor developed by Fermi *et al* was called 'Pile' as it had a pile-like structure of graphite with pigeon holes for keeping fissile material, enriched uranium. There were also spaces for cadmium arrester rods which could be inserted or withdrawn to regulate fission rate. The first reactor named Chicago Pile-1 was put in operation on 2 December 1942 underneath the University of Chicago's Stagg Field football stadium (figure 1.8).

Figure 1.9 shows the build-up of a self-sustained fission chain reaction with the use of moderators that reduce the energy of fast fission neutrons to thermal energies. Different kinds of moderators like graphite, heavy water, light water etc have been used for moderation. The fission fragments; a pair of nuclei, each about half in size of the initial nucleus, like ${}^{84}_{36}Kr$ and ${}^{137}_{56}Ba$ or ${}^{96}_{43}Tc$ & ${}^{115}_{49}In$ etc., may be produced in fission.

Figure 1.8. The Chicago Pile-1 (CP-1), first man-made nuclear fission reactor.

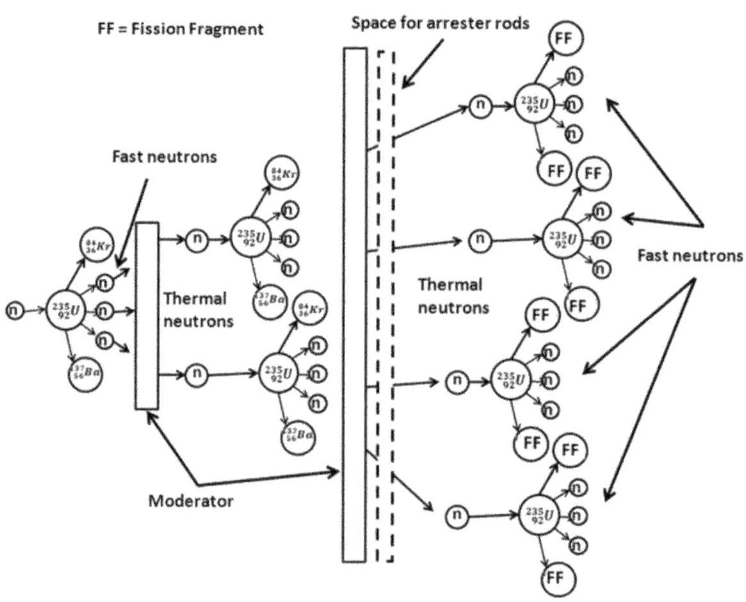

Figure 1.9. Build-up of fission chain reaction.

Unfortunately, the discovery of the fission reaction took place in the backdrop of World War II, and, therefore, the destructive power of the sustained chain reaction showed up first when on August 6 and 9, 1945 America dropped two atomic bombs, using an uncontrolled fission chain reaction, on the cities of Hiroshima and

Nagasaki in Japan. Around 130 000 civilians were killed and a very large number suffered radiation damage. Figure 1.10 represents pictures of atom bomb mushroom clouds at Hiroshima and Nagasaki.

It took almost 12 years for the peaceful use of fission chain reaction to appear when on December 2, 1957, the World's first large scale nuclear power plant begun operation in Shippingport, Pennsylvania, USA. At the time of writing this text, almost 468 fission-based nuclear power plants are operational in the World, producing almost 2500 TWh electric power which accounts for roughly 11.7% of World energy production.

Long-lived radioactive materials are produced in large amounts as a by-product of fission, along with a considerable amount of weapon grade fissionable material. The problem of safe disposal of radioactive waste and management of weapon grade material are two big problems associated with peaceful use of the fission reaction. Safe running of reactors, without any leak of radioactive material and radiation is also a problem, which showed its dirty face at Chernobyl, USSR, on 26 April, 1986 and at Fukushima Daiichi, Japan on 11 March 2011. In Russia, the fourth unit of a reactor complex housing four reactors each producing 1000 MW of power, became super critical, partly because of faulty design, and partly because of poor maintenance and upkeep. Uncontrolled build-up of the fission chain reaction resulted in core meltdown; some 32 lives were lost on the first day of the disaster and many more suffered radiation injuries because of excessive radiation doses.

The Japan nuclear disaster was basically an outcome of Nature's fury; first the area was hit by an earthquake of intensity 9.0 on the Richter scale, which was followed by a huge tsunami of more than 12 m high waves. Eleven reactors operating in the region all shut down automatically when the earthquake hit. The reactors proved robust seismically but were damaged by the tsunami which flooded the cooling system of the reactors. There was some radioactive leak but no one lost their life or was injured by radiation. Despite nearly 19 000 people losing their lives by drowning in Tsunami waves, however, no deaths were on account of nuclear

Figure 1.10. Pictures of atom bomb mushroom clouds at Hiroshima and Nagasaki.

disaster. These recent nuclear mishaps once again put the question of safety of nuclear reactors in the forefront.

During the last decade and a half, development of accelerator driven energy amplifiers has attracted considerable attention which not only claims a very safe and foolproof working of nuclear fission reactors but has the potential of incinerating radioactive waste and all accumulated weapon grade materials in a few years' time. These energy amplifiers use a high energy (around 800 MeV) and high beam current proton accelerator to initiate spallation reaction in a heavy metal like lead that generates high flux of neutrons in a wide energy band. Fissionable materials are then distributed in the system at locations where the flux of neutrons in the required energy band is high so as to fission them out. Long-life radioactive waste from other reactors may also be used as fuel for these energy amplifiers. Similarly, weapon grade material may also be incinerated to produce additional heat. A part of the energy generated may be used to run the system while the rest may be used for commercial purpose. Since the system runs on proton beams, the energy amplifier may be stopped just by switching off the accelerator (figure 1.11).

1.10 Fusion reaction

Fusion reactions in which two light nuclei are made to fuse are promising candidates for safe and clean generation of energy. Fusion has the advantage over fission that no radioactive waste is generated in this process and that it has no carbon footprint. However, in order to fuse light nuclei, they must first be given large kinetic energy to reach within the range of the nuclear field to fuse and release energy. At required

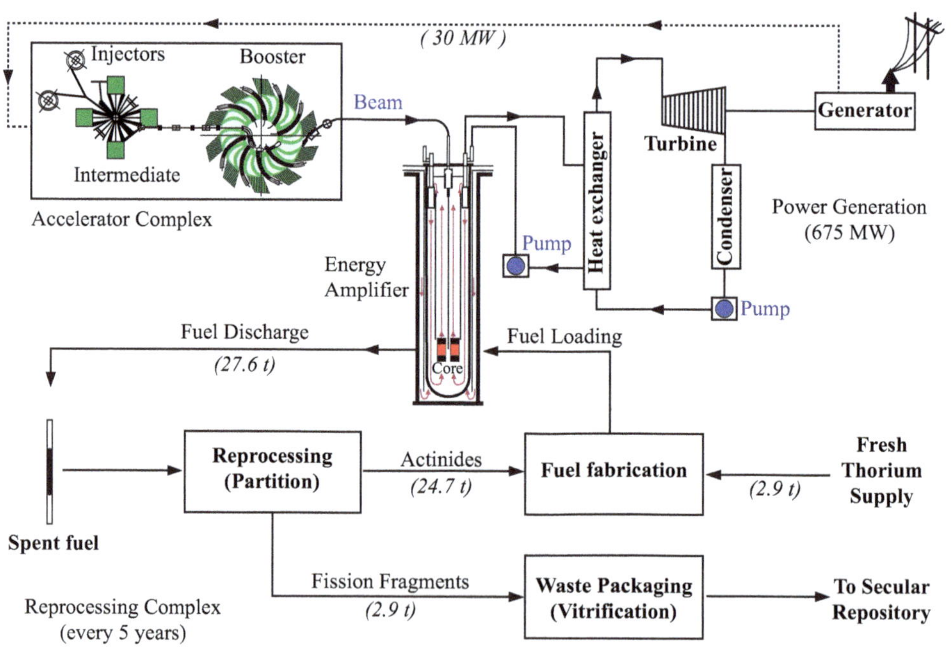

Figure 1.11. An artist's view of an accelerator driven energy amplifier complex/unit.

high kinetic energies matter acquires the fourth state—the plasma state. Confinement of plasma, to keep it concentrated to the desired level for sufficient time for fusion to take place, is the biggest obstacle. Apart from that, finding materials for a fusion reactor that may sustain such high temperatures and development of the mechanism for extracting heat energy are also major challenges. Although scientists working on fusion projects are unanimous that commercially viable fusion energy is not likely for at least more than two decades or so, two major projects on fusion are underway. These are the International Thermonuclear Experimental Reactor (ITER), a 35 nation collaboration to design, build and operate a fusion reaction for a short period of time, also called the largest tokamak project in France, and the other with the name Wendelstein $7-X$ ($W7-x$) at Max Planck Institute for Plasma Physics (IPP) in Germany, also termed as world's largest Stellarator. In a recently held meeting of the ITER council it was stated that about 55% of the work of fabrication and installation of technical components like vacuum vessel sectors, toroidal field magnets, the cryoplant, site service building and magnet power supply etc are in place. The construction cost of the facility is expected to be of the order of 20 billion Euros. The first experiment with ITER is planned by 2025.

1.11 Accelerators

Since the discovery of neutrons, if any other event made a big impact on the study of nuclear reactions, it was the accelerator machines developed to impart large kinetic energies to charged nuclear and other particles and ions. The centre or the place from where the thought of having accelerator machines originated was, beyond any doubt, the Cavendish Laboratory at Cambridge university in England. After carrying out two landmark experiments, the large angle alpha scattering experiment establishing the presence of the nucleus at the centre of each atom and the experiment on the transmutation of nitrogen hitting them by alpha particles obtained from a radioactive source, Rutherford, in his 1927 address to the Royal Society London publicly expressed the desire of the scientific community to have a copious supply of charged particles with energies far more than the energies of alpha and beta particles available from radioactive sources. These accelerated particles may be used to disintegrate and smash nuclei with higher charge and mass. Rutherford's appeal received immediate attention of national press and several proposals were submitted.

The first particle accelerator was built at Cavendish Lab by John D Cockcroft and E T S Walton in 1930, based on the principle of electrostatics that a charged particle of charge Q in traversing a potential gradient ΔV will gain an energy ΔE given by,

$$\Delta E = Q\Delta V$$

Knowing well that with the available maximum voltage it may not be possible to accelerate alpha particles to energies higher than what is available from radioactive sources, they tried to accelerate protons instead of alphas. They also estimated, based on the theoretical work recently published by a Russian scientist George

Gamow, working at Neil Bohr's institute for theoretical physics, Copenhagen, Denmark that lighter charged particles have a better chance of quantum mechanical tunnelling through the Coulomb barrier and reach the nuclear field of the target nucleus. Cockcroft and Walton initially tried and accelerated protons to the energy of 2 keV using directly a 2000 Volt transformer available in the lab. However, with such low energy protons they could not produce any nuclear reaction and, therefore, they used the technique of voltage multiplication to accelerate protons in a 2.6 m long evacuated tube to energy 500 keV. Accelerated protons were then made to hit a beryllium target which was found to disintegrate into two alphas.

Robert J Van de Graaff, a post-doctoral fellow at the University of Princeton in a speech before American Physical Society in 1931 outlined how he was able to generate high voltage of the order of 1 MV by transporting charges to big metallic spheres insulated from the ground by glass rods. The charge was transported through silk belts run by motors. Charges were sprayed on moving belts by needle-comb arrangement and were picked up at the other end by a similar arrangement. Using this method, it was possible to build a large potential difference between the ground and metallic sphere which was restricted by the spark between the ground and the high voltage terminal. Because of its simplicity and the fact that Van de Graaff had already produced around 750 kV by his machine, many universities and institutes started building such machines. By the year of 1933 the first practical Van de Graaff generator delivering a 0.6 MeV proton beam became operational. Modern day Van de Graaff machines can sustain up to 15 MV of voltage, the high voltage terminal being kept in a high-pressure Freon gas insulator tank. Another modification, called Tandem Van de Graaff, uses the same potential difference twice, once accelerating the negative ions and then stripping negative ions to positive ions the same potential difference may again be used to accelerate positive ions.

A Swedish scientist Gustav Ising in 1924 gave the idea of accelerating heavy ions in a linear path by applying different voltages. Rolf Wideroe, a PhD student at Technische Hochshule Karlshruhe, Germany designed a heavy ion accelerator using drift tubes of increasing lengths (see figure 1.12), which were fed RF voltage such that accelerated ions reached the next gap at the right moment to get an energy push.

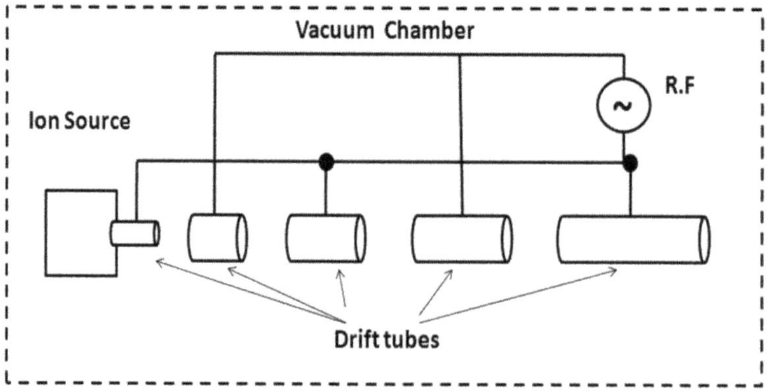

Figure 1.12. Principle of operation of heavy ion Linac designed by Rolf Wideroe.

He was able to accelerate potassium ions to 50 keV using an RF of 1 MHz, 25 kV supply. Ernest Orlando Lawrence, working at Berkeley, USA picked the idea of cyclotron seeing the figures of the heavy ion Linac of Rolf Wideroe. Lawrence realised that if a particle of mass m and charge q is constrained to move in a circular path by applying a magnetic field B normal to the velocity v of the particle, then the cyclic frequency ω of the particle may be given by;

$$\omega = \frac{v}{r} = \frac{qB}{m}$$

As may be observed, the cyclic frequency is independent of the radius of the path. If, therefore, an injection of energy is given to the charged particle at a fixed interval of time, its radius of path will increase but it will remain synchronised and will gain energy in discrete steps as has happened in the heavy ion Linac of Wideroe. Lawrence with his PhD student Stanley M Livingston, using a 13 000 Gauss, 4 inch size pole piece magnet, 10 MHz, 1000 V RF oscillator were able to accelerate protons to about 0.6 MeV in January 1931 in their first cyclotron. Later, with improved geometry of magnetic pole pieces that provided focusing, by increasing the size of the magnetic pole pieces and with the availability of deuterium ions, Lawrence obtained 5 MeV deuterium ions by 1934, 8 MeV by 1936 and 20 MeV by 1939. Within a span of about ten years almost 60 cyclotrons were built at various universities and research institutes and by 1960 the number crossed the 100 mark.

By 1945 the maximum energy obtained by different types of accelerators approached limiting values, because of insulation break down in electrostatic accelerators like Van de Graff, relativistic increase of mass of the particles which de-tuned the synchronization between the cyclic frequency and the frequency of the applied RF and excessive energy loss in the case of electrons.

In 1945, V Veksler [6] in Russia and E M McMillan [7] at University of California, USA almost simultaneously published the principle of phase stability to maintain synchronization of RF frequency ω_{rf} and the cyclic frequency of motion ω. If it is assumed that at each turn in motion the particle gains energy equal to its rest mass m_0, then at the end of the $(n-1)^{\text{th}}$ turn the frequency of the motion will be,

$$\omega_{n-1} = \frac{Bq}{nm_0}$$

Thus, if ω_{rf} is an integer multiple of a particle's revolution frequency, the particle acceleration will remain synchronized. The principle of phase stability was soon applied to electron accelerators and also to other cyclotrons used to accelerate deuteron etc, such that 184 inch Berkeley cyclotron, which on modification to confirm to the principle of phase stability is called synchrocyclotron produced 190 MeV deuterons and 380 MeV He^{2+} ions by 1947. Further progress in particle accelerators was achieved by better focusing of accelerated beams in cyclic machines using the principle of alternate gradient or strong focusing.

Progress in particle accelerators is measured by the acceleration of particles to higher energies, utilization of new techniques and absorption of new ideas. Starting from electrostatic accelerators that could deliver particles up to a few MeV, the concept of resonant acceleration using RF fields extended the energy limit to around 25 MeV. Relativistic increase in mass of the particle to be accelerated imposed this upper limit. Then came the principle of phase stability that resulted in the construction of Synchrocyclotron and Synchrotron that pushed the energy limit to around 2 GeV by the early 1950s. In 1950 came the principle of alternate gradient focusing that dramatically reduced the size of magnets and the energy barrier moved to 400 GeV. In 1960 came the concept of colliding beams and the energy frontier dramatically moved further. Today the World is limited only by the prohibitive cost of building a new accelerator and where to build it.

1.12 Particle accelerators in India

The first particle accelerator, a 1 MeV proton accelerator was installed at Tata Institute of Fundamental Research (TIFR), India in the year 1950. Cockcroft Walton type DC machines primarily for producing 14 MeV neutrons were installed at several universities and research institutes during 1955 and 1960, including at: (i) Aligarh Muslim University, Aligarh (AMU); (ii) Saha Institute for Nuclear Physics (SINP), Kolkata; (iii) Bose Institute, Kolkata; (iv) Banaras Hindu University, Varanasi (BHU); (v) Andhra University, Andhra Pradesh; (vi) Indra Gandhi Centre for Atomic Research (IGCAR), Kalapakkam; (vii) Bhabha Atomic research centre, (BARC), Mumbai, India etc. A 38 inch Cyclotron was installed at SINP, Kolkata in 1960. A 2 MV Van de Graaff generator at IIT Kanpur, India was installed in 1961 to study fission. A 5.5 MeV Van de Graaf accelerator at Bhabha Atomic Research Centre (BARC), Mumbai, was operational in 1962 which has since been retrofitted as a Folded Tandem Ion Accelerator. A cyclotron facility capable of delivering 5 MeV protons, gifted by Rochester University, USA was installed at Punjab University, Chandigarh, Punjab, India in the early seventies. A variable energy 224 cm cyclotron became operational at VECC, Kolkata in 1978. A race track microtron started working at University of Pune, Pune in 1981. A 14 UD Pelletron accelerator was commissioned at TIFR, Mumbai in 1988 which was coupled to a LINAC booster in 2007. A 15 UD Pelletron started operating at the Inter University Accelerator centre (IUAC), New Delhi, India in 1992. In 1995, a 3 MeV Pelletron accelerator was installed at Institute of Physics, Bhubaneswar, Orissa. Centre of Advanced Technology (CAT) developed a variable energy microtron which was installed at the University of Mangalore in 1995. A 3 MeV tandem accelerator was installed at National Centre for Compositional Characterization of Materials and Guru Ghasidas University, Bilaspur around 2014. 1.7 MeV Tendetron accelerators have been installed at (IGCAR) Kalpakkam and IIT, Kanpur recently. Indus-1 and Indus-2, the two synchrotron radiation sources, respectively, capable of giving electrons of energy 459 MeV and 2.5 GeV have been working since 1999 and 2005, respectively.

Some other accelerators like a low energy high intensity proton accelerator (LEHIPA) which will be coupled to a 3 MeV Radio Frequency Quadrupole (RFQ) and a 20 MeV drift tube LINAC (DLT) and to some other accelerating systems to get about 1 GeV proton beam to be used for accelerator driven energy amplifier systems is under construction at BARC. Institute for Plasma Research is also developing a high flux neutron source capable of giving 1012 neutrons per second using T(D,n) reaction and they are in process of developing a suitable accelerator. Several other accelerators are going to be installed at different places for the study of astrophysics and for the development of radioactive ion beams facilities etc.

1.13 Radiation detectors

Detectors are essential to detect, identify and count the number of radiations involved in a nuclear reaction to establish the reaction. Particles or radiation involved in nuclear reactions are either charged like protons, beta particles, alpha particles or other heavy ions, or they may be electrically neutral radiation like neutrons, gamma rays and other neutral mesons etc. Charged radiation directly produces ionization when it passes through air or any other gas while electrically neutral particles produce ionization through secondary reactions like gamma rays via photo-electric effect and neutrons through scattering of charged nuclei etc. Thus, ionization produced by radiation has been used for the first time to detect the presence of radiation. Starting from the gold leaf electroscope, invented by Abraham Bennet in 1876, photographic films (invented by George Eastman in 1889) and photographic emulsion plates used for recording of cosmic ray and other high-energy interactions were developed in 1871 by Richard Leach Maddox. These detectors were capable of detecting the presence of radiation but, in general, were not able to pick up each individual radiation. A breakthrough towards detecting each radiation was made by Rutherford when he found that tiny flashes of visible light are emitted when charged radiation falls on a fluorescent screen like that of ZnS. He used this technique for detecting alpha particles in his famous large angle alpha scattering experiments. However, counting of light flashes by eye was very difficult and tiring. It required training and acclimatization of the scanner in a dark room and required so much concentration that even the best trained person could do it only for few minutes. It is amazing that in spite of these handicaps Rutherford and his group could carry of experiments in which alpha particles of the order of a few billions were recorded. Another instrument that recorded the path of individual radiation was the cloud chamber invented by C T R Wilson in 1911. A cloud chamber consists of an expandable glass chamber which is filled with steam. On sudden expansion, temperature falls and vapours become supersaturated resulting in drop formation along the path of the ionizing radiation. A camera and flash light system, activated by the expansion mechanism, records the droplet path of the radiation. When this instrument is put in a magnetic field, the curvature of the radiation path may be used to determine the energy of the radiation. The biggest disadvantage of the cloud chamber was its size and other associated systems that made it unwieldy for easy transportation. Later, since the 1950s the bubble chamber based on super-heated steam that makes particle trails visible in liquid medium, has been in use, particularly in high energy experiments.

In the year 1908, a German Scientist, Hans Wilhelm Geiger, developed a tube filled with some low-pressure inert gas and fitted with electrodes for detecting ionizing radiation through the charge collected at the electrodes. Later, in 1928 in collaboration with Walther Muller he produced a practical tube that was capable of detecting and counting the number of different types of charged particles. The big advantage of the GM tube or counter was the size of the output pulse, in the case of a counter of normal size pulses of a few hundred micro volts may be produced, which may be easy to record. The GM counter, however, could not determine the energy of the particle and could also not handle high counting rates.

In 1940, Frisch developed a gridded ionization chamber, which still finds limited use in alpha spectroscopy. In late 1940 a third type of gas filled detector called a proportional counter was developed. Proportional counters produce electrical pulses the size of which is proportional to the energy lost by the radiation in the active volume of the detector. Gas filled detectors were extensively used for the study of nuclear reactions until the 1940s (figure 1.13).

The method of scintillations for detecting and counting radiation started by Rutherford was almost overtaken by gas filled detectors till the development of photomultiplier tubes that were capable of amplifying by a large factor the weak electronic pulses formed by faint flashes of visible light. Photomultiplier tubes (PMTs) (figure 1.14) made rapid progress after the development of efficient photo cathodes and secondary emission multipliers, called dynodes. In 1935 Lams and Salzberg [8] succeeded in making a triode photomultiplier tube with a photocathode combined with a single dynode for use in movie sound pick up. In 1939, Zworykin and Rajchman [9] developed a photomultiplier tube employing electrostatic focusing of secondary electrons, which has since been further improved to have a Sb–Cs photocathode. Further improvements in dynode structures were made by Morton *et al* [10].

The crucial component of a scintillation spectrometer is the scintillating material where tiny sparks of visible light are produced. A good scintillator should convert almost all the energy deposited in it by the impinging radiation into prompt fluorescence, the light yield should be proportional to energy deposited, the luminescence should be of short duration, the material should be homogeneous, and must be transparent to its own radiation. It is a fact that no material simultaneously meets all these requirements. Therefore, choice of a particular scintillator is always a compromise among these and other factors. For example, NaI(Tl), sodium iodide crystal that has been activated with thallium, developed in 1958, is very efficient for gamma rays but not for other charged particles. Organic scintillators like anthracine are good for charged particles like beta particles.

The detectors that measure the ionization produced in solid dielectrics and which are a solid state analogue of gas filled counters, emerged in the early 1960s. These solid state detectors developed rapidly since then. Diffused junction and surface barrier detectors found widespread use in the detection of charged particles. High purity germanium detectors HP(Ge) are now very frequently used for gamma ray spectroscopy. HP(Ge) detectors are basically semiconductor diodes with p–i–n structure and the middle intrinsic region is sensitive to ionizing radiation like x-rays and gamma rays. A special feature of this detector is its energy resolution which is a measure of the capability of the detector to distinguish two gamma rays of nearly the same energy as two different peaks. The percent resolution of a NaI(Tl)

Figure 1.13. Photo of Frisch grid ionization chamber developed at Oak Ridge Institute of Nuclear Studies in 1959.

spectrometer is typically of the order of 5% while the resolution of an HP(Ge) detector is almost 30 times better.

Figure 1.15 shows the basic structure and circuit diagram of an HP(Ge) detector while figures 1.16 and 1.17 respectively, show the complete detector setup along with source–detector arrangement and the typical gamma ray spectra of standard ^{152}Eu gamma source recorded using such a setup.

With the availability of accelerated beams of all ions from helium to uranium and of a detecting system that could handle large count rates, it became possible to carry

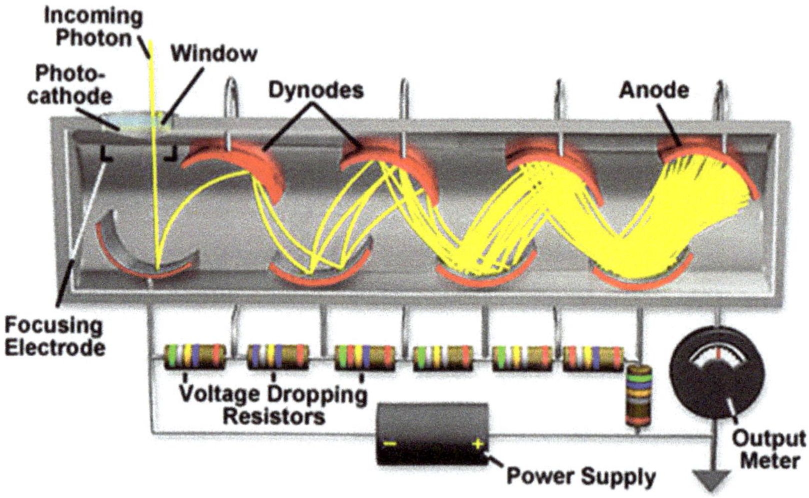

Figure 1.14. Schematic diagram of a photomultiplier tube with arrangement of power supply and voltage dropping resistors.

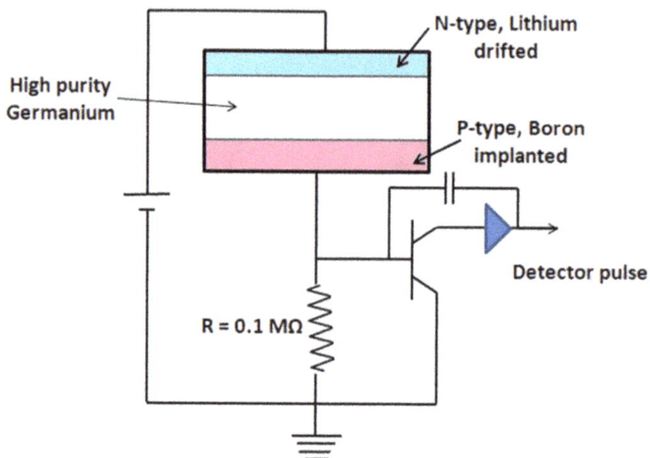

Figure 1.15. Circuit diagram of a typical high purity germanium, HP(Ge) detector.

out in-beam experiments. In-beam experiments provide real time information about reaction products. It is now possible to record time coincident events, the charge and the energy of reaction products and even the point of origin of the product using suitable focal plane detectors like multiple wire proportional counter and micro-channel detectors. Computer based fast data recording and retrieval facilities have made it possible to study nuclear reactions in much detail and with great accuracy.

1.14 Study of reaction mechanism

As has already been mentioned, Fermi [4] for the first time made a detailed study of reactions induced by newly discovered neutrons and measured (n,γ) cross-sections

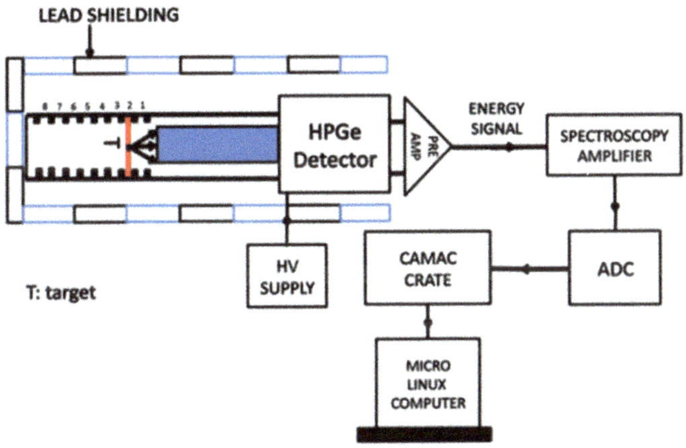

Figure 1.16. A typical HP(Ge) spectrometer setup along with source–detector arrangement.

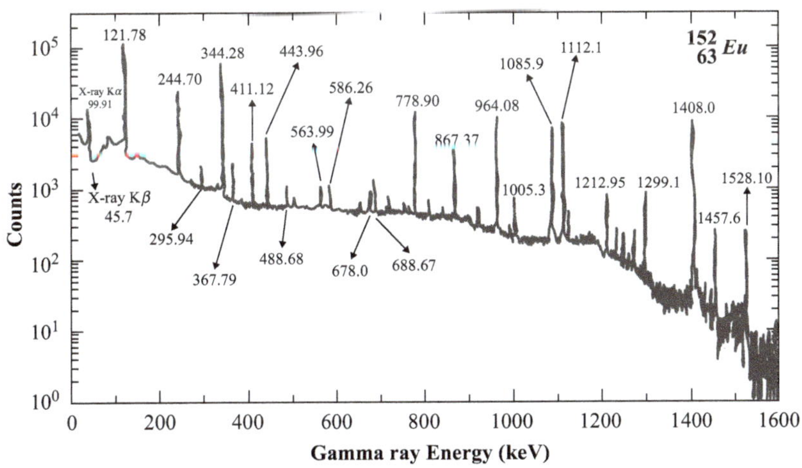

Figure 1.17. Gamma ray spectrum of standard ^{152}Eu source taken by a 100 c.c. HP(Ge) detector.

for about 18 reactions using the activation technique, which involves identification of reaction residues by their characteristic induced radioactivity. A major conclusion that Fermi drew from his measurements was that for most of the cases the (n,γ) cross-section was very large and for some targets at low energies there were sharp peaks which was not so for other targets. From his observation of large value of (n,γ) cross-sections Fermi inferred that the system formed by the absorption of the incident neutron stays in that state for a long time before emitting the gamma ray. Fermi argued that this is possible if the width of the state formed by the absorption of neutron was sufficiently large so that its lifetime is large on a nuclear scale. He estimated the lifetime of these states to be of the order of or even larger than 10^{-16} s, which is much larger than the nuclear timescale of interaction time $\approx 10^{-22}$ s for the neutron with the nucleus.

Within two years of Fermi's measurements, Bohr [5] and Breit and Wigner [11] proposed a theoretical model for the reaction mechanism called compound reaction mechanism. The basic assumption of the model was that the incident particle (neutron in this case) on entering the target nucleus forms a composite system which lives for a while during which the projectile shares its energy and momentum with other nucleons of the target leading to the formation of a compound nucleus with thermodynamic equilibrium. However, when by chance enough energy is concentrated on a nucleon or on a bunch of nucleons, they are emitted from the compound nucleus leading to its decay. The compound nucleus is treated as a quasi-bound state with discrete energy eigenvalues. However, these eigenvalues, denoted by W_λ, are complex as they have a finite lifetime and may be given as;

$$W_\lambda = E_\lambda - i\frac{\Gamma_\lambda}{2} \tag{1.1}$$

Here, the level width Γ_λ is related to the mean life τ_λ of the compound nuclear state λ by the relation,

$$\tau_\lambda = \frac{\hbar}{\Gamma_\lambda} \tag{1.2}$$

The reaction proceeds only if incident energy matches one of the energies E_λ within the width Γ_λ, giving rise to the resonance structure. If incident energy does not match any E_λ within the width Γ_λ, only elastic scattering of the projectile will take place. Thus, the long life of the compound nuclear state explains the large values of (n,γ) cross-sections observed by Fermi. Further, if the width of the compound nuclear state Γ_λ is of the order of few tens of keV, which is quite often the case, the mean life τ_λ turns out to be of the order of 10^{-16} s, as estimated by Fermi.

Breit and Wigner derived their famous Breit–Wigner [11] formula for cross-sections at energies near to the resonance energy, using the analogy of atomic resonance fluorescence and treating the problem through Wigner–Weisskopf perturbation theory. Both the comparison of the decay of the compound nucleus with atomic resonance fluorescence was not valid since the incident neutron in this case does not disappear in the intermediate state, while a photon disappears in the resonance fluorescence and also treating the nuclear force with perturbation theory is not correct because of its strong nature. A valid quantum mechanical description, free from such discrepancies was later presented by Kapur and Peierls [12] in the framework of R-matrix.

R-matrix formalism is characterised by the division of coordinate space into an internal region I, where all nucleons are close together and interact with each other with nuclear force. A channel γ is where the system is partitioned into two particles, say C and c, whose centres of mass are separated by $r_\gamma \geqslant a_\gamma$, where a_γ is the channel radius and is larger than the range of the nuclear force between c and C. The boundary between channel γ and region I is defined by $r_\gamma a_\gamma$. The compound nucleus state wave function X_λ is defined to obey the Schrödinger equation in region I,

$HX_\lambda = W_\lambda X_\lambda$ (in region I) Here H is the total Hamiltonian and W_λ the energy eigenvalue.

If one neglects all charges and angular momentum of interacting nucleons, then X_λ has the form,

$$X_\lambda = \sum A_{\lambda\gamma} \frac{e^{ik_\gamma r_\gamma}}{r_\gamma} \phi_\gamma(\varepsilon_\gamma)(r_\gamma = a_\gamma \text{ for all } \gamma) \tag{1.3}$$

Here, $A_{\lambda\gamma}$ is the amplitude of X_λ at the boundary of channel γ, $\phi_\gamma(\varepsilon_\gamma)$ is the internal wave function, and k_γ is the wave number of relative motions at the experimental total energy E. A decaying compound state is supposed to represent a compound state long after it is formed with no incident wave anymore, it only decays through all open channels into outgoing waves, as given by equation (1.3).

The total wave function, Ψ_α^+, which describes a reaction initiated in channel α at energy E, satisfies the Schrodinger equation, $H\ \Psi_\alpha^+ = E\ \Psi_\alpha^+$ and an asymptotic boundary condition:

$$\left(\Psi_\alpha^+\right)_\infty = \left[\frac{e^{ik_\alpha r_\alpha}}{r_\alpha} - \sum_\beta \sqrt{\frac{v_\alpha}{v_\beta}} S_{\beta\alpha} \frac{e^{ik_\beta r_\beta}}{r_\beta}\right] \tag{1.4}$$

where $S_{\beta\alpha}$ is the S-matrix element of transition $\alpha \rightarrow \beta$, and the v are the relative velocities in the channels. If one expands $\Psi_\alpha^+ in\{X_\lambda\}$ in region I as

$$\Psi_\alpha^+ = \sum_\lambda a_\lambda X_\lambda \tag{1.5}$$

one can calculate the coefficients, a_λ, using the Schrödinger equations and the boundary conditions for Ψ_α^+ and $\{X_\lambda\}$. One can then obtain $S_{\beta\alpha}$, if one substitutes (1.3) and (1.4) into (1.5) and compares the amplitudes of the outgoing wave in channel β on both sides of the equation. The result is

$$S_{\beta\alpha} = e^{-2ik_\alpha a_\alpha}\delta_{\beta\alpha} - i\sum_\lambda \frac{g_{\lambda\beta}g_{\lambda\alpha}}{E - W_\lambda} \tag{1.6}$$

where

$$g_{\lambda\gamma} = \sqrt{\frac{\hbar v_\gamma}{(2\pi)^3}} N_\lambda A_{\lambda\gamma}$$

The first term in equation (1.6) corresponds to the elastic scattering by a hard sphere of radius a_α and is the only process if E is away from one of W_λ. If E approaches closer to W_λ the second term of equation (1.6) produces resonance.

The total cross-section of the reaction $\alpha \rightarrow \beta$ is given by

$$\sigma_{\beta\alpha} = \frac{\pi}{k_\alpha^2} \mid \delta_{\beta\alpha} - S_{\beta\alpha} \mid^2 \tag{1.7}$$

If E is within Γ_λ from E_λ, and all other resonances are so far away that they can be neglected in the S-matrix element, one obtains the Breit–Wigner one-level formula.

$$\sigma_{\beta\alpha} = \frac{\pi}{k^2} \frac{\Gamma_{\lambda\beta}\Gamma_{\lambda\alpha}}{(E - E_\lambda)^2 + \Gamma_\lambda^2/4} \quad (\text{for} \neq \alpha) \tag{1.8}$$

where

$$\Gamma_{\lambda\gamma} = | g_{\lambda\gamma} |^2$$

In s-matrix description one makes use of channel radius which is not unique. Resonance parameters depend on the choice. Normally choice of the channel radius is made such that it represents the experimental data.

On the other hand, Feshbach [13] used the formalism in which there is no channel radius as he made a division of the Hilbert space wave functions, instead of the coordinate space. The Hilbert space is divided into two components: P-space that has all open channels and the complimentary Q-space. A compound nuclear state is defined as an eigenstate of QHQ. In this formalism one may define a compound nuclear state as an eigenstate of QHQ,

$$QHQ\Phi_s = E_s\Phi_s \tag{1.9}$$

The resonance energy W_s may be written as,

$$W_s = E_s + \langle \Phi_s \mid QHP \frac{1}{E + i\varepsilon - P\bar{H}P} QHP \mid \Phi_s \rangle \tag{1.10}$$

where P and Q are projection operators onto P and Q spaces, respectively. Feshbach's formalism has no ambiguity related to channel radius. It, however, has its own non-uniqueness related to the partition of the Hilbert space. Q-space is often defined as a space in which all nucleons are bound in a single-particle potential. This definition naturally depends on the single-particle potential.

1.15 Statistical reactions

With the increase of excitation energy, the spacing between compound nuclear states decreases and their level widths increase, therefore, there is considerable overlap between compound nuclear states. If one considers only those compound states that fall within the energy spread ΔI of the incident energy and leaves out all other nuclear states, then the average cross-section $\langle \sigma_{\beta\alpha} \rangle$ for the reaction $\alpha \rightarrow \beta$ may be given as,

$$\left\langle \sigma_{\beta\alpha} \right\rangle = \frac{\pi}{k_\alpha^2} \left\langle \left| \sum\nolimits_\lambda \frac{g_{\lambda\beta} g_{\lambda\alpha}}{(E - E_\lambda + i^{\Gamma_\lambda}/_2)} \right|^2 \right\rangle \tag{1.11}$$

Assuming that the parameters of the remaining resonances are statistically random, the average of the cross terms may be neglected and equation (1.11) reduces to,

$$\left\langle \sigma_{\beta\alpha} \right\rangle = \frac{\pi}{k_\alpha^2} \sum_{\lambda \Delta I} \frac{2\pi}{\Delta I} \frac{\Gamma_{\lambda\beta}\Gamma_{\lambda\alpha}}{\Gamma_\lambda} \approx \frac{\pi}{k_\alpha^2} \frac{2\pi}{<D>} \left\langle \frac{\Gamma_{\lambda\beta}\Gamma_{\lambda\alpha}}{\Gamma_\lambda} \right\rangle \tag{1.12}$$

Here $<D>$ is the average level separation and $\left\langle \frac{\Gamma_{\lambda\beta}\Gamma_{\lambda\alpha}}{\Gamma_\lambda} \right\rangle$ is the average of their ratio in energy interval ΔI. If level widths are un-correlated to each other then,

$$\langle \sigma_{\beta\alpha} \rangle = \frac{\pi}{k_\alpha^2} \frac{T_\beta T_\alpha}{\sum_\gamma T_\gamma} \tag{1.13}$$

Here,

$$T_\gamma = 2\pi \quad < \frac{\Gamma_{\lambda\gamma}}{D} > \tag{1.14}$$

Equation (1.14) holds if D and Γ are unrelated to each other.

Equation (1.13) represents the well-known Hauser–Feshbach formula for the decay to a discrete state of the residual nucleus. T_γ is called transmission coefficient.

The cross-section for the formation of the compound nucleus through channel α, $\sigma_{\alpha\iota}^{CN}$ may be obtained by summing the expression given by equation (1.13) over all open channel β;

$$\sigma_{\alpha\iota}^{CN} = \sum_\beta \langle \sigma_{\beta\alpha} \rangle = \frac{\pi}{k_\alpha^2} T_\alpha \tag{1.15}$$

Equation (1.15) may be rewritten as,

$$\langle \sigma_{\beta\alpha} \rangle = \sigma_{\alpha\iota}^{CN} \,.\, R_\beta \tag{1.16}$$

where, $R_\beta = \frac{T_\beta}{\sum_\gamma T_\gamma}$ is the branching ration of the decay from channel β.

Equation (1.16) shows the statistical independence of the formation of the compound nucleus and its subsequent decay. Therefore, in the regime of statistical nuclear reactions, the nuclear reaction is a two-step process: in the first step a compound nucleus is formed and in step two it decays. The formation and decay steps are independent of each other, because before its decay the compound nucleus forgets its history of formation. This essentially means that if the same compound nucleus in a given state is formed by two different incident channels then the mode of decay of the compound nucleus will not depend on the properties of the incident channels. This essentially means that a compound nucleus in a given quantum state decays to a given final state with a fixed probability that does not depend on the specific way it was formed.

1.16 Experimental verification of independence of formation and decay of compound nucleus

Experimental verification of this independence of decay from the incidence channel was first explored by Ghoshal [14] in 1950. He produced the same compound nucleus (CN)^{64}Zn at the same excitation energy via two different channels (^{63}Cu + p) and (^{60}Ni + α). The incident energies of proton and alpha particles were so adjusted that the compound nucleus ^{63}Zn has the same excitation energy. Ghoshal concentrated on three decay channel: (i) (^{62}Zn + 2n); (ii) (^{63}Zn + n); and (iii) (^{62}Cu + np) since the residual nuclei of these channels can be easily identified through the characteristic half-lives and decay modes, respectively, of: (^{62}Zn) 9.18 h, β^+; (^{63}Zn) 38.5 min, β^+; and (^{62}Cu) 9.7 min, β^+. Ghoshal studied the excitation functions for these three induced activities produced via the two incident channels. The schematic representation of Ghoshals's experiment is given in figure 1.18 while the measured excitation functions for the three decay channels populated via two different incident channels are shown in figure 1.19.

It may be observed in figure 1.19, that the two excitation functions corresponding to the same decay channel almost match with each other, indicating that the incident channels have no influence on the strength of the decay channels. The small discrepancies in the two excitation functions for the same decay channel may be attributed to the fact that in the experiment only the excitation energy of the compound nucleus was kept the same by appropriately choosing the incident energies of the proton and alpha particles. However, the angular momentum distribution, isospin distribution, etc, of the compound nucleus formed by the two incident channels were different. Several attempts to verify the independence hypothesis were made carrying out experiments similar to that of Ghoshal but choosing different combinations of incident particles. The main problem with such experiments remains the same, the compound nuclei formed via different incident channels may have the same excitation energy but their other quantum mechanical parameters are different. As such, the CN formed through different channels are not exactly identical in quantum mechanical terms. Since the independence hypothesis is applicable only to quantum mechanically identical compound nuclei, there is always some discrepancy between the observed and expected results. Better agreement between the experimental and expected results may be achieved in some cases by applying corrections, calculated theoretically, for the differences in the compound nuclei.

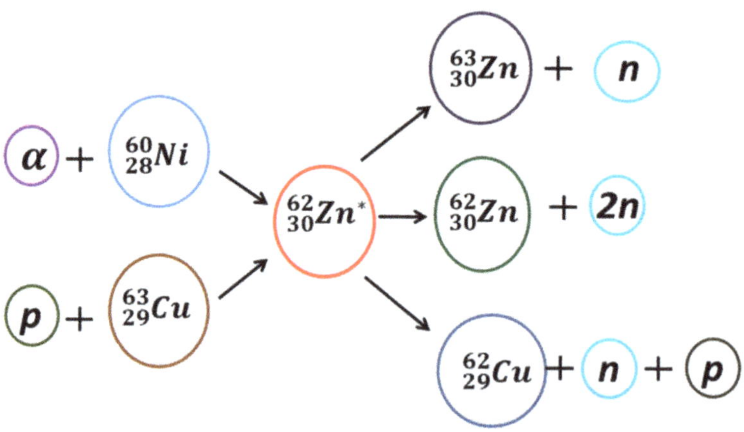

Figure 1.18. Schematic representation of Ghoshal's experiment.

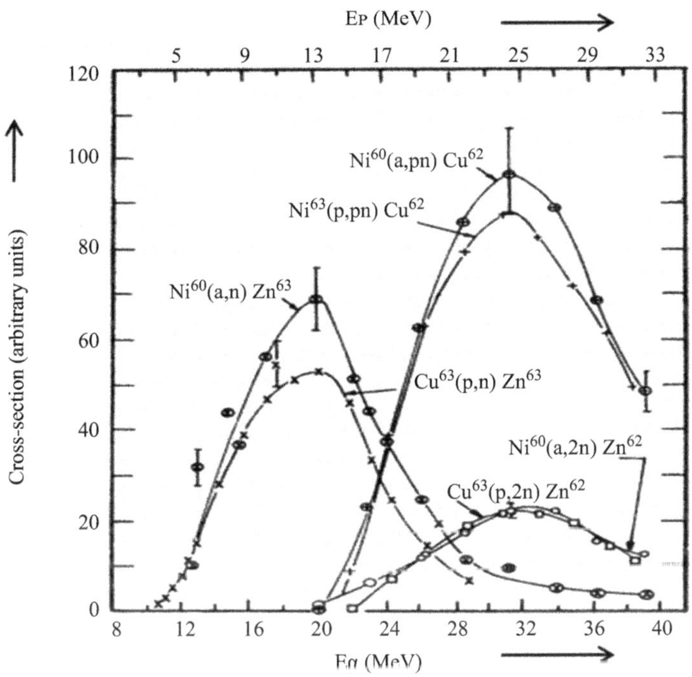

Figure 1.19. Measured excitation functions for three decay channels (adapted from [14] with permission from American Physical Society).

1.17 Particle spectra

Let us consider a reaction where a target nucleus 'A' is hit by an energetic projectile 'a' such that the compound nucleus 'C' formed by the absorption of projectile 'a' by target 'A' is excited to the continuum of excited levels. If an ejectile 'b' is now emitted from the excited compound nucleus that it leaves the residual nucleus 'B' in the continuum of excitation, then it can be shown that the cross-section per unit energy interval of the outgoing particle is given as,

$$\frac{\partial \sigma_{\beta\alpha}}{\partial E_\beta} = \sigma_\alpha^{CN} R_\beta(E_\beta) w_B(U_B) \tag{1.17}$$

Here, $w_B(U_B)$ is the level density of the residual nucleus 'B' at excitation energy U_B, such that,

$$U_B = U_B^{\text{Max}} - E_\beta \text{ where } U_B^{\text{Max}} \text{ is the maximum value of } U_B \tag{1.18}$$

The energy spectrum of particle 'b' is determined by the factor $R_\beta(E_\beta)w_B(U_B) = \frac{T_\beta}{\sum_\gamma T_\gamma} w_B(U_B)$. Since $\sum_\gamma T_\gamma$ does not depend on E_β, The energy spectrum of the ejectile is determined by $T_\beta w_B(U_B)$.

Since $T_\beta \sim E_\beta \sigma_\beta^{CN}(E_\beta)$, the energy spectrum of the ejectile is determined by

$$\frac{\partial \sigma_{\beta\alpha}}{\partial E_\beta} = \sigma_\alpha^{CN} R_\beta(E_\beta) w_B(U_B) \sim T_\beta w_B(U_B) \sim E_\beta \sigma_\beta^{CN}(E_\beta) w_B(U_B) \quad (1.19)$$

Here, $\sigma_\beta^{CN}(E_\beta)$ is the cross-section for the formation of the compound state through channel β. Equation (1.19) tells that the energy spectrum is determined by the cross-section $\sigma_\beta^{CN}(E_\beta)$ for the inverse process and the level density $w_B(U_B)$ of the residual nucleus B, (not of the CN).

The level density $w_B(U_B)$ is an exponential function of U_B. Expanding its logarithm around $U_B^{\max}$ and retaining only the first order terms, one gets,

$$w_B(U_B) = w_B(U_B^{\max})\left(e^{-\frac{E_\beta}{T_B}}\right) \quad (1.20)$$

Substituting this value of $w_B(U_B)$ from equation (1.20) back in equation (1.19) one gets;

$$\frac{\partial \sigma_{\beta\alpha}}{\partial E_\beta} = \sigma_\alpha^{CN} R_\beta(E_\beta) w_B(U_B) \sim E_\beta \sigma_\beta^{CN}(E_\beta) w_B(U_B) \sim E_\beta \sigma_\beta^{CN}(E_\beta)\left(e^{-\frac{E_\beta}{T_B}}\right) \quad (1.21)$$

Since $\sigma_\beta^{CN}(E_\beta)$ is a slowly varying function of energy E_β, the energy spectrum of the outgoing particle essentially depends on factor $E_\beta\left(e^{-\frac{E_\beta}{T_B}}\right)$. A distribution represented by $E_\beta\left(e^{-\frac{E_\beta}{T_B}}\right)$ is a typical Maxwellian distribution, as shown in figure 1.20. Energy distribution of molecules evaporated from a liquid at temperature T_B also has the same shape. An excited CN may, therefore, be compared to a liquid at temperature T_B and particles emitted from it may be said to have evaporated from it. This is called evaporation model. The temperature T_B may be determined from the analysis of the tail of the curve as it goes as $e^{-\frac{E}{T}}$.

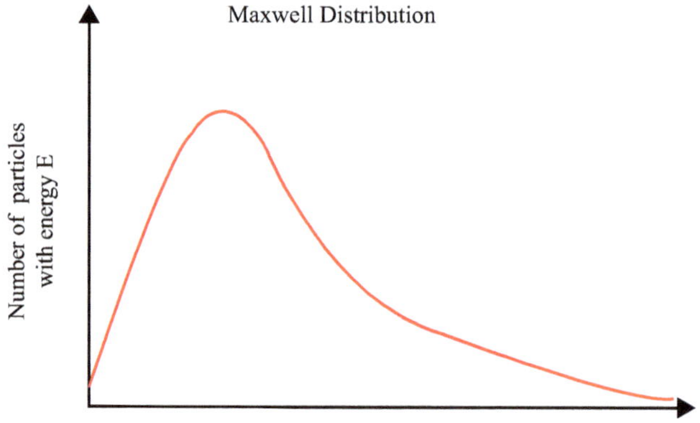

Figure 1.20. Maxwell distribution curve.

1.18 Experimentally measured energy spectrum of ejectiles

With the availability of accelerated beams of nuclear ions and high-resolution fast detectors with associated computerised data accusation, it become possible to carry out in-beam experiments to record the energy spectrum of emitted charged particles in nuclear reactions. As a result a large amount of experimental data [15, 16] on ejectile energy spectra accumulated during the last two decades or so. The general profile of measured ejectile energy spectrum is shown in figure 1.21.

The energy spectrum may be divided into three distinct parts: (i) a Maxwellian peak at low energy corresponding to the compound nuclear process with interaction time of the order of 10^{-16} s; (ii) the high energy side of the spectrum consisting of closely packed sharp peaks that may be assigned to the isolated levels of the residual nucleus and produced via direct reaction between the target and the projectile. The interaction time in this region is very small, of the order of 10^{-22} s, the time that the projectile takes to cross the target nucleus; (iii) the region lying between the compound and the direct reaction component encompassing interaction times from 10^{-22} to 10^{-16} s is termed as the pre-equilibrium or pre-compound reaction component. It is believed that this part of the spectrum originates from the processes that occur between the absorption of the projectile by the target and the establishment of thermodynamic equilibrium in the composite system. Both intuition and these experimental observations suggest that nucleons and or clusters of nucleons may be emitted from the equilibrating composite system at each step of equilibration, assumed to take place through two body residual interactions. Several theories, mostly semi-classical and a few totally quantum mechanical, have been proposed to explain pre-equilibrium component. Recently, pre-equilibrium emission of light particles has also been observed in reactions induced by heavy ions [17, 18].

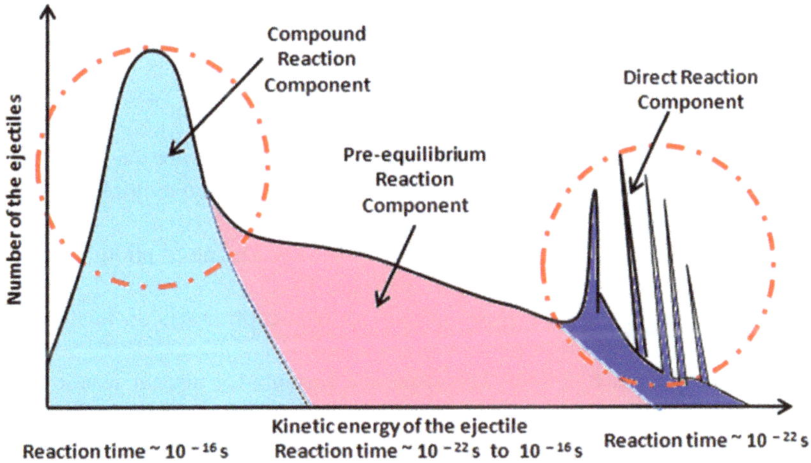

Figure 1.21. A typical energy spectrum of ejectiles emitted in statistical nuclear reactions.

1.19 Motivation

Interest in pre-equilibrium emission studies has been renewed recently when light particle emission prior to the establishment of equilibrium, in light and heavy ion induced reactions was found at considerably low energies where the compound reaction mechanism is expected to dominate. Generally, the total reaction cross-section in such reactions is taken to be the sum of the compound, pre-equilibrium and direct reaction cross-sections, neglecting all possible interference between them. Further, it is possible to reproduce the experimental data for an individual reaction or group of reactions by choosing free parameters of different codes used to describe the reaction mechanism, but no attempt has so far been made to use the same set of free parameters to analyse a large body of data obtained by different groups of experimentalists or that has been obtained through different experimental methods. Another important aspect is to compare the relative strengths and weaknesses of semi- aclassical and quantum mechanical theories of pre-equilibrium emission. With this view, a project to analyse a large body of existing data on pre-equilibrium/pre-compound emission using different codes describing various semi-classical and quantum mechanical reaction mechanisms in a consistent way has been undertaken. Results of this analysis with necessary background theories are presented in this text.

Bibliography

[1] Rutherford E 1911 The scattering of α and β particles by matter and structure of atom *Philos. Mag.* **6** 21

[2] Rutherford E 1919 Collision of a particle with light atoms. IV. An annomalous effect in nitrogen; *Lond. Edinb. Dublin Philos. Mag. J. Sci., 6th series* **37** 581

[3] Chadwick J 1932 Possible existence of a neutron *Nature* **129** 312
Chadwick J 1932 The existence of a neutron *Proc. Roy. Soc.* A **138** 692

[4] Fermi E, Amaldi E, D'Agostino O, Rasetti F and Segre E 1934 Artificial radioactivity produced by neutron bombardment *Proc. Roy. Soc. Ser.* A **146** 483

[5] Bohr N 1936 Neutron capture and nuclear constitution *Nature* **137** 344

[6] Veksler V 1945 A new method of accelerating of relativistic particles a new method of accelerating of relativistic particles *J. Phys. (USSR)* **9** 153

[7] McMillan E M 1945 The synchrotron—a proposed high energy particle accelerator *Phys. Rev.* **68** 143

[8] Lams H E and Salzberg B 1935 The secondary emission phototube *Proc. IRE* **23** 55

[9] Zworykin V K and Rajchman J A 1939 The electrostatic electron multiplier the electrostatic electron multiplier *Proc. IRE* **27** 558

[10] Morton G A, Mathesan R M and Greenblatt M H 1956 Design of photomultipliers for the sub-millimicrosecond region *IRE. Trans. Nucl. Sci* **3** 122

[11] Breit G and Wigner E P 1936 Capture of slow neutronscapture of slow neutrons *Phys. Rev.* **49** 519

[12] Kapur P L and Peierls R E 1938 The dispersion formula for nuclear reactions *Proc. Roy. Soc. Sec.* A **166** 277

[13] Feshbach H 1962 A unified theory of nuclear reactions *Ann. Phys.* **19** 287
Feshbach H 1958 Unified theory of nuclear reactions **5** 357

[14] Ghoshal S N 1950 An experimental verification of the theory of compound nucleus *Phys. Rev.* **80** 939

[15] Ahmad M, Brient C E, Egun P M, Grimes S M, Saraf S and Satyanarayana H 2017 Charged-particle emission in reactions of 9- and 11-MeV neutrons with ^{63,65}Cu *Nucl. Sci. Eng.* **95** 296

[16] Grimes S M, Haight R C, Alvar K R, Barschall H H and Borchers R R 1979 Charged-particle emission in reactions of 15-MeV neutrons with isotopes of chromium, iron, nickel, and copper *Phys. Rev.* C **19** 2127

[17] Sharma M K, Singh P P, Singh D P, Yadav A, Raj Sharma V, Bala I, Kumar R, Unnati, Singh B P and Prasad R 2015 Systematic study of pre-equilibrium emission at low energies in ^{12}C- and ^{16}O-induced reactions, *Phys. Rev.* C **19** 014603

[18] Abhishek Y *et al* 2011 Signature of pre-equilibrium-emission in forward-to-backward yield ratio measurement *Int. J. Mod. Phys.* E **20** 2133–42

B P Singh, Manoj K Sharma and R Prasad

Chapter 2

Characteristics of binary nuclear reactions

Large numbers of nuclear reactions are binary in nature with two nuclear entities interacting through nuclear force in the entrance channel resulting in two entities in the final channel. In most cases parameters of the entrance channel are decided by the experimenter, however, some parameters of the final channel may be obtained using the conservation laws. Classification of binary reactions, conservation laws obeyed in binary nuclear reactions and their application in obtaining exit channel parameters are discussed here.

2.1 Introduction

A nuclear reaction occurs when two or more nuclei or some other nuclear entities interact through nuclear forces. In general, composition and energies of interacting entities change during a nuclear reaction. Most nuclear reactions are binary in nature, i.e., there are two entities in the entrance channel and two in the exit channel. Since nuclei and most other nuclear entities, (except neutrons which were discovered by Chadwick [1] in 1932) are charged, kinetic energy needs to be imparted to at least one of the interacting entities so as to bring it within the range (of the order of few Fermi) of nuclear force of the target system. However, for neutrons, a nuclear reaction may take place even with thermal neutrons [2] (0.025 eV). It may, however, be added that gamma rays, that are electromagnetic radiations may also initiate nuclear reactions. In experimental study of nuclear reactions the projectiles are often obtained either from some radioactive sources or from the particle accelerators. Accelerators have the advantage that they provide mono-energetic beams of high flux and of desired energies. In spite of the fact that tremendous efforts have been made during the last almost 100 years or so to understand the properties of nuclear force, the fact is that exact and complete knowledge of nuclear force is still missing. Therefore, theoretical models of nucleus and nuclear processes are made and experiments are done to compare the prediction of theoretical models with the results of experiments to develop insight into nuclear reaction dynamics. Further,

doi:10.1088/978-0-7503-5077-8ch2

the predictive power of the theoretical models may also be tested, so as to refine the theory. The study of accelerator-based nuclear reactions has become a powerful tool to study the properties and structure of nuclei. The nuclear reactions may be classified in a variety of ways. A description of such classification is presented here.

2.2 Classification of nuclear reactions

Nuclear reactions [3] may be classified in different ways. It is, however common to classify reactions on the basis of the kinetic energy E of the incident particle. The typical classification may be given as:

(a) Very low energy reactions; $0< E < 1$ keV;
(b) Low energy reactions; 1 keV $< E <$ 500 keV;
(c) Intermediate energy reactions; 500 keV $< E <$ 10 MeV;
(d) High energy reactions; 10 MeV $< E <$ 50 MeV;
(e) Very high energy reactions; $E>$ 50 MeV.

The above classification is, generally, given if the incident particles involved are gamma radiation, neutrons, protons, deuterons or alpha particles.

It is also possible to classify nuclear reactions on the basis of the mass (or mass number A) of either the incident or the target ion.

(i) Light nuclei ; $1 < A < 25$;
(ii) Intermediate nuclei; $25 < A< 80$;
(iii) Heavy nuclei; $80 < A < 250$, or so.

As may be observed, the above-mentioned classifications give very broad categorization of different types of reactions without any clear cut demarcation. These classifications are generally used for managing a large amount of global data in various groups and for studying systematic trends etc.

A typical experimental set-up for the study of a nuclear reaction consists of an energetic nucleus/ion/particle in the incident channel with a stationary target. However, in the outgoing or exit channel there may be one, two, three or more nuclei/particles. Since most of the nuclear reactions are binary in nature, we assume only two particles in the exit channel. A typical binary nuclear reaction is represented as;

$$a + {}^{A}_{Z}X \rightarrow {}^{A'}_{Z'}Y + b$$

where, 'a' is incident nuclear particle, ${}^{A}_{Z}X$ the target nucleus, while 'b' is the emitted particle or 'ejectile' and ${}^{A'}_{Z'}Z$ is the residual nucleus. The symbols X and Y are the abbreviations for the chemical symbols for atoms the nuclei of which take part in the reaction. The above reaction may be represented in abbreviated form as ${}^{A}_{Z}X(a, b){}^{A'}_{Z'}Z$. The residual nucleus Y, in general, is a heavy nucleus and emitted particle b is a lighter nucleus/nuclear particle or gamma ray. Since, the chemical symbol (X) has inherent definite atomic number, therefore, the atomic numbers (Z) of interacting nuclei are, generally, not written in nuclear reactions. Thus nuclear reaction plays an important role in artificial transmutation [4, 5]. In most of the experiments the lighter particles of type '*b*' are detected with the help of a suitable detector. Some of the common types of the nuclear reactions are given below;

(i) **Elastic scattering:** In such a reaction, the incident particle '*a*' and the target nucleus '*X*' remain in their initial quantum states, the incident particle, however, may change its direction of motion as a result of interaction. Such a reaction may be represented as;

$$a + {}^{A}_{Z}X \rightarrow {}^{A}_{Z}X + a$$

and may be symbolically represented as $X(a,a)X$.

(ii) **Inelastic scattering:** In inelastic scattering, though the ejectile and the residual nucleus are the same, respectively, as the incident particle and the target nucleus, just like in elastic scattering, the energy states of both of them get altered due to interaction between them. The incident particle may give some of its energy to the target leaving it in an excited state or may gain some energy from the target to bring it down to a state of lower excitation. Such a reaction may be represented as;

$$a + {}^{A}_{Z}X \rightarrow {}^{A}_{Z}X^{*} + a'$$

In the above equation, the ${}^{A}_{Z}X^{*}$ represents the target nucleus in an excited state while a' represents the outgoing particle which is the same as the incident particle but with different energy. The above reaction may be symbolically represented as $X(a,a)X^{*}$. An example of such a reaction is ${}^{16}O(p,p){}^{16}O^{*}$.

(iii) **Direct reactions:** The direct reactions [6] generally take place at higher energies and are further classified as stripping and pick-up reactions. In the case of stripping reaction, a part of the incident particle gets stripped off and fuses with the target nucleus, while the remaining part goes on moving in the forward direction. An example of such a reaction is given below;

$$d + {}^{39}K \rightarrow {}^{40}K + p$$

Since, the incident deuteron is a combination of a neutron and a proton, the neutron gets stripped off and fuses with target nucleus ${}^{39}K$ forming ${}^{40}K$, while the remaining part, i.e., proton goes on moving in the forward direction. Such a reaction may be represented as ${}^{39}K(d,p){}^{40}K$.

On the other hand, in the case of pick-up reaction the incident particle picks up a nucleon of a group of nucleons from the target nucleus and moves in the forward direction. An example of such a reaction may be given by;

$$p + {}^{40}K \rightarrow {}^{39}K + d$$

Such a reaction may be represented as, ${}^{40}K(p,d){}^{39}K$. It may be remarked that the direct reactions are likely to occur at higher energies where interaction time is very short, typically of the order of the time taken by a nucleon to traverse the nuclear diameter, which is $\approx 10^{-21}$ s.

(iv) **Radiative capture reaction:** In such reactions [7] the incident particle fuses with the target nucleus forming an excited composite nucleus, which then decays by the emission of a gamma ray. An example of such a reaction may be given as;

$$p + {}^{27}_{13}\mathrm{Al} \rightarrow \left({}^{28}_{14}\mathrm{Si}\right)^{*} \rightarrow {}^{28}_{14}\mathrm{Si} + \gamma$$

Such a reaction may be represented as ${}^{27}\mathrm{Al}(p,\gamma){}^{28}\mathrm{S}$.

(v) **Photo disintegration reaction:** In such a reaction the incident particle is a photon that transfers energy to the target nucleus which finally emits a nuclear particle. An example of such a reaction is given below;

$$\gamma + {}^{2}_{1}\mathrm{H} \rightarrow {}^{1}_{1}\mathrm{H} + n$$

Such a reaction may be represented as ${}^{2}\mathrm{H}(\gamma,n){}^{1}\mathrm{H}$. Since the binding energies of nucleons in the target nuclei are rather large, only high energy photons may initiate photo disintegration.

(vi) **Compound nucleus reactions**: In such reactions [8] the incident particle is absorbed by the target nucleus forming a compound nucleus; the energy and momentum of the incident particle are shared among all the nucleons of the composite system till thermal equilibrium is established. The compound nucleus is usually formed in an excited state. In this compound system, by chance, energy more than the separation energy is concentrated on one or a group of nucleons which then get ejected. An example of such a reaction may be given as,

$$p + {}^{63}_{29}\mathrm{Cu} \rightarrow \left({}^{64}_{30}\mathrm{Zn}\right)^{*} \rightarrow {}^{68}_{30}\mathrm{Zn} + n$$

Such a reaction may be represented as ${}^{63}\mathrm{Cu}(p,n){}^{63}\mathrm{Zn}$. In compound nucleus reactions, since the sharing of energy takes place among a large number of nucleons, the establishment of equilibrium requires a sufficiently long time, hence these reactions are said to involve timescales $\approx 10^{-16}$ s.

(vii) **Many body reactions:** In such reactions more than two particles are observed in the outgoing channel. An example of such a reaction may be given as;

$$p + {}^{18}_{8}\mathrm{O} \rightarrow {}^{17}_{8}\mathrm{O} + p + n$$

The above reaction may be abbreviated ${}^{18}\mathrm{O}(p,pn){}^{17}\mathrm{O}$.

(viii) **Nuclear fission reaction:** These reactions are interesting, as in the case of fission [9] the incident particle is absorbed by the target nucleus forming a composite system which breaks into nearly two equal or comparable fragments, emitting a few nucleons and energy. An example of such a reaction is the following thermal neutron induced fission,

$${}^{1}_{0}n + {}^{235}_{92}\mathrm{U} \rightarrow {}^{136}\mathrm{Te} + {}^{98}\mathrm{Zr} + 2n + \mathrm{Energy}$$

The above fission reaction may be represented as $^{235}U(n,f)$ in short. An interesting property of fission is that the exit channel is not fixed, that is, several combinations of fission fragments with a different number of neutrons may be emitted as a result of fission of a given nucleus. Although fission in heavy nuclei may be initiated by high energy charged particles, in some cases it may also be initiated by thermal neutrons as well. Fission reactions are of importance, as a large amount of energy is released in such reactions and, therefore, these reactions are used in nuclear reactors for energy generation.

(ix) **Heavy ion complete fusion reactions:** Nuclear reactions initiated by ions heavier than alpha particle are termed as heavy ion reactions. Heavy ion reactions, in general require higher incident energy because of the large Coulomb barrier between the target and incident heavy ions. Broadly speaking, heavy ion reactions may be classified [10] into two types: the complete fusion and the incomplete fusion types. In complete fusion reactions the incident heavy ion merges with the target nucleus forming a composite nucleus which may decay by emitting nuclear particles. An example of such a reaction is given below;

$$^{16}_{8}O + ^{159}_{65}Tb \rightarrow ^{173}_{73}Ta + 2n$$

which may be represented as, $^{159}Tb(^{16}O,2n)^{173}Ta$. Since the projectile completely fuses with the target nucleus, these reactions are said to lead to the complete momentum transfer of incident ion to the composite nucleus.

(x) **Heavy ion incomplete fusion reaction:** In the case of heavy ion induced reactions, when the projectile reaches within the target nuclear field, it may break up into two parts, one of the fragments fuses with the target nucleus, forming a composite system. The other fragment may move, as a spectator without any further interaction. The composite system, formed by the incomplete fusion [10] of the incident ion, then decays by usual modes of de-excitation. For example, if the projectile is $^{16}_{8}O$, it may break up into $\alpha + ^{12}_{6}C$, and one of these fragments e.g., $^{12}_{6}C$ may fuse with the target nucleus. Such an incomplete fusion reaction may be represented as;

$$\left(\alpha + ^{12}_{6}C\right) \text{ or } \left(^{16}_{8}O\right) + ^{159}_{65}Tb \rightarrow ^{167}_{69}Tm + \alpha 2n$$

The short notation for the above reaction is $^{159}Tb(^{16}O,\alpha\, 2n)^{173}Ta$. Since in these reactions only a part of the projectile (and not the complete projectile) fuses with the target nucleus, these reactions lead to incomplete momentum transfer.

(xi) **Multi-fragmentation reaction:** In such reactions [11], very high energy nuclei collide with other nuclei creating nuclear matter under unusual conditions of density and excitation energy. The states so obtained may be in a different phase from normal nuclei which may lead to the characteristic of matter during early stages of the creation of the Universe.

(xii) **Spallation reaction:** Spallation reactions [12] are generally initiated by light projectiles (proton, neutron, or light nucleus) with the high kinetic energy from several hundreds of MeV to several GeV. When such high energy incident nuclear particles impinge on some heavy material like lead (Pb), a large number of neutrons and secondary charged particles are emitted.These reactions have become of interest in recent years due to their relevance in recently proposed accelerator driven subcritical reactors.

(xiii) **Coulomb excitation:** Since both the incident charged ion and the target ion have long range electric field, it is possible for the incident particle to excite the target nucleus electromagnetically. This is referred to as Coulomb excitation [6], or in short as Coulex. Such reactions are important when the incident particle is a heavy ion. In such a process the rotational bands in deformed target nuclei may be excited. The cross-sections for such reactions are generally large, as such these are suitable for studying the structure of exotic nuclei with radioactive ion beams where the intensities are quite low.

2.3 Conservation laws for nuclear reactions

Let us consider a general binary nuclear reaction represented by

$$^{A_a}_{Z_a}a + ^{A_x}_{Z_x}X = ^{A_b}_{Z_b}b + ^{A_Y}_{Z_Y}Y$$

where a projectile 'a' with kinetic energy 'E_a' 'mass' 'M_a' atomic number Z_a, atomic mass number A_a and intrinsic spin s_a impinges on a target nucleus X (mass M_X, atomic number Z_X, atomic mass number A_X and intrinsic spin s_X). For generality, it may be assumed that the target nucleus is also moving with a kinetic energy E_X. The interaction results in the residual nucleus Y (of kinetic energy E_Y, mass M_Y, atomic number Z, atomic mass number A, and intrinsic spin s) and the ejectile b having kinetic energy E_b, mass M_b, atomic number Z_b, atomic mass number A_b and intrinsic spin s_b. It may, however, be noted that in most of the experimental set-ups $E_X = 0$ as the target is stationary, however, as already mentioned it is assumed for generality that the target also has a kinetic energy E_X.

Some of the important conservation laws which are obeyed by binary nuclear reactions are listed below;

(i) **Conservation of total energy:** In a nuclear reaction the total energy of the entrance channel and the outgoing channels should remain conserved, i.e., total energy of reactants minus the total energy of product particles must be equal to zero. Thus, the sum of kinetic energies and the rest mass energies of the nuclei involved in the entrance channel should be equal to the sum of kinetic energies and rest mass energies of the exit channel nuclei. The above energy conservation rule may be written as;

$$(E_a + M_a c^2) + (E_X + M_X c^2) = (E_Y + M_Y c^2) + (E_b + M_b c^2) \tag{2.1}$$

where, E_a, E_X, E_Y, E_b are the kinetic energies of respective nuclei/particles while $M_a c^2$, $M_X c^2$, $M_Y c^2$, $M_b c^2$ are the rest mass energies, respectively, of particle, target X, residual nucleus Y and the ejectile b.

(ii) **Conservation of linear momentum:** In a nuclear reaction the total linear momentum of the nuclei/particles in the entrance channel before collision and of the nuclei/particles in the exit channel observed after the collision are conserved. This conservation means that;

$$M_a V_a + M_X V_X = M_Y V_Y + M_b V_b \tag{2.2}$$

Here $M_a V_a$, $M_b V_b$ are the linear momentum of the incident and the outgoing particles while $M_X V_X$ and $M_Y M_X$ are the linear momentum of the target and the residual nucleus. It is needless to say that the linear momentum (MV) of a particle of mass M and velocity V is related to the kinetic energy E through the relation, MV = $\sqrt{2ME}$.

(iii) **Conservation of atomic mass number:** The sum of atomic mass numbers of nuclei/particles involved in the entrance channel should be equal to the sum of atomic mass numbers of nuclei/particles in the exit channel. Since, the atomic mass number (A) gives information about the number of nucleons; it is also referred to as the conservation of nucleon number. As such, ΣA is conserved meaning thereby that the value of ΣA is the same before and after the reaction. Mathematically one may write,

$$A_a + A_X = A_b + A_Y \tag{2.3}$$

(iv) **Conservation of charge:** The sum of atomic numbers of nuclei/particles involved in the entrance channel should be equal to the sum of atomic numbers of nuclei/particles in the exit channel. Since the atomic number (Z) is related to the charge of the nucleus ($+Ze$), this law is also referred to as the law of charge conservation. As such, the total charge is also conserved before and after the reaction, i.e.,

$$Z_a + Z_X = Z_b + Z_Y \tag{2.4}$$

(v) **Conservation of total angular momentum:** The law of the conservation of total angular momentum demands that the sum of the intrinsic spins of the reactants plus the angular momentum of their relative motion in entrance channel should be equal to the sum of the intrinsic spins of the final products plus the angular momentum of their relative motion in exit channel.

The total angular momentum of the incident channel J_i may be written as,

$$J_i = s_a + s_X + \ell_{aX} \tag{2.5}$$

where $\ell_{aX}$is the angular momentum of the relative motion of a and X. Similarly, the total angular momentum, $J_f$of the exit channel may be written as,

$$J_f = s_b + s_Y + \ell_{bY} \tag{2.6}$$

Here ℓ_{bY} is the angular momentum corresponding to the relative motion of b and Y.

According to the law of total angular momentum conservation,

$$J_i = J_f \text{ or } s_a + s_X + \ell_{aX} = s_b + s_Y + \ell_{bY} \tag{2.7}$$

(vi) **Conservation of parity:** According to quantum mechanics all subatomic particles, systems and processes may be assigned a quantum number, called parity and denoted by π, that is related to the mirror symmetry of the system. Quantum mechanically, a system is defined in terms of its wave function. If a wave function and its mirror image are identical, the wave function or the system is said to have positive or even parity, if not then negative or odd parity. Parity may have only two values; +1 or −1, i.e., either even or odd. The total parity of a system is equal to the multiplication of the parities of individual particles in the system and the parity of the process involved. If for example, particle a with parity π^a and nucleus X with parity π^X are in a state of relative motion of angular momentum ℓ_{aX}, in the entrance channel, then the parity of the entrance channel π^i may be given as,

$$\pi^i = \pi^a \pi^X \pi^{\ell_{aX}} = \pi^a \pi^b (-1)^{\ell_{aX}} \tag{2.8}$$

Here $(-1)^{\ell_{aX}}$ is the parity of the relative motion of the two particles a and X. It may be observed that for states of even angular momentum values ($0\hbar$, $2\hbar$, $4\hbar$, …) of relative motion, the parity of the relative motion will be even and for odd values of relative angular momentum ($1\hbar$, $3\hbar$, $5\hbar$, …) parity will be odd.

Therefore, the law of conservation of parity in strong nuclear reactions demands that:

$$\pi^i = \pi^f \text{ or } \pi^a \pi^X (-1)^{\ell_{aX}} = \pi^b \pi^Y (-1)^{\ell_{bY}} \tag{2.9}$$

Let us now consider a hypothetical reaction where both the target and the projectile have even parity and the angular momentum of their relative motion is, say, $3\hbar$. Further, the residual nucleus has even parity and the ejectile odd parity. Therefore, the parity of the entrance channel $\pi^i = (+1)(+1)(-1)^3 = -1$ or odd. Similarly, the parity of the exit channel $\pi^f = (+1)(-1)(-1)^{\ell_{Yb}} = (-1)(-1)^{\ell_{Yb}}$. In order to conserve parity the angular momentum of relative motion in exit channel ℓ_{Yb} should be even, it may have values $0\hbar$, $2\hbar$, … and so on. Thus the law of conservation of parity often decides the state of relative motion of particles in the exit channel. It may, however, be remarked that parity is conserved only in strong nuclear interactions. It is not conserved in weak nuclear interactions like the beta decay.

(vii) **Conservation of isotopic spin:** Isospin or isobaric spin or isotopic spin is a property that is characteristic of families of related subatomic particles differing essentially in values of their electrical charge. Prior to the discovery of quarks, the concept of isospin originated from the fact that both the neutron and the proton behave exactly in identical manner in a nucleus, except for the charge. It appears as if the neutron and the proton are the two states of the same particle, called nucleon, which split into two when an electromagnetic field is switched on. The families of similar particles are known as isospin multiplets; doublets have two members, triplets three and so on. Neutron and proton form an isospin doublet. Like intrinsic spin, neutrons and protons have half integer isospin; the difference is while intrinsic spins are measured in units of $\hbar$ the isospin is dimensionless. Since the discovery of quarks it has been established that the isospin of a nuclear particle depends on its quark structure, quarks have an isospin vector T that has a value 1/2. The third component T_3 for an up-quark is taken as +1/2 and for a down-quark as −1/2. Therefore, in general for hadrons,

$$T_3 = \frac{1}{2}\left(N_{\text{up}} - N_{\text{down}}\right) \tag{2.10}$$

Here N_{up} and N_{down} are, respectively, the number of up-quarks and the down-quarks in the hadron. As such, the values of T_3 assigned to proton and neutron are, respectively, +1/2 and −1/2, while for a typical nucleus ${}^{A}_{Z}\mathrm{X}_N$, it is $\frac{1}{2}(Z - N)$. It may be realized that the assignment of the positive sign to the isotopic spin projection T_3 of a proton is quite arbitrary, but now it has become a convention. A given Z and T_3 combination identifies a specific isotope associated with Z, while a given A and T_3 combination identifies a specific isobar associated with A. The third component T_3 of isospin is conserved in an interaction initiated by strong nuclear force.

Let us consider the following binary nuclear reaction,

$${}^{16}_{8}\mathrm{O} + {}^{159}_{65}\mathrm{Tb} \rightarrow {}^{173}_{73}\mathrm{Ta} + 2n$$

It is easy to calculate the isospin projection T_3 for each individual nucleus as;

$$T_3\left({}^{16}_{8}\mathrm{O}\right) = 0;\; T_3\left({}^{159}_{65}\mathrm{Tb}\right) = -14.5;\; T_3\left({}^{173}_{73}\mathrm{Ta}\right) = -13.5 \text{ and } T_3(2n) = -1$$

Conservation of T_3 in the reaction can be easily verified from the above data.

It may be remarked that some of these conservation laws may assume different forms, particularly in very high energy reactions where new elementary particles are likely to be produced. The expression for kinetic energy $\frac{1}{2}mv^2$ was deduced assuming that mass of a particle is independent of its velocity. But according to Einstein's special theory of relativity we have seen that it is not true. As an example, the relativistic increase in mass of a proton is about 1% for kinetic energy 10 MeV. In fact, for accurate analysis with moving particles one must use the relativistic mass rather than rest mass. This is particularly important in view of the high-energy reactions where particles are accelerated to high energies using the accelerators.

2.4 Quantities not conserved in nuclear reactions

Two important quantities that are not conserved in nuclear reactions are the magnetic moment (μ) and the electric quadrupole moment (q). In fact, these quantities depend on the mass, charge and current distributions inside the nuclei involved in the reactions and therefore are not conserved. Further, conservation of total number of nucleons in nuclear reactions does not necessarily imply that mass is conserved.

In the following table some of the nuclear reactions are given along with what kind of information is obtained by studying them.

Type of nuclear reaction	What is studied?
Elastic scattering	Size of nucleus and interaction potential, energy levels, quantum numbers, giant resonances in nuclei
Direct reactions	Importance and details of shell model
Fusion reactions	Astrophysical process, generation of nuclear data relevant to reactor technology
Fission reactions	Nuclear energy generation, suitability of liquid drop model
Compound nucleus reactions	Properties of statistical nuclear reactions
Multi-fragmentation	To investigate the phase transformation in nuclear matter, and study the collective model of nuclei
Spallation reactions	Used to produce high energy neutrons that may be used in accelerator driven subcritical reactors

2.5 Nuclear reactions versus chemical reactions

As we are familiar, chemical reactions involve atoms of elements and are brought about by the exchange of valence electrons through electromagnetic interactions. On the other hand, nuclear reactions are initiated by nuclear forces. During a chemical reaction there may be transfer, loss, gain, and sharing of electrons and at the same time there is no effect on the nucleus of the atom.

In the case of nuclear reactions, some properties of the interacting nuclei, like their energies or their compositions or both undergo change. Change of the composition of the nuclei results in the formation of new elements. The nuclear reactions may also involve the emission of alpha, beta and gamma radiation. On the other hand in the case of chemical reactions there is only a rearrangement of electrons and they do not involve any changes in the structure of the nuclei of the interacting atoms. Further, the rates of chemical reactions are generally influenced by external factors like pressure, temperature and sometimes due to the presence of catalysts used. On the other hand, nuclear reactions are not affected by these factors.

A very major difference between chemical and the nuclear reactions is the conversion of mass into energy and vice versa in nuclear reactions. Moreover, the binding energies of nucleons in atomic nuclei are a few orders of magnitude larger than the binding energies of electrons in atoms. Therefore, energy changes involved in the case of nuclear reactions are many times larger than those in chemical reactions. These energy changes come from the inter conversion of mass and energy.

2.6 Energetics of nuclear reactions

It is obvious that only those nuclear reactions that obey all the above-mentioned conservation laws may take place in Nature. However, adherence to the conservation laws does not guarantee that a particular reaction will take place. Apart from the conservation laws, whether a reaction will actually take place or not is also decided by the balance of total energy (mass energy + kinetic energy) between the entrance and the exit channels, i.e., the energetic of the reaction. To explain this, for the sake of simplicity let us again consider the binary reaction of the form;

$$^{A_a}_{Z_a}a + ^{A_X}_{Z_X}X \rightarrow ^{A_Y}_{Z_Y}Y + ^{A_b}_{Z_b}b \tag{2.11}$$

where, a is incident particle, X the target nucleus, b the emitted particle and Y is the residual nucleus. Applying the law of conservation of total energy, we may write;

$$(M_ac^2 + E_a) + (M_xc^2 + E_X) = (M_Yc^2 + E_Y) + (M_bc^2 + E_b) \tag{2.12}$$

Here, M_a, M_X, M_Y, M_b represent the masses while, E_a, E_X, E_Y, E_b represent the kinetic energies, respectively, of the incident particle a, target nucleus X, recoiling residual nucleus Y and the outgoing particle b. Collecting terms containing mass energy on the left-hand side and terms containing kinetic energies on the right-hand side, one gets;

$$(M_ac^2 + M_xc^2) - (M_Yc^2 + M_bc^2) = (E_Y + E_b) - (E_a + E_X) \tag{2.13}$$

The left-hand side of equation (2.13) represents the difference of rest mass energies of the entrance channel and the exit channel. On the other hand the right-hand side represents the difference of kinetic energies in the exit channel and the incident channel. The right-hand side in equation (2.13) defines the Q-value of the reaction. The Q-value represents the 'energy balance' of the reaction and is given by;

$$Q \equiv \left[(M_ac^2 + M_Xc^2) - (M_bc^2 + M_Yc^2)\right] = [(E_Y + E_b) - (E_a + E_X)] \tag{2.14}$$

Depending on whether the mass difference of the entrance channel and the exit channel partners is positive or negative, the Q-value will be positive or negative. Since the Q-value is the difference of rest mass energies, it has the same value in both the laboratory and the centre-of-mass frames.

(a) **Reactions with positive Q-value**
In the case, when Q is positive, $(M_ac^2 + E_Xc^2) > (E_Yc^2 + M_bc^2)$ and $(E_Y + E_b) > (E_a + E_X)$, the reaction is energetically allowed. Energy is liberated in the reaction and the liberated energy appears in the form of the kinetic energies of the reaction products

in the exit channel. Such reactions are called **exoergic** reactions. It may be observed from equation (2.14) that the sum of the kinetic energies of the residual nucleus and the emitted particle in the exit channel is larger than the sum of the kinetic energies of the target nucleus and the incident particle in the entrance channel. This excess kinetic energy is produced by the conversion of mass into energy in the nuclear reaction. In most experiments carried out using accelerated beams, the target is at rest and as such only the kinetic energy of incident particle matters and the energy of the target nucleus $E_Y$is taken as zero. The above condition, $(M_a c^2 + M_X c^2) > (M_Y c^2 + M_b c^2)$, means that on the basis of energy considerations the reaction can occur. But it does not necessarily mean that reaction will definitely take place. For the reaction to occur, an additional condition, not contained in Q-value, is the need to bring the two nuclei in the entrance channel close to each other so that they come within the range of each other's nuclear forces. Needless to say, that for a nuclear reaction to occur, the following conditions must be fulfilled:

(i) It should be energetically possible, i.e., either the Q-value of the reaction is positive or at least a little more than zero or if Q-value is negative, then the sum of the kinetic energies of interacting partners in the entrance channel must be larger than the magnitude of the negative Q-value.

(ii) The interacting partners in the entrance channel must be within the range of mutual nuclear force.

(iii) The reaction conforms to all conservation laws.

(b) **Reactions with negative Q-value**

In the case where the Q-value of the reaction is negative, $(M_a c^2 + M_X c^2) < (M_Y c^2 + M_b c^2)$ and $(E_Y + E_b) < (E_a + E_X)$, the reaction is **endoergic**. In an endoergic reaction the sum of mass energies of the residual nucleus and the emitted ejectile is more than the sum of the mass energies of the target and the projectile, which means that mass or matter is created in such reactions. Therefore, endoergic reactions can take place only if the total kinetic energy in the entrance channel (KE of projectile + KE of target) provides for this additional mass created in the reaction and compensates for deficit of mass difference. At first sight it appears that in the case of endoergic reaction, total kinetic energy equal to $|-Q|$ (which is the mass energy difference) is enough to initiate the reaction. However, it is not true. The reason is simple, the total input kinetic energy of the incident channel is not available for creation of additional mass; a part of this energy is used up in imparting linear momentum to the products in the exit channel which is essential for the conservation of linear momentum. The minimum energy required to initiate an endoergic reaction is called threshold energy of the reaction and can be calculated using simple energy and momentum conservation laws.

Let us assume that the reaction shown by equation (2.11) has a negative Q-value of magnitude, say, Q_1. In order to calculate the threshold energy for this endoergic reaction, let us further assume that the target X is stationary and the projectile a impinges on X with velocity V_a. Let the reaction proceed in two steps. In the first step the target X and projectile a fuse to form a composite nucleus say, C, of mass $M = (M_a + M_X)$ that moves with velocity V. In the second step, the composite nucleus C decays by emitting residual nucleus Y and ejectile b, as indicated in equation (2.15)

$$a + X \rightarrow C \rightarrow Y + b \tag{2.15}$$

Since the target is stationary the linear momentum of the entrance channel is simply (M_a, V_a). This should be equal to the linear momentum of the composite nucleus C, i.e.;

$$(M_a V_a) = MV \tag{2.16}$$

From equation (2.16) the velocity V of the composite nucleus C is given as,

$$V = \frac{M_a}{M} V_a \tag{2.17}$$

Now the kinetic energy E_c of the composite nucleus C may be given by,

$$E_c = \frac{1}{2}MV^2 = \frac{1}{2}M\left(\frac{M_a}{M}V_a\right)^2 = \frac{1}{2}\frac{M_a^2 V_a^2}{M} \tag{2.18}$$

It may be noted that the kinetic energy available in the entrance channel is only $(E_a - E_c)$, not E_a. The surplus kinetic energy available in the entrance channel $(E_a - E_c)$ should be equal to the mass deficit between the entrance and exit channels, i.e., $(E_a - E_c)$ must be equal to the magnitude of the negative Q-value, for the reaction to proceed. Hence,

$$(E_a - E_c) = |-Q_1| \tag{2.19}$$

Or

$$\left[\frac{1}{2}M_a V_a^2 - \frac{1}{2}\frac{M_a^2 V_a^2}{M}\right] = |-Q_1| \tag{2.20}$$

Or

$$\frac{1}{2}M_a V_a^2\left(1 - \frac{M_a}{M}\right) = |-Q_1| \tag{2.21}$$

Putting $M = (M_X + M_a)$ in equation (2.21), one gets

$$\begin{aligned}\frac{1}{2}M_a V_a^2 &= |-Q_1|\left(1 - \frac{M_a}{M}\right)^{-1} = |-Q_1|\left(\frac{M_X}{M_a + M_X}\right)^{-1} \\ &= |-Q_1|\left(\frac{M_a + M_X}{M_X}\right)\end{aligned} \tag{2.22}$$

Equation (2.22) gives the minimum value of the kinetic energy of the projectile a in laboratory frame (where target is stationary) that will just be sufficient to compensate for the generation of additional mass in the exit channel and will also provide for the conservation of linear momentum. The kinetic energy of projectile given by equation (2.12) is called the **threshold energy**.

$$\text{Threshold energy} = |-Q_1|\left(\frac{M_a + M_X}{M_X}\right) = |-Q_1|\left(1 + \frac{M_a}{M_X}\right) \tag{2.23}$$

It may be observed in equation (2.23) that the threshold energy may be split into two parts:

$$\text{Threshold energy} = |-Q_1| + |-Q_1|(M_a/M_X)$$

The component $|-Q_1|$ of the threshold energy is used up in generating additional mass in the exit channel, while the part $[|-Q_1|(M_a/M_X)]$ is consumed in providing linear momentum to the reaction product so as to conserve the input linear momentum.The magnitude of the threshold energy is slightly more than the magnitude of the negative Q-value. It is expected also as a part of the projectile energy gets used up in conserving the input linear momentum and the remaining energy component gets converted into mass. It is also obvious that in the case where a negative Q-value reaction is initiated by its threshold energy, the final products, residual nucleus Y and ejectile b, will be emitted with zero kinetic energy in laboratory frame. As such, it is possible to calculate the threshold energy for a given reaction if its Q-value and masses of target and projectile are known. It may further be noted that nuclear masses are used in derivations of these equations. However, in actual practice one may use masses of neutral atoms, or even atomic mass number A of nuclear particles, instead of nuclear masses without any significant error.

(c) ***Q*-equation**

Having defined the Q-value for a nuclear reaction, the next question is how to determine experimentally the Q-value for a specific reaction. The experimental set-up for studying nuclear reactions in most cases consists of an accelerator that provides the accelerated collimated beam of desired ion of desired energy. The accelerated ion beam is made to hit a stationary target of desired material. Since the energy of incident ions is sufficient to overcome the coulomb barrier between the target and the incident ion, the projectile ion reaches close enough to experience the nuclear field of the target nucleus and, therefore, a nuclear interaction takes place. As a result of nuclear interaction between the nucleons of the projectile ion and the target nucleus, rearrangement of nucleons both of the projectile and the target takes place along with redistribution of their energies. This rearrangement of nucleons and redistribution of energies etc, i.e., the nuclear reaction, occurs in extremely short time, estimated to be of the order of 10^{-21} s to 10^{-16} s. Since, with present day technology it is not possible to obtain time resolution of the order of the nuclear interaction times; it is experimentally not possible to decipher the nuclear reaction process. However, it is possible to detect and record the energy, type, mass etc of the final products of the reaction, i.e., of the residual nucleus and the ejectile using appropriate detectors.

In order to experimentally determine the Q-value for a reaction, let us consider a nuclear reaction of the type shown by equation (2.15), where a projectile a of mass M_a, velocity V_a (kinetic energy $E_a = \frac{1}{2}M_a v^2$) is made to hit a target X of mass M_X. Although in actual laboratory experiment the target nucleus is stationary, for the sake of generality, let us assume that the target nucleus is also moving in the incident direction with velocity V_X. One can always revert back to real laboratory experiment by putting $V_X = 0$. It may be noted that all parameters of the interacting system $(a + X)$ are quite well known. As already mentioned, it is possible to determine experimentally, using suitable detectors and fixing their angular positions with respect to the incident ion beam, the energies of the final products, Y and the ejectile b. Let us assume that a suitable detector is placed at an angle ϕ with respect to the beam direction that may detect and determine the kinetic energy of the residual nucleus Y. Further, at some instant t this detector records a nucleus Y of energy E_Y at angle ϕ. At the same instant t, another detector, placed at angle θ, detects an ejectile of energy E_b. Now, experimentally we know the parameters (a, M_a, V_a, E_a; X, M_X, V_X, E_X) of the entrance channel and (b, b, V_b, E_b, θ; Y, M_Y, V_Y, E_Y, and ϕ) for the exit channel. We now proceed to correlate these experimental parameters to the Q-value of the reaction. The analytical relation that correlates the measured quantities with Q-vale of a reaction is termed as the Q-equation.

As has been mentioned, what really happens during the course of a nuclear reaction is totally unknown, and, therefore, it is customary to assume a mechanism for nuclear reaction. One may assume that a composite nucleus C of mass M ($=M_a + M_X$) is formed by the fusion of the target and the projectile as an intermediate step. This composite nucleus C will be moving in the incidence direction, say, with velocity V, to conserve the input linear momentum. Further, the intermediate system C may decay within a very short time ($\approx 10^{-16}$ s) emitting the residual nucleus Y in direction ϕ and ejectile b in direction θ, as shown in figure 2.1. This type of mechanism for nuclear reaction is called cthompound reaction

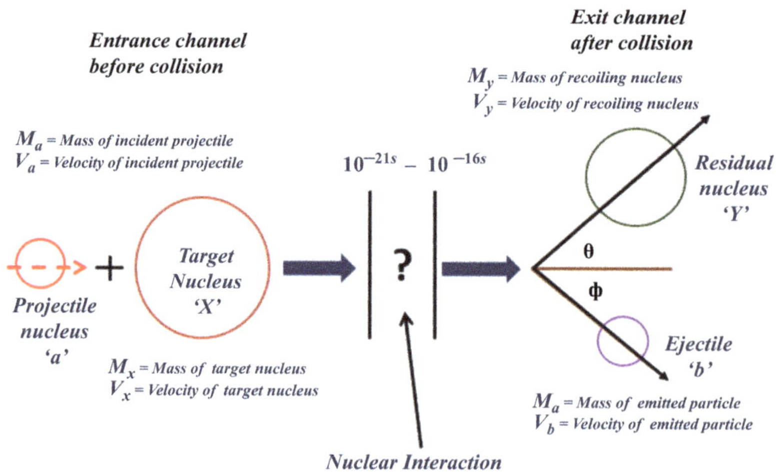

Figure 2.1. Binary nuclear reaction in laboratory frame.

mechanism. On the other hand, it may also be assumed that the incident ion does not interact with the whole of the target, it hits a few nucleons of the target and ultimately the residual nucleus Y and ejectile b are emitted at specified angles, this broadly refers to direct reaction mechanism. However, the energetic of a reaction is not affected by the assumed mechanism for the nuclear reaction, because it is governed by the laws of conservation of total energy and the total linear momentum in the entrance and the exit channels.

One may now use simple laws of conservation of total energy and linear momentum to obtain the Q-equation as worked out in the following.

Equating the total energies in the incident and exit channels one gets,

$$M_a c^2 + M_x c^2 + E_a + E_X = M_b c^2 + M_Y c^2 + E_b + E_Y$$

Or

$$[(M_a c^2 + M_x c^2) - (M_b c^2 + M_Y c^2)] + E_a + E_X = E_b + E_Y \tag{2.24}$$

But,

$$[(M_a c^2 + M_x c^2) - (M_b c^2 + M_Y c^2)] = Q, \tag{2.25}$$

Hence,

$$Q + E_a + E_X = E_b + E_Y \tag{2.26}$$

Or

$$E_b + E_Y) - (E_a + E_X) = Q \tag{2.26a}$$

For the actual laboratory experiment in which $E_X = 0$, equation (2.26) reduces to;

$$Q = E_b + E_Y - E_a \tag{2.27}$$

The linear momentum p of a particle of mass m, velocity v and kinetic energy E_m may be written as,

$$p = mv = \sqrt{2mE_m} \tag{2.28}$$

Using the above relation, linear momentum of projectile a, target X, residual nucleus Y and the ejectile b may be written, respectively, as;

$$p_a = \sqrt{2M_a E_a},\ p_X = \sqrt{2M_X E_X},\ p_Y = \sqrt{2M_Y E_Y},\ \text{and}\ p_b = \sqrt{2M_b E_b}, \tag{2.29}$$

In laboratory frame, $E_X$is zero so $p_X = 0$.

Conservation of linear momentum components in X-direction and Y-direction leads to;

$$\sqrt{2M_a E_a} = \sqrt{2M_b E_b}\cos\theta + \sqrt{2M_Y E_Y}\cos\phi \tag{2.30}$$

And

$$0 = \sqrt{2M_b E_b}\sin\theta - \sqrt{2M_Y E_Y}\sin\Phi \tag{2.31}$$

Separating the sin ϕ and cos ϕ terms from the above two equations, one may get,

$$\sqrt{2M_YE_Y}\cos\phi = \sqrt{2M_aE_a} - \sqrt{2M_bE_b}\cos\theta \tag{2.32}$$

And

$$\sqrt{2M_YE_Y}\sin\phi = \sqrt{2M_bE_b}\sin\theta \tag{2.33}$$

Equations (2.32) and (2.33) are quite general but are applicable only to the system where the target is at rest, which means for laboratory frame only. Further, all nuclear experiments are carried out in vacuum so that the incident beam ions and also the reaction products, the residual nucleus and ejectile, do not suffer scattering by air molecules.

In principle in an ideal experiment, one may pick up both the residual nucleus Y at angle ϕ and the ejectile b at angle θ in time coincidence with each other. However, practically it is quite difficult. It is because the residual nucleus is, in general, heavy and has small range; therefore, in most cases it is absorbed within the thickness of the target material. Moreover, detection of heavy charged particles requires special detectors (like fission fragment detectors). On the other hand, detection of lighter ejectiles, like proton, deuteron, alpha particles etc is much easier, simple surface barrier detectors may be used for detecting them. In most experiments a partial telescope at angle θ is employed that detects the ejectile and also measures its energy. This avoids the need for detecting the residual nucleus and measuring of its energy. It is, therefore, why the Q-equation is generally written in a form which does not include parameters (E_Y and ϕ) associated with residual nucleus.

Squaring and adding equations (2.32) and (2.33) to eliminate ϕ gives,

$$2M_YE_Y = 2M_aE_a + 2M_bE_b - 4\sqrt{M_aE_aM_bE_b}\cos\theta \tag{2.34}$$

Hence,

$$E_Y = \frac{M_a}{M_Y}E_a + \frac{M_b}{M_Y}E_b - 2\frac{\sqrt{M_aE_aM_bE_b}}{M_Y}\cos\theta \tag{2.35}$$

Substituting this value of E_Y in equation (2.27) gives,

$$Q = E_b + \left[\frac{M_a}{M_Y}E_a + \frac{M_b}{M_Y}E_b - 2\frac{\sqrt{M_aE_aM_bE_b}}{M_Y}\cos\theta\right] - E_a \tag{2.36}$$

Or

$$Q = \left(1 + \frac{M_b}{M_Y}\right)E_b - \left(1 - \frac{M_a}{M_Y}\right)E_a - 2\frac{\sqrt{M_aE_aM_bE_b}}{M_Y}\cos\theta \tag{2.37}$$

Equation (2.37) is called the Q-equation. The Q-value for a specific reaction can be computed by measuring the energy E_b of the ejectile at angle θ in a laboratory experiment. Masses of different nuclei/particles contained in the Q-equation are actually nuclear masses, but for all practical purposes one may use either masses of corresponding neutral atoms or even atomic mass number A, without any significant errors. It is because of the appearances of the ratios of masses in the Q-equation. It may further be observed that if the ejectile is recorded at $\theta = 90^0$ in laboratory, then the last term of equation (2.37) vanishes and calculation of Q becomes still easier.

Q-equation may also be written as;

$$\left(\frac{M_Y + M_b}{M_Y}\right)E_b - \left(\frac{M_Y - M_a}{M_Y}\right)E_a - 2\frac{(M_a M_b E_a E_b)^{1/2}}{M_Y}\cos\theta - Q = 0 \qquad (2.38)$$

Or

$$E_b - \left(\frac{M_Y - M_a}{M_Y + M_b}\right)E_a - 2\frac{(M_a M_b E_a E_b)^{1/2}}{M_Y + M_b}\cos\theta - \left(\frac{M_Y}{M_Y + M_a}\right)Q = 0 \qquad (2.39)$$

Or

$$\left(\sqrt{E_b}\right)^2 - \left\{\frac{2(M_a M_b E_a)^{1/2}\cos\theta}{M_Y + M_b}\right\}\sqrt{E_b} - \left[\left(\frac{M_Y Q + \{M_Y - M_a\}E_a}{M_Y + M_a}\right)\right] = 0 \quad (2.40)$$

We make the following substitutions in equation (2.40)

$$\upsilon = \left\{\frac{(M_a M_b E_a)^{\frac{1}{2}}\cos\theta}{M_Y + M_b}\right\} \quad \text{and} \quad \omega = \left[\left(\frac{M_Y Q + \{M_Y - M_a\}E_a}{M_Y + M_a}\right)\right] \qquad (2.41)$$

So that equation (2.40) becomes the following quadric equation in $\sqrt{E_b}$;

$$\left(\sqrt{E_b}\right)^2 - \sqrt{2}\,\upsilon\sqrt{E_b} - \omega = 0 \qquad (2.42)$$

The roots of the above equations may be given as,

$$\sqrt{E_b} = \upsilon \mp \sqrt{\upsilon^2 + \omega} \qquad (2.43)$$

It is worth noting that for energetically possible reactions the value of $\sqrt{E_b}$ should be real and positive. If both υ and ω become negative, $\sqrt{E_b}$ will become either negative or imaginary and reaction will not be possible. Now υ may become negative when angle θ (the angle at which ejectile detector is set) is very large so that $\cos\theta$ becomes negative. Similarly, ω may become negative (i) when Q-value is negative and $|M_Y Q| > |\{M_Y - M_a\}E_a|$, (ii) when $\{M_Y - M_a\}E_a$ is negative and $|\{M_Y - M_a\}E_a| > |M_Y Q|$. In these cases the reaction will become energetically impossible.

When the Q-value of a reaction is known, equation (2.40) may be used to find out the energy of ejectile b at any desired angle θ and one may also study the variation in E_b with the incident energy E_a of the projectile. For such a study it is convenient to consider the case for positive Q value (exoergic) and negative Q-value (endoergic) reactions separately.

Case-I: Exoergic reactions, i.e., $Q > 0$ (or positive Q-value cases)

(a) ***Case of $Q > 0$ and the kinetic energy of the projectile very nearly zero.***

This happens in the case of nuclear reactions initiated by thermal neutrons. Thermal neutrons are the neutrons that are in thermal equilibrium with the molecules of gases at room temperature, the kinetic energy of thermal neutrons is ≈0.025 eV, which for all practical purposes may be taken as zero. Since neutrons have no charge, thermal neutrons may enter a target nucleus without encountering any Coulomb barrier, and may start a nuclear reaction. Thermal neutron fission of uranium is an example of such a reaction.

For any reaction initiated by thermal neutron, E_a may be taken as zero, hence putting $E_a = 0$ in equtaion (2.41) one gets:

$$v = \left\{\frac{(M_a M_b E_a)^{\frac{1}{2}} \cos\theta}{M_Y + M_b}\right\} = 0 \text{ and } w = \left[\left(\frac{M_Y Q + \{M_Y - M_a\} E_a}{M_Y + M_a}\right)\right]$$

$$= \left[\left(\frac{M_Y Q}{M_Y + M_a}\right)\right]$$

And from equation (2.43)

$$E_b = w = \left[\left(\frac{M_Y Q}{M_Y + M_a}\right)\right] \tag{2.44}$$

Equation (2.44) does not contain any θ term, and therefore, energy of the ejectile will be the same at all angles. Further, since the linear momentum in X-direction is zero in the entrance channel, the linear momentum in X-direction must also be zero in the exit channel, which means that the reaction products, Y and b will be emitted in opposite direction or $(\theta + \phi) = 180°$.

(b) ***Case $Q > 0$, kinetic energy of projectile E_a non-zero***

In this case $w = \left[\left(\frac{M_Y Q + \{M_Y - M_a\} E_a}{M_Y + M_a}\right)\right]$ is always positive as M_Y in most cases is larger than M_a.

However, $v = \left\{\frac{(M_a M_b E_a)^{\frac{1}{2}} \cos\theta}{M_Y + M_b}\right\}$, remains positive until $\cos\theta$ is positive, but may become negative for large values of θ. Therefore, from equation (2.43) E_b may have both positive and negative values. Negative values of E_b are unrealistic.

Let us assume that v is positive, then $\sqrt{v^2 + w}$ will always be larger than v and therefore, the root $\sqrt{E_b} = v + \sqrt{v^2 + w}$ will give a finite value to E_b, the other root $\sqrt{E_b} = v - \sqrt{v^2 + w}$ will give a −ve value and will not be physically justified. Therefore, for positive values of v, the emitted particle b will have a single value of energy at each angle θ. On the other hand, for large values of θ when $\cos\theta$ is negative, the roots will be given by $\sqrt{E_b} = -v \mp \sqrt{v^2 + w}$: In the case $\sqrt{E_b} = -v + \sqrt{v^2 + w}$, the root will be real and giving a finite energy of emitted b particle having different values at different angles. The other root $\sqrt{E_b} = -v - \sqrt{v^2 + w}$ is physically not justified.

It may thus be observed that in the case of $Q > 0$, $E_a > 0$ and $M_Y > M_a$, E is single valued at each angle and its magnitude depends on θ. The least value of E_b will occur for $\theta = 180°$ (back emission) where $\cos\theta = -1$. Also, at $\theta = 90°$, $E_b = w$.

(c) ***Case when $Q > 0$, $E_a > 0$ and $M_Y < M_a$***
Such situations may arise in reactions like $X(\alpha, n)Y$ or $X(^{12}\mathrm{C}, \alpha)\,Y$ when residue Y is treated as ejectile. The magnitude of w from equation (2.41) is given as;

$$w = \left[\left(\frac{M_Y Q + \{M_Y - M_a\}E_a}{M_Y + M_a}\right)\right]$$

If in the above equation $M_Y < M_a$, then for incident energies $E_Y > \frac{M_Y Q}{(M_Y - M_a)}$, the w will become negative and roots given by equation (2.43) will become,

$$\sqrt{E_b} = v \mp \sqrt{v^2 - w} \tag{2.45}$$

$\sqrt{E_b}$ will have two positive roots; $\sqrt{E_b} = v + \sqrt{v^2 - w}$ and $\sqrt{E_b} = v - \sqrt{v^2 - w}$. Since both these roots are positive, there will be two mono-energetic groups of particle b, at each angle, one with $E_b = \left(v + \sqrt{v^2 - w}\right)^2$ and the other with energy $E_b = \left(v - \sqrt{v^2 - w}\right)^2$.

Case-II: Endoergic reactions, i.e., $Q < 0$ (or negative Q-value case)

(a) ***Case of $Q < 0$ and the kinetic energy of the projectile very nearly zero***
In this case when E_a approaches zero, from equation (2.41) it follows that;

$$v = \left\{\frac{(M_a M_b E_a)^{\frac{1}{2}}\cos\theta}{M_Y + M_b}\right\} = 0$$

and

$$w = \left[\left(\frac{M_Y Q + \{M_Y - M_a\}E_a}{M_Y + M_a}\right)\right] = \left\{\frac{M_Y(-Q)}{M_Y + M_a}\right\} = \textit{A negative quantity}.$$

Therefore, $\sqrt{E_b} = 0 \mp \sqrt{-w}$

That means that the case is physically impossible. It is expected also, since the total sum of mass energies in the exit channel is more than the total sum of mass energies in the entrance channel and there is no kinetic energy available in the entrance channel that may provide for the excess of mass energy.

(b) ***Case $Q < 0$, kinetic energy of projectile E_a non-zero, $M_Y > M_a$***
In the case of endoergic reactions the projectile particle must have a minimum energy E_{thr}, called threshold energy, to initiate the reaction. If E_a is less than E_{thr}, the reaction will not proceed. If the target is hit with incident particles of threshold energy $E_a = E_{\mathrm{thr}}$, the reaction will proceed but the ejectile will have minimum energy and it will first appear in the direction of the incident beam, i.e., at $\theta = 0$. In order to calculate E_{thr}, first one must

find the condition when $\sqrt{E_b}$ will have a minimum real and positive value in direction $\theta = 0$. According to equation (2.41),

$$\sqrt{E_b} = \upsilon \mp \sqrt{\upsilon^2 + \omega},$$

that means the real values of E_b will occur only when $\sqrt{E_b} = \upsilon + \sqrt{\upsilon^2 + \omega}$. Also, if $E_a = E_{\text{thr}}$ the minimum real value of E_b will happen for $(\upsilon^2 + \omega = 0)$, calculated at $\theta = 0$, which means when,

$$(\upsilon^2)_{(\theta=0)} + \omega = 0$$

Or

$$\frac{M_a M_b E_{\text{thr}}}{(M_Y + M_b)^2} = -\left[\frac{M_Y Q + (M_Y - M_a)E_{\text{thr}}}{(M_Y + M_b)}\right] \tag{2.46}$$

On collecting the terms containing E_{thr} on the RHS one gets;

$$E_{\text{thr}} = -Q\left[\frac{M_Y + M_b}{M_Y + M_b - M_a}\right] \tag{2.47}$$

But from equation (2.25)

$$[(M_a c^2 + M_x c^2) - (M_b c^2 + M_Y c^2)] = Q$$

Or

$$[(M_a + M_x) - (M_b + M_Y)] = \frac{Q}{c^2}$$

Or

$$(M_b + M_Y) = \left(M_a + M_x - \frac{Q}{c^2}\right) \tag{2.48}$$

In general $M_x \gg \frac{Q}{c^2}$ and therefore $\frac{Q}{c^2}$ may be neglected in equation (2.48), that gives;

$$(M_b + M_Y) = (M_a + M_x) \tag{2.49}$$

Substituting, $M_Y + M_b = M_a + M_x$ in equation (2.47), one gets,

$$E_{\text{thr}} = -Q\left[\frac{M_a + M_X}{M_X}\right] \tag{2.50}$$

(c) ***Double values of the ejectile kinetic energy E_b***

When projectile a hits the target with $E_a = E_{\text{thr}}$, the ejectile b and the residual nucleus Y both first appear in the forward direction $\theta = \phi = 0$ with $E_b = (\upsilon^2)_{(\theta=0)} = E_{\text{thr}}\left\{\frac{M_a M_b}{(M_Y + M_b)}\right\}$. As the projectile energy E_a is increased beyond E_{thr}, the ejectile starts appearing for other values of θ. The sign of factor $\omega\left[= \left(\frac{M_Y Q + \{M_Y - M_a\}E_a}{M_Y + M_a}\right)\right]$ for negative Q-value reactions is decided by the factor

$M_Y(-Q) + \{M_Y - M_a\}E_a$; for $E_a < \frac{M_Y(|-Q|)}{(M_Y - M_a)}$, ω remains negative.

Let us denote $E_a^1 = \frac{M_Y(|-Q|)}{(M_Y - M_a)}$. Therefore, for the incident energy E_a in the range from E_{thr} to E_a^1 where ω has a negative value, for each value of angle θ two mono-energetic groups of ejectiles b having energies $(E_b)_1^\theta$ and $(E_b)_2^\theta$ given respectively by,

$$(E_b)_1^\theta = \left\{(\upsilon)_\theta + \sqrt{[(\upsilon)_\theta^2 - \omega]}\right\}^2 \text{and } (E_b)_2^\theta = \left\{(\upsilon)_\theta - \sqrt{[(\upsilon)_\theta^2 - \omega]}\right\}^2$$

appear in the exit channel. Here $(\upsilon)_\theta$ stands for the value of υ at angle θ as defined by equation (2.41). On increasing the incident energy E_a beyond E_a^1 factor ω becomes positive and $\sqrt{[(\upsilon)_\theta^2 + \omega]}$ becomes larger than $(\upsilon)_\theta$. As a result only one root of $\sqrt{E_b}$ given by $\sqrt{E_b} = (\upsilon)_\theta + \sqrt{[(\upsilon)_0^2 - \omega]}$ has a positive real value for $E_a > E_a^1$. Hence ejectiles having only a single value of energy (that depends on the value of θ) are emitted at each angle.

Case-III: Elastic and inelastic scattering

Interaction between any two bodies, in general, may result in one of the three processes, namely: (i) reaction; (ii) elastic scattering; or (iii) inelastic scattering. In the case of a reaction, the sum of the total mass energy, the total kinetic energy and any other kind of internal energies (like rotational, vibrational or excitation) is conserved along with the conservation of total linear momentum. These conservation laws, in the case of nuclear reactions result in the Q-equation that reflects the total energy balance. As has been indicated, in a reaction mass energy may be converted into kinetic energy or into any other form of energy and vice versa.

In the case of elastic scattering, the total mass energy and total kinetic energies are separately conserved, along with the conservation of linear momentum. This essentially means that the masses and internal energies of the interacting bodies do not change resulting in having same projectile and target pairs as partners in the exit channel. In elastic scattering, therefore, the projectile and the target appear in the exit channel with changes only in their directions of motion with changed values of their kinetic energies such that the sum of kinetic energies remain constant. The Q-equation given by (2.37) may also be used to describe the energetics of both the elastic and inelastic scattering processes. The elastic scattering process in laboratory frame may be indicated by figure 2.2. In the case of elastic scattering, the composition and the internal states of both the target and the projectile remain unaltered before and after scattering and, therefore, the Q-value, which is a measure of the change in mass or kinetic energies, of the process is zero. In this case mass and kinetic energies are separately conserved and no new particles appear in exit channel. The projectile 'a' in the exit channel plays the role of ejectile 'b' and target 'X' behaves like the residual nucleus Y. Thus, on substitution of $M_b = M_a$ and $M_Y = X$, with $Q = 0$, one may write the Q-equation for elastic scattering as;

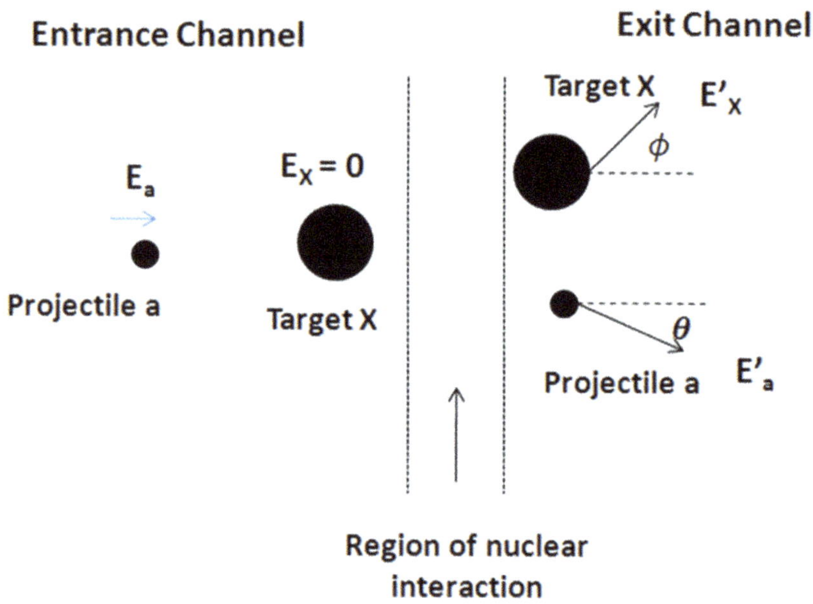

Figure 2.2. Pictorial representation of elastic scattering in laboratory frame.

$$0 = \left(1 + \frac{M_a}{M_X}\right)E_a' - \left(1 - \frac{M_a}{M_X}\right)E_a - 2\frac{\sqrt{M_a E_a M_a E_a'}}{M_X}\cos\theta \tag{2.51}$$

The values of parameters v and w for the case of elastic scattering are given as;

$$v = \frac{M_a E_a^{1/2}\cos\theta}{M_X + M_a} \quad \text{and} \quad w = \frac{(M_X - M_a)E_a}{M_X + M_a} \tag{2.52}$$

Since, w, in case of elastic scattering, can never be negative, there is only one positive real root for $\sqrt{E_b} = \sqrt{E_a'}$. Therefore, E_a' will be single valued at each angle θ.

In inelastic scattering mass energies of interacting particles are conserved but the sum of kinetic energies of the target and the projectile is not conserved. A part of the kinetic energy in the incident channel may convert into some other form like excitation energy or rotational energy (but not mass energy) in the exit channel. Suppose a part U of the total input kinetic energy $(E_a + E_X)$ gets converted into some other form, say in the excitation energy of the target, then;

$$E_a + E_X = E_a' + E_X' + U$$

Or

$$(E_a' + E_X') - (E_a + E_X) = -U \tag{2.53}$$

Comparing equation (2.53) with equation (2.26a) one gets, Q for inelastic scattering equals to the $-U$.

Energy balance or Q-equation for inelastic scattering may be obtained by making the following substitutions in equation (2.37.);

$$M_Y = M_X;\ M_b = M_a;\ Q = -U;\ \text{and}\ E_b = E_a^{'} \tag{2.54}$$

Derivation of Q-equation and other calculations carried out so far were non-relativistic. It is because in all reactions of interest in low and medium energy physics, the kinetic energies of interacting ions are always much smaller than their rest masses. It may however suffice to say that relativistic effects may be taken care of by including a correction term in the Q-equation.

2.7 Centre-of-mass frame of reference

Let us consider an experiment in which one studies the angular distribution of some ejectile b in a given reaction $a + X \rightarrow Y + b$. This experiment may be performed in the laboratory in several different ways; one may place the target X in a direction normal to the incident beam and record the angular distribution or place the target at 45° to the beam and record angular distribution and so on. Each record of angular distribution so obtained will have a different structure. This difference in the structure of angular distributions arises because of the different angles of incidence, but inherently all these different structures refer to the same interaction in each case. It may therefore be concluded that in laboratory frame of reference the same interaction may give different outcomes depending on the experimental set-up. This ambiguity in results obtained in laboratory frame of reference, arising due to different experimental set-ups may be resolved if the interaction is studied in the centre-of-mass (CM) reference frame. Whatever may be the experimental set-up, if transformed to CM frame all angular distributions will show the same structure. It is, therefore, evident that study of an interaction in CM frame reveals its true dynamics (figure 2.3).

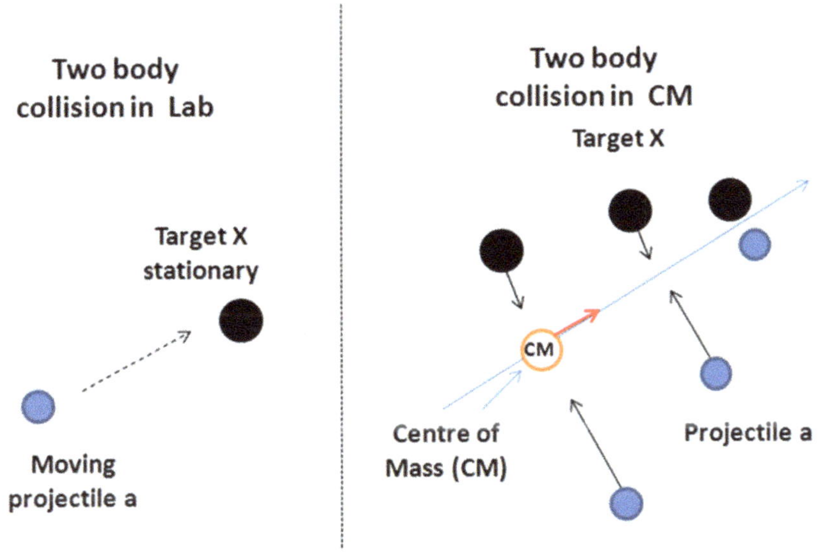

Figure 2.3. Two-body collision in laboratory and CM frames.

CM frame of reference is an inertial frame of reference in which the observer is travelling along with the CM of the system. The total momentum in CM frame is zero and the reactants (target and projectile) appear to undergo a head-on collision at the position of their CM. The CM reference frame is independent of experimental geometry, allowing results from different types of laboratory experiments to be compared providing a much more intuitive picture of the interaction. For simplicity we only consider calculations involving elastic collision between two bodies of mass M_a and M_X moving with velocity vectors v_a and y_X. v_a and v_X are velocities in laboratory frame. Let the velocities of these particles in CM frame be denoted by u_a and u_X, then;

$$u_a = v_a - v_{\mathrm{CM}} \text{ and } u_X = v_X - v_{\mathrm{CM}} \tag{2.55}$$

The velocity of the CM may be determined using the fact that the total momentum in CM is zero. If M denotes the mass of the CM, then momentum of CM is;

$$Mv_{\mathrm{CM}} = M_a v_a + M_X v_X \tag{2.56}$$

Or

$$v_{\mathrm{CM}} = \frac{(M_a v_a + M_X v_X)}{M} \tag{2.57}$$

Equation (2.56) defines the momentum of the CM, similarly, the kinetic energy E_{CM} may be defined as,

$$E_{\mathrm{CM}} = \frac{1}{2} M v_{\mathrm{CM}}^2 \tag{2.58}$$

It is important to note that because the total momentum of the system has to be conserved, the velocity, momentum and kinetic energy of the CM is conserved throughout the interaction. Energy 'tied-up' in the motion of the CM is, therefore, not available for interaction. The remaining kinetic energy is the kinetic energy associated with the relative motion of '*a*' and '*X*'. This energy, denoted by E_{rel},is sometimes called the 'collision energy' or 'CM frame kinetic energy' and is given by;

$$E_{\mathrm{rel}} = \frac{1}{2} \mu v_{\mathrm{rel}}^2 \tag{2.59}$$

Here μ stands for the reduced mass given by;

$$\mu = \frac{M_a M_X}{M_a + M_X} \tag{2.60}$$

2.8 Cross-section

When a target nucleus is bombarded by a beam of mono-energetic ions, a variety of nuclear interactions may take place simultaneously, each with its own characteristic probability. Probabilities of different interactions are measured in terms of their cross-sections. The cross-section for an interaction is a quantity that may be experimentally measured in a given experimental set-up and on the other hand it may also be calculated theoretically, assuming models for the nucleus and for the interaction mechanism.

When a beam of incident particles hits a target foil containing a large number of nuclei, each nucleus of the foil may be assumed to expose an area σ_t to the beam particles, such that incident particles when hitting this area definitely produce one of the many possible interactions. If the incident beam particles hit outside this area no interaction takes place and the particle passes without any interaction with the target nucleus. The area σ_t exposed by each nucleus, represents the total reaction or interaction cross-section. It may be further subdivided into smaller areas $\sigma_1, \sigma_2, \sigma_3, \ldots, \sigma_n$, such that if incident particles hit area σ_1 only, interaction or reaction type-1 takes place; if it hits area σ_2 only interaction type-2 occurs; if it hits area σ_3 only interaction type-3 occurs and so on. In this scheme it is obvious that these partial areas will be proportional to the probability of occurrence of their specific interaction type. These partial areas are designated as the cross-section for each specific process or interaction. Further, $\sigma_t = \sum_i \sigma_i$. As is obvious, interaction cross-sections have units and dimensions of area. Figure 2.4 gives a pictorial view of the total and partial cross-sections for different interactions. The incident ions will fall on different areas corresponding to different interactions and will initiate different interactions simultaneously. In a given time those interactions that have larger cross-section area will happen in larger number.

2.8.1 Nuclear scattering cross-section

Scattering of incident ions always takes place whenever a beam of projectiles is made to hit a target. The incident ions on hitting the target may initiate different nuclear reactions but scattering is bound to happen. It is said that there can be no reaction without scattering, though there may be scattering without any reaction.

Like any other interaction/reaction, one may assign a cross-section σ_{SC} with the scattering of incident particles by the target. A schematic of nuclear scattering is shown in figure 2.5. This figure shows an experimental set-up where a beam of

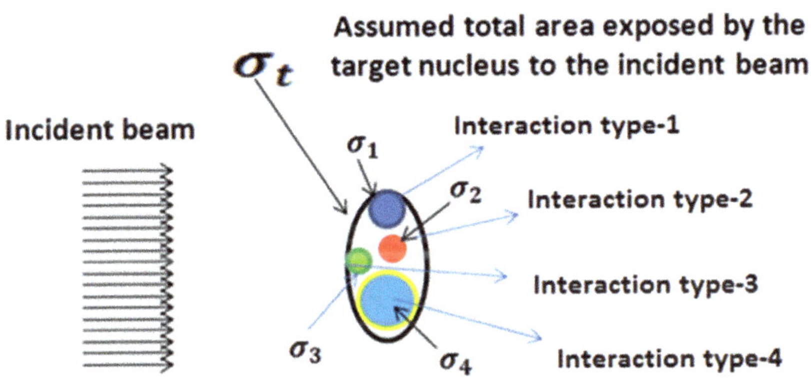

Figure 2.4. Pictorial representation of interaction cross-section.

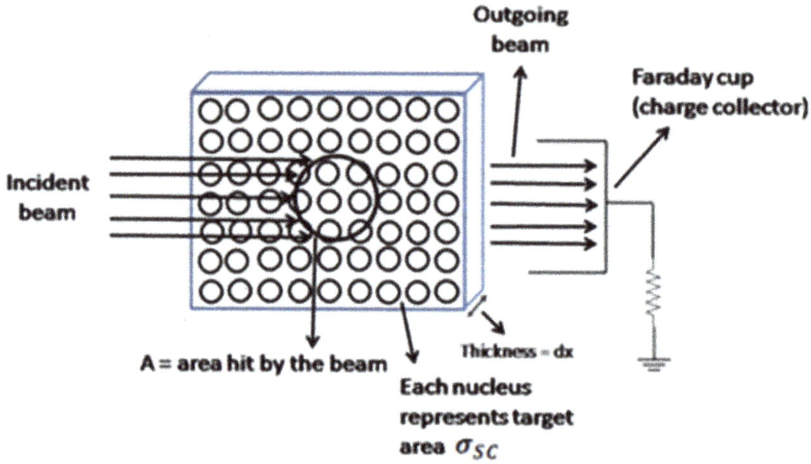

Figure 2.5. Illustration of scattering cross-section.

mono-energetic projectile '*a*' is made to hit a rectangular foil of thickness *dx*. The foil contains '*n*' number of target nuclei *X* per unit volume. Each of the target nuclei in the foil will expose an area σ_{SC} to the incident ion beam such that if any beam particle hits within this area, it definitely gets scattered. The smaller circles in the rectangular foil represent areas (of magnitude σ_{SC}) projected by target nuclei to the incident beam. The thick dotted circle represents the area *A* of the foil hit by the beam of incoming particles. It is obvious that those incident ions that hit within an area σ_{SC} associated with each target nucleus are scattered or are removed from the beam. It may, however, be noted that the cross-section, σ_{SC} represented by small circles does not represent the actual size of the nucleus and should not be considered to be a sharply defined area. It may be pointed out that a scattering event takes place only if the centres of the projectile and the target nucleus come as close to each other as the radius of the circle. In other words one can say that if the incident particle hits within this projected area it will definitely get scattered, this particular particle will not initiate any other reaction. On the other hand, if the incident particle does not hit this area, it will pass through the foil either without any interaction or may initiate some other interaction except scattering.

Let, *A* be the area of cross-section of the incident ion beam and *N* the number of incident ions (particle *a*) hitting the target foil per unit time. Further, let us assume that the target foil contains '*n*' nuclei (nucleus *X*) per unit volume and that the thickness of the target foil is *dx*.

Then, number of target nuclei involved in any kind of interaction with the beam = Number of target nuclei contained in a cylinder of base *A* and height $dx = n{\cdot}A{\cdot}\mathrm{d}x$.

The probability of the scattering will be proportional to the area of the incident beam blocked by the projected areas of target nuclei. Therefore, the fraction of the incident beam particles that undergoes scattering will be given by;

$$\frac{dN}{N} = \frac{\text{Blocked area of the incident beam by the target nuclei}}{\text{Total area of the incident beam}} = \frac{\sigma_{SC}\bullet n\bullet A\bullet dx}{A} \tag{2.61}$$

In equation (2.61) $\frac{dN}{N}$ represents the fraction of the beam that has been lost due to scattering, hence it is a negative quantity. Therefore,

$$-\frac{dN}{N} = \sigma_{SC}\bullet n\bullet dx \tag{2.62}$$

Integration of the two sides of equation (2.62) gives

$$-\int_{N_0}^{N} \frac{dN}{N} = \int_0^x \sigma_{SC}\bullet n\bullet dx$$

Or

$$-[\ln N]_{N_0}^{N} = \sigma_{SC}\bullet n\bullet x$$

Or

$$\ln(N/N_0) = -\sigma_{SC}\bullet n\bullet x$$

Or

$$N = N_0 e^{-\sigma_{SC} n\bullet x} \tag{2.63}$$

Equation (2.63) tells us that if 'N_0' number of incident beam particles per unit time is made to hit an area A of the target foil that contains 'n' number of target nuclei per unit volume, then 'N' incident beam particles will pass through the foil without undergoing scattering. ($N_0 - N$) number of projectiles will be lost from the beam as they will initiate nuclear scattering with the cross-section σ_{SC}. For positive values of σ_{SC}, x and n, the factor $e^{-\sigma_{ab} nx}$ will have a value less than 1 and, therefore, $N < N_0$. Further, in equation (2.63) both N_0 and N have the same dimensions, therefore, the factor $e^{-\sigma_{SC} nx}$ should be dimensionless, In other words the dimensions of cross-section σ_{SC} should be reciprocal of the dimensions of ($n{\cdot}x$). Now 'n' that represents the number density of target nuclei in the foil has dimensions of [L^{-3}] while 'x' that represent the thickness of the foil has dimension of [L]. As such, the dimension of σ_{SC} turns out to be [L^2]; i.e., of area. In the MKS system units of area are m^2 which is a very big unit for nuclear processes. Nuclear cross-sections are, therefore, measured in units of 'barn', where 1 barn = 10^{-28} m^2 = 100 fm^2. In equation (2.63) N_0, n, and x are known for a given experimental set-up, if one can record the number of particles scattered per unit time or the number of particles N passed through the target foil without scattering, then one may calculate scattering cross-section. The total cross-section σ_{SC} determines how many particles are scattered from the beam while passing through the material. It may be remarked that from this cross-section, it is not possible to get the direction of scattered particles. In order to get directional information one uses the differential cross-section, which defines the directional distribution of scattered particles. Depending on the type of interaction the scattered particles may be distributed either isotropically or may have some skewed directional distribution. The differential cross-section is defined as the number of particles scattered per unit time per unit solid angle at some angle θ with respect to the direction of the incident beam. The differential cross-section is represented as $\frac{d\sigma(\theta,\Phi)}{d\Omega}$. The differential cross-section is measured in units of barn/steradian. The differential cross-section may be used to obtain the total cross-section, as;

$$\sigma = \int_0^{2\pi} d\Phi \int_0^{\pi} \frac{d\sigma(\theta, \Phi)}{d\Omega} \sin\theta \, d\theta \tag{2.64}$$

It may be mentioned that the cross-section depends not only on the projectile–target combination but also on the incident energy as well. The measured cross-section, for a given interaction, as a function of beam energy, is generally referred to as the excitation function. Since, different types of interactions dominate at different energies, from the features of excitation function at low, medium and high energy, one can get information of considerable value about the reaction mechanism.

2.8.2 Experimental determination of nuclear reaction cross-section

As already mentioned, nuclear reaction cross-section is a quantity that may be calculated theoretically assuming a model for the nucleus and a mechanism for the reaction. Formulations that may be used to determine the cross-section in a typical experiment are described in the following.

Let us consider a binary nuclear reaction of the type;

$$^{A_a}_{Z_a}a + ^{A_x}_{Z_x}X \rightarrow ^{A_y}_{Z_y}Y + ^{A_b}_{Z_b}b$$

Here, $^{A_a}_{Z_a}a$ is the incident particle hitting target nucleus $^{A_x}_{Z_x}X$. After the reaction, light particle of type $^{A_b}_{Z_b}b$ is emitted leaving behind the residual nucleus $^{A_y}_{Z_y}Y$.

The cross-section σ_{ab} for the binary nuclear reaction $X(a,b)Y$, may be experimentally determined using the following formulation. Since cross-section is the probability of occurrence of an event of the type $X(a,b)$ when a target foil or sample having target nuclei 'X' is hit by a beam of incident particle 'a', the cross-section σ_{ab} or the probability of the reaction may be written as the number of favorable events of the type $X(a,b)Y$ divided by total number of possible events of all types, i.e.;

$$\sigma_{ab}(E_B) = \frac{\text{Number of events of the type } X(a, b)Y/\text{area}}{N_T \Phi_B t} \tag{2.65}$$

where, N_T is the number of target nuclei (X) in the sample, Φ_B is the flux of incident beam of ion 'a' of energy E_B and 't' the time for which the sample is bombarded with the energetic beam. Cross-section for a reaction in general depends on the energy E_B of the incident ion. For the computation of the cross-section, equation (2.65) given above may be used provided quantities appearing on the right-hand side of the equation are known. Quantities, like N_T, Φ_B and t appearing in the denominator are all those which may be set to desired values in an experiment and are, therefore, known. However, the quantity in the numerator, i.e., number of events of type $X(a, b)$ are required to be determined experimentally. This number of events of the type X

(a,b) may be obtained from two different types of experiments, i.e., (i) in-beam, and (ii) off-beam measurements. In case of an in-beam measurement one uses detectors that may identify either the emitted particles of the type ${}^{A_b}_{Z_b}b$ or the residual nucleus ${}^{A_y}_{Z_y}Y$ or both produced in the reaction. In a good experiment and depending on the requirements, detectors that may measure the energy and mass of the emitted particles may also be used. From the analysis of the experimental data taken over a period of time during the irradiation of the sample, (and therefore, called in-beam experiment), cross-sections for specific reactions may be determined. On the other hand the off-beam experiments may be used only for those reactions in which the residual nucleus Y is radioactive. In such cases, irradiation of the sample by the incident beam induces radio-activity in target samples which may be studied after the stop of irradiation. This technique of determining reaction cross-section is called activation technique [12].

A brief description of the formulations used to determine the cross-section in off-beam experiments is presented here. As has already been mentioned, the irradiation of a sample by the particle beam may initiate various reactions in it, and many radioactive isotopes are likely to be formed by the process of transmutation. The rate of formation N, of a particular reaction product/residue may be given by the expression;

$$N = N_T \Phi_B \sigma_{(ab)} \tag{2.66}$$

where, Φ_B is the flux of incident beam, N_T is the initial number of nuclei in the target sample, and $\sigma_{X(a,b)Y}$ is the cross-section for the particular reaction channel. We are interested in determining the quantity $\sigma_{(ab)}$ of the reaction.

The disintegration rate of the radio-activity induced in the target sample (due to radioactive residue ${}^{A_y}_{Z_y}Y$) after a time gap 't' from the stop of irradiation is given as;

$$\left(\frac{dN}{dt}\right) = N\frac{\{1 - e^{(-\lambda t_1)}\}}{e^{(\lambda t)}} \tag{2.67}$$

where, t_1 is the time of irradiation and λ is the decay constant of the induced activity which is related to half-life $t_{1/2}$ of the residues as,

$$\lambda = (\ln 2 / t_{1/2}) \tag{2.68}$$

Further, the number of decays, 'dN', of the induced activity in a very small time interval 'dt' may be given as;

$$dN = N\frac{\{1 - e^{(-\lambda t_1)}\}}{e^{(\lambda t)}}dt \tag{2.69}$$

In the case where the induced activity in the sample irradiated by the beam is recorded for a time duration t_3, after a time gap t_2 between the stop of irradiation and the start of counting, then the number of nuclei decayed in the time gap between t_2 to (t_2+t_3) may be given as;

$$C = N\frac{\{1 - e^{(-\lambda t_1)}\}\{1 - e^{(-\lambda t_3)}\}}{\lambda e^{(\lambda t_2)}} \tag{2.70}$$

The radio-activity induced in the sample is generally measured using a radiation detector. If the activity induced in the sample is recorded by employing a γ-ray spectrometer set-up of efficiency $G\varepsilon$, then absolute counts 'C' and observed counts 'A' may be related as;

$$C = \frac{A}{\theta_\gamma K_\gamma G_\epsilon} \tag{2.71}$$

where, A is the total counts recorded during the accumulation time t_3 of the induced activity of decay constant λ, the term θ_γ is the branching ratio of the characteristic γ-ray and $K\gamma = [\{1-\exp(-\mu d)\}/\mu d]$ is the self-absorption correction factor of gamma rays for the material of the sample of thickness d (g cm^{-2}) and of absorption coefficient μ(cm^2 g^{-1}). This correction factor can be calculated by knowing the thickness 'd'(gm cm^{-2}), of the sample and the self-absorption coefficient for the gamma ray of given energy.

Now comparing equations (2.70) and (2.71) and substituting the value of N from equation (2.66), the expression for cross-section $\sigma_{(ab)}$ can be obtained as,

$$\sigma_{(,b)}(E_B) = \frac{A\lambda e^{(\lambda t_2)}}{N_T \Phi_B \theta_\gamma K_\gamma G_\epsilon \{1 - e^{(-\lambda t_1)}\}\{1 - e^{(-\lambda t_3)}\}} \tag{2.72}$$

In the above expression, on the left-hand side the cross-section is represented as $\sigma_{(ab)}(E_B)$, that indicates that the cross-section is a function of beam energy. As may be observed in expression (2.72) all quantities on the right-hand side are either known or can be measured experimentally. For example the quantities like, half-life of the residues of interest or decay constant λ, the gamma ray branching ratio θ_γ, and the efficiency of the detector G_ϵ for the gamma ray energy of interest etc, are all known beforehand, apart from the irradiation time, time lapse and counting time etc. As such, at each incidence energy one can measure the cross-section.

The value of cross-section for a particular reaction is independent of the flux of incident beam particles and the density of the target material used. The measured cross-sections at various energies may be used to get an idea of at what energy the population of a particular radioisotope is a maximum; as such it gives an idea of optimum irradiation conditions required for production of specific artificial radio-isotope or radionuclide. This information is useful from the point of view of production of medically important radioactive isotopes. These isotopes may be of interest not only in medical field but also in agricultural, industrial and other applications. From the basic physics point of view, the measured cross-section for a particular type of reaction may also be used to study the underlying reaction mechanism. Since the cross-section may also be calculated theoretically based on some model predictions, therefore, it may be used to test the predictive power of the theoretical model codes.

2.8.3 Experimental requirements for nuclear reaction studies

Information of considerable value regarding the reaction mechanism is obtained by measuring reaction cross-sections and comparing them with the predictions of the theoretical models. Nuclear physics experiments for measuring cross-sections are generally quite involved and require auxiliary measurements of high accuracy. In most nuclear physics experiments where accelerated ion beams are used, the following four components are essential.

(i) **Target or sample:** In order to study nuclear reactions, a proper choice of target is essential. Most accelerator-based nuclear reaction experiments require thin targets. There are basically two important methods for preparing thin targets. If the required target material is available in the form of a thick metallic sheet, then thin targets may be made by rolling the thick foils using a target rolling machine. A picture of the rolling machine at the target laboratory of the Inter University Accelerator Centre (IUAC), New Delhi, India is shown in figure 2.6.

However, if the target/sample material is in powder form or in the form of ingots then the vacuum evaporation method is used to prepare the target for experiments (figure 2.7). Here, the material is heated at very high temperature in an evacuated chamber so that it vaporizes. Thin foils of backing material like carbon or aluminum are kept inside the vacuum chamber so that the evaporated material settles on these foils in a thin layer.

Further, in order to calculate the number of target nuclei in the sample, precise knowledge of the thickness of the sample/target is important. An approximate method of determining sample thickness is to weigh by a microbalance a piece of the sample of known dimensions and then divide the measured weight by the area of the sample to get sample thickness in

Figure 2.6. Target rolling machine at the target laboratory of Inter University Accelerator Centre, IUAC, New Delhi, India.

Figure 2.7. Vacuum evaporator set-up at the target laboratory of Inter University Accelerator Centre (IUAC), New Delhi, India.

units of kg m^{-2} or in g cm^{-2}. However, a more precise method involves measuring the energy lost by alpha particles of known energy in the thickness of the target in a transmission experiment. In a transmission experiment a beam of nearly mono-energetic alpha particles of known energy, say E, from some standard radioactive source is made to pass through the sample and the energy, E', of the alpha particles coming out on the other side after passing through the sample is measured using an alpha spectrometer. Once the energy lost $\Delta E = E - E'$ (MeV) is determined, it is easy to calculate sample thickness using the known value of stopping power, available in tabular form in literature.

Experimentalists also check the purity of the target, otherwise interfering reactions may be produced. Generally, spectroscopically pure targets are used in nuclear physics experiments to avoid masking of desired activity by other activities induced in the sample. In order to check the purity of the sample one may carry out the x-ray fluorescence (XRF) analysis or the

Rutherford backscattering (RBS) study. In actual experiments small samples of target material, typically of 1×1 cm^{2}size are generally used and carefully put on a target ladder for placing in front of the beam of accelerated nuclear particles.

(ii) **Beam of nuclear particles:** The incident particles with suitable energy in the form of a beam are required to initiate desired reactions. This is done by the use of either a particle from a radioactive source or from a particle accelerator or from a nuclear reactor. However, for precise measurements mono-energetic beams from the particle accelerators are generally preferred. Some of the frequently used accelerators are: cyclotrons [13], linear accelerators, Van de Graff [14], and Pelletron accelerators [15]. A schematic representation of a pelletron accelerator at Inter University accelerator Centre (IUAC), New Delhi, India is shown in figure 2.8. The advantage of a particle accelerator is that accelerators deliver a mono-energetic focused beam of desired ion. Precise knowledge of beam energy is important for proper selection of threshold energy and also to understand energy dependence of the reaction yield. In order to study the energy dependence of cross-section, samples may be bombarded at different beam energies.

(iii) **Detectors:** In order to identify the reaction, either the emitted particles or the residual nuclei or both needs to be detected. For detecting the heavier nuclei one can use the recoil separators coupled with special kinds of multiwire detectors [16]. However, the emitted light particles or gamma rays are generally detected with the help of radiation (particles or gamma ray)

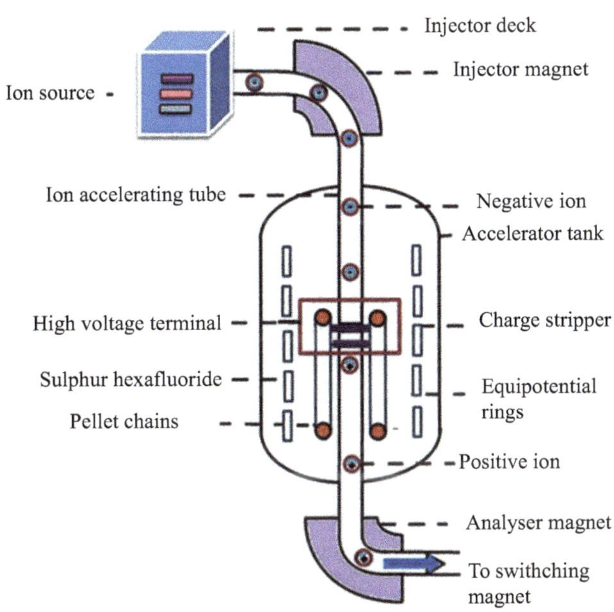

Figure 2.8. Schematic representation of the 15 UD Pelletron Accelerator at IUAC, New Delhi, India.

detectors. Depending on the requirements it can be a set-up of charged particle detectors, neutron or gamma ray detector, as the case may be. It is important that the detector should have high resolution. One can either use a single detector or a group of detectors may also be used to get detailed information about the reaction kinematics and other parameters. Highly sophisticated detectors with high resolution are now available for precise measurements. Use of position sensitive detectors has further enhanced the analyzing power of experiments.

(iv) **Data acquisition and analysis:** Modern day advanced experiments use multi-parameter detectors that generate a large number of signals when reaction products pass through them. The signal may carry information of many parameters like energy, time, mass, charge, trajectory etc, of the reaction products. Detector signals are to be acquired and analyzed. The data or signals acquired from the detector have to be amplified and collected in some useful form e.g., in the form of energy, a mass spectrum etc. The data acquisition (DAQ) system digitizes, formats, and stores this information in a way which can be retrieved and analyzed later. This is an important aspect for any experiment as the data generated has to be collected in real time for later analysis. In the modern DAQ systems the data collection is done with the help of dedicated computer software, especially for accelerator-based experiments. Careful analysis of the recorded data helps in understanding the underlying physics.

Bibliography

[1] Chadwick J 1932 Possible existence of a neutron *Nature* **129** 312

[2] Encyclopaedia Britannica 2018 Thermal neutron

[3] Nuclear Science—a guide to the nuclear science wall chart ©2018 Contemporary Physics Education Project (CPEP), ch 7

[4] Rutherford E 1919 Collisions of alpha particles with light atoms. IV. An anomalous effect in nitrogen, *Lond. Edinb. Dublin Philos. Mag. J. Sci., 6th series* **37** 581–7

[5] Blackett P M S 1925 The ejection of protons from nitrogen nuclei, photographed by the Wilson method *Proc. R. Soc. Lond.* A **1925** 349–60

[6] Butler S T 1950 On angular distributions from *(d,p)* and *(d,n)* nuclear reactions *Phys. Rev.* **80** 1095.
Butler S T 1951 *Proc R. Soc. Lond.* A **208** 36

[7] Brune C S and Davids B 2015 Radiative capture reactions in astrophysics *Annu. Rev. Nucl. Part. Sci.* **65** 87–112

[8] Bohr N 1936 Neutron capture and nuclear constitution *Nature* **137** 344–48

[9] Hahn O and Strassmann F 1939 The detection and behavior of the alkaline earth metals which result from the irradiation of uranium by neutron *Naturwissenschaften* **27** 11

[10] Prasad R and Singh B P 2018 *Fundamentals and Applications of Heavy Ion Collisions: Below 10 MeV/Nucleon Energies* (Cambridge: Cambridge University Press)

[11] Moretto L G and Wozniak G J 1993 Multifragmentation in heavy-ion processes *Ann Rev. Nucl. Part. Sci* **43** 379–455

[12] Russell G J 1990 Spallation physics—an overview *ICANS-XI International Collaboration on Advanced Neutron Sources (KEK, Tsukuba, October 22–26, 1990)*

[13] Lawrence E O and Stanley Livingston M 1932 The production of high speed light ions without the use of high voltages *Phys. Rev.* **40** 19

[14] Van De Graaff R J 1933 The electrostatic production of high voltage for nuclear investigations *Phys. Rev.* **43** 149

[15] Herb R G 1971 The pelletron accelerator *IEEE Trans. Nucl. Sci.* **18** 71–5

[16] Charpak G, Bouclier R, Bressani T, Favier J and Zupancic C 1968 The use of multiwire proportional counters to select and localize charged particles *Nucl. Instrum. Methods* **62** 235

Chapter 3

Theoretical formulation of compound and pre-compound emission

The first theory for the nuclear reaction mechanism was proposed by Bohr in 1936, which is known as the compound reaction mechanism. The theory revolves round the independence hypothesis which assumes that a nuclear reaction takes place in two mutually independent steps: formation of the compound nucleus (CN) and its subsequent decay. Experimental verification of the independence hypothesis became a challenge; Hauser–Feshbach developed a theoretical framework based on compound reaction mechanism for statistical nuclear reactions. Details of experimental verification of the independence hypothesis are discussed in the following. Hauser–Feshbach formalism and the role played by nuclear level density in compound nuclear reactions are also presented here. Some early experimental evidence of pre-equilibrium emission is also discussed in this chapter.

3.1 Introduction

The history of nuclear reaction studies goes back to the time when during 1909 to 1911, Rutherford and his colleagues observed the large angle elastic scattering of alpha particles from Au (gold) and Pt (platinum) foils, thereby discovering the atomic nucleus [1, 2]. In order to have systematic studies on the nuclear reaction mechanism, Enrico Fermi [3] made some measurements on (n,γ) reactions. By that time Chadwick had already established the presence of the neutron. Fermi observed that the measured (n,γ) reaction cross-sections for a large number of nuclei were quite large, which appeared as a puzzle. Further, for some nuclides (n,γ) cross-sections at very low energies were found to be unusually large. The observation of large (n,γ) cross-sections suggested that the system, formed as a result of neutron capture by the target nucleus, stayed in a highly excited state for a considerably long time which was estimated to be $\approx 10^{-16}$ s. Later, the large (n,γ) cross-sections at low energies were assigned to the phenomenon of 'resonances'.

doi:10.1088/978-0-7503-5077-8ch3

The resonances were found to occur in direct, intermediate as well as in compound reactions. A long-lived compound system is likely to have narrow resonance structure described by the characteristic Breit–Wigner formula. Each of the observed resonances was found to correspond to an excited nuclear state in the continuum, generally with a relatively long lifetime. Neutron resonances in heavy nuclei were some of the significant examples. At higher excitations resonances become significantly large and can be described only with statistical methods presented by Hauser-Feshbach theory. The broad features of the neutron induced reactions were explained by the CN model proposed by Bohr. The details of the model are described in this chapter.

As mentioned in chapter 2, the time scales involved in the nuclear reactions are typically in the range $\approx 10^{-16} - 10^{-21}$ s, which cannot be measured experimentally; therefore, it is not possible to investigate directly as to what transpired during the time of nuclear interaction. In the absence of direct evidence, theoretical models for reaction mechanism were proposed. Simultaneously, attempts were made to correlate the experimentally measured properties of reaction products with the predictions of theoretical models. To some extent it is similar to deciphering the reasons for a plane crash by looking at its debris. The basic quantity that may be measured experimentally on one hand and may be calculated theoretically assuming a specific reaction mechanism on the other is the cross-section for a particular reaction of interest. In the following, the theoretical approach within the framework of CN model is presented.

3.2 Compound reaction mechanism

As already mentioned, one of the earliest approaches to understand the reaction dynamics of binary nuclear reactions is that of the compound reaction (CN) mechanism proposed by Danish physicist Niels Bohr [4] way back in 1936. According to Bohr's CN theory a nuclear reaction proceeds in two steps. The first step of a binary reaction is the formation of a CN as a result of the absorption of the incident particle by the target nucleus. The CN so formed is in excited state, i.e.,

$$\text{Step-1}\ {}^{A_a}_{Z_a}a + {}^{A_x}_{Z_x}X \rightarrow {}^{A_c}_{Z_c}\mathrm{C}^* \tag{3.1}$$

Here ${}^{A_a}_{Z_a}a$ is the projectile nucleus, ${}^{A_x}_{Z_x}X$ the target nucleus and ${}^{A_c}_{Z_c}\mathrm{C}^*$ the excited CN. Further, A_a is the atomic mass number and Z_a is the atomic number of incident particle while A_x is the atomic mass number and Z_x is the atomic number of the target nucleus. A_c is the atomic mass number and Z_c is the atomic number of the CN. Here, $A_c = A_a + A_x$, while $Z_c = Z_a + Z_x$.

The second step is the decay of the CN. Once the projectile is absorbed by the target, the theory assumes that the energy and the momentum of the incident particle are shared rapidly among all the nucleons of the compound system and soon a thermodynamic equilibrium gets established. Now by chance, amount of energy more than separation energy may get accumulated on one or a group of nucleons (light cluster) which may be emitted leading to the decay of the CN in step-2, i.e.,

$$\text{Step-2}\ {}^{A_c}_{Z_c}\mathrm{C}^* \rightarrow {}^{A_y}_{Z_y}Y + {}^{A_b}_{Z_b}b \tag{3.2}$$

In the above equation, ${}^{A_Y}_{Z_Y}Y$ denotes the residual nucleus and ${}^{A_b}_{Z_b}b$ the outgoing lighter nuclear particle, also called ejectile or emitted particle. An essential assumption of Bohr's theory was the independence of step-1 and step-2 of the reaction process, called 'independence hypothesis'. The independence hypothesis is the result of the establishment of thermodynamic equilibrium as the last sequence of thorough mixing. As a result of thermodynamic equilibrium the equilibrated CN forgets about the history of its formation and the probability of its decay into a specific set of products becomes independent of its mode of formation. In other words, the decay of the equilibrated CN depends only on the set of its good quantum numbers and on nothing else. It may be mentioned that a given equilibrated CN may decay via several channels each with a given probability. Schematically, the complete binary nuclear reactions may be written as shown in figure 3.1. Here, A_i, Z_i are respectively the atomic mass number and atomic number of the ith entity (particle, cluster or nucleus). The basic conservation laws of nuclear physics require that the total energy (sum of the kinetic, potential and mass energies) in the entrance channel must be equal to the total energy in each possible exit channel and that;

$$A_a + A_X = A_C = A_{yi} + A_{bi} \text{ and } Z_a + Z_X = Z_c = Z_{Yi} + Z_{bi} \tag{3.3}$$

Also, other good quantum numbers like the angular momentum and isospin etc, must also be conserved. Each equation on the right hand side of the nuclear reaction represents an exit channel. Each energetically allowed exit channel that fulfils the conservation laws, occurs with a finite probability. Cross-section (denoted by the Greek alphabet sigma σ) for any given exit channel, which is proportional to the probability of occurrence of that channel, may be measured in an experiment. For example cross-section $\sigma_{X(a,\ b_1)Y_1}$ for the exit channel $\left({}^{A_{Y_1}}_{Z_{Y_1}}Y_1 + {}^{A_{b_1}}_{Z_{b_1}}b_1\right)$ may be written as,

$$\sigma_{X(a,\ b_1)Y_1} = \frac{\text{Number of events of type } X(a,\ b_1)Y_1 \text{ in a given time } t}{\text{(No. of target nuclei)(Flux of the incident particle)(time)}} \tag{3.4}$$

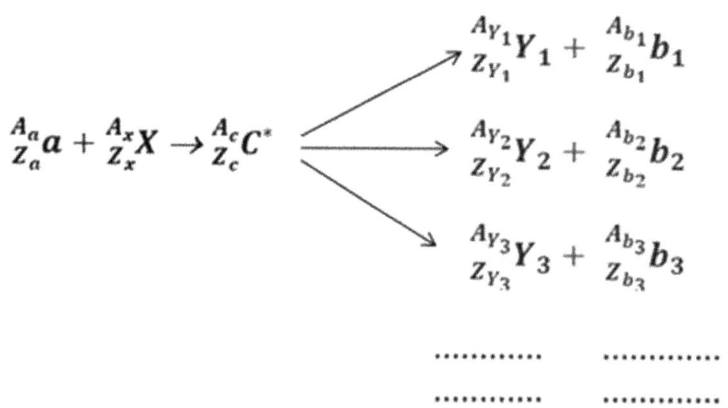

Figure 3.1. A Schematic description of complete binary nuclear reaction.

As may be verified from the above definition of the cross-section, the dimensions of the cross-section are that of area (m^2). Nuclear reaction cross-sections are measured in a unit called 'barn', denoted by 'b'; where 1 barn $= 10^{-24}$ cm^2. Since all quantities appearing in the denominator of the expression for cross-section can be directly determined in an actual experiment beforehand, the only quantity to be determined during the experiment is the number of events of the specific type in a given time. There are several possible ways in which the number of events may be counted: (i) by identifying both the ejectile ${}^{A_{b_1}}_{Z_{b_1}}b_1$ and the residual nucleus ${}^{A_{Y_1}}_{Z_{Y_1}}\mathrm{Y}_1$ and counting the coincidence rate between them in an in-beam experiment; (ii) counting the coincidence rate between projectile ${}^{A_a}_{Z_a}a$ and ejectile ${}^{A_{b_1}}_{Z_{b_1}}b_1$ in an in-beam experiment; (iii) in the case where the residual nucleus ${}^{A_{Y_1}}_{Z_{Y_1}}\mathrm{Y}_1$ is radioactive with measurable half-life, then the number of residual nuclei produced in a given time may also be determined by measuring the intensity of the radioactivity induced in the target sample as a result of irradiation after the experiment in an off-beam experiment. In such, experiments activity induced is measured after the irradiation of the sample is stopped. A detailed formulation for measuring cross-sections in an experiment has already been given in chapter 2.

For example, the following exit channels may be populated in the reaction,

$$
{}^{1}_{1}p + {}^{27}_{13}Al \rightarrow
\begin{cases}
{}^{24}_{12}Mg + {}^{4}_{2}He \\
{}^{27}_{14}Si + {}^{1}_{0}n \\
{}^{24}_{11}Na + (3\,{}^{1}_{1}p\;1\,{}^{1}_{0}n)
\end{cases}
$$

The cross-section, as indicated above, is a quantity that may be measured in an experiment on one hand, that may also be calculated theoretically assuming (i) a certain mechanism for the nuclear reaction, and/or (ii) a certain structure for the nucleus. The degree of agreement between the experimental and the theoretical values of cross-section for a given reaction may give information regarding the validity of assumptions about reaction mechanism and about the nuclear structure.

In Bohr's theory of compound reaction mechanism it is assumed that the composite system formed as the result of absorption of the projectile by the target nucleus does not decay till thermodynamic equilibrium is established. This will be possible only if the mean-free path of the projectile in the composite system is much shorter than the nuclear radius, so that the projectile undergoes rapid collisions with other nucleons of the composite system sharing its energy and momentum at each collision and rapidly distributing the available excitation energy among other nucleons. Once the equilibrium is established, the excitation energy per nucleon U_C^*/A in the CN (C) may be less than the separation energy of a nucleon (S_n or S_p) i.e.,

$$U_C^*/A < S_n \text{ or } S_p \text{ or } S_a \tag{3.5}$$

where, S_n, S_p andS_a are respectively, the separation energy of neutron, proton and of incident particle from the CN.

In Bohr's CN theory the reaction mechanism is regarded as the two-step process: (i) formation of the equilibrated CN via entrance channel in the first step; and (ii) subsequent decay of the CN via the exit channel in the second step;

$$\begin{aligned} &{}^{A_a}_{Z_a}a + {}^{A_x}_{Z_x}X \rightarrow {}^{A_c}_{Z_c}\mathrm{C}^* && \text{(step-1)} \\ &{}^{A_c}_{Z_c}\mathrm{C}^* \rightarrow {}^{A_Y}_{Z_Y}Y + {}^{A_b}_{Z_b}b && \text{(step-2)} \end{aligned} \tag{3.6}$$

Both the entrance channel $\left({}^{A_a}_{Z_a}a + {}^{A_x}_{Z_x}X\right)$ and the exit channel $\left({}^{A_y}_{Z_y}Y + {}^{A_b}_{Z_b}b\right)$ are defined, respectively, through the quantum numbers of projectile ${}^{A_a}_{Z_a}a$ and of target ${}^{A_x}_{Z_x}X$ and of residual nucleus ${}^{A_Y}_{Z_Y}Y$ and the ejectile ${}^{A_b}_{Z_b}b$. If one lumps all the quantum numbers of projectile ${}^{A_a}_{Z_a}a$ and target ${}^{A_x}_{Z_x}X$ together and denotes them by a composite symbol α, and that of the residual nucleus ${}^{A_Y}_{Z_Y}Y$ and ejectile ${}^{A_b}_{Z_b}b$ by a composite symbol β, the cross-section for the above reaction may be written as;

$$\sigma(\alpha, \beta) = \sigma_C(\alpha) \bullet G_C(\beta) \tag{3.7}$$

where, $\sigma_C(\alpha)$ is the cross-section for the formation of a CN at an excitation energy U_C through entrance channel α, while the term $G_C(\beta)$ is the relative probability for the CN at an excitation energy U_C to break up via channel β emitting particle of type ${}^{A_b}_{Z_b}b$ and the residual nucleus ${}^{A_Y}_{Z_Y}Y$. As already mentioned, in a nuclear reaction all good quantum numbers like the parity as well as total angular momentum are conserved.

To calculate the cross-section for the formation of CN $\sigma_C(\alpha)$ at a given excitation energy, the semi-classical approach may be employed. In the simplest approach the wave properties of the incident particles in the beam are neglected. If the incident particle is a neutron, then classical target area gives the formation cross-section, while in the case of a charged projectile it is given by;

$$\sigma_C(\alpha) = \pi R^2\left(1 - \frac{V_{\text{Coulomb}}}{E_\alpha}\right); \text{ if } E_\alpha \gg V_{\text{Coulomb}} \tag{3.8}$$

Here, V_{Coulomb} is the Coulomb barrier between the interacting partners given by;

$$V_{\text{Coulomb}} = \frac{1}{4\pi\varepsilon_0} \cdot \frac{(Z_a e)(Z_X e)}{R};$$

$$V_{\text{Coulomb}} = \frac{1}{4\pi\varepsilon_0} \cdot \frac{(Z_a)(Z_X)e^2}{R}; \tag{3.9}$$

$Z_a e$, and $Z_X e$ being the charges of the projectile nucleus ${}^{A_a}_{Z_a}a$ and the target nucleus ${}^{A_x}_{Z_x}X$,respectively, while R is the separation between the centres of projectile and target nuclei when they touch each other. Here R is taken as the sum of radii of projectile and target nuclei when they touch each other, i.e., $R = R_a + R_X$.

The radius of individual nuclei may be taken by the expression $R = R_0A^{1/3}$ Fermi. Where R_0 is taken as a constant and its typical value ranges from 1.3 to 1.5 Fermi. The term A is the atomic mass number of the nuclei under consideration.

For the values of $E_\alpha < V_{\text{Coulomb}}$, the $\sigma_C(\alpha) = 0$, because the projectile cannot reach to the surface of the target nucleus. The above classical expression can be modified in light of the typical wave phenomenon of the incident particle with de Broglie wavelength λ_α, in the entrance channel, i.e., since the position of the incident particle a is uncertain by λ_α, we may replace R by $(R + \lambda_\alpha)$, and to take into account the quantum mechanical tunnelling even at energies $E_\alpha < V_{\text{Coulomb}}$, one can write the formation cross-section as;

$$\sigma_C(\alpha) = \pi(R + \lambda_\alpha)^2.\ T_\alpha \tag{3.10}$$

where, T_α is the transmission coefficient of incident particle ${}^{A_a}_{Z_a}a$ in the incident or the entrance channel α.

Here, for the sake of simplicity in calculations it may be assumed that the nucleus has a well-defined boundary with radius R, and when the incident particle enters the nuclear boundary it experiences the attractive negative potential, moving with increased energy represented with wavenumber K. Further, when the incident particle enters the nucleus, it shares the energy and momentum with other nucleons of the target nucleus so that for values of $r < R$ the probability of emission of particles with the same incident energy is negligible. To calculate the transmission coefficient T_α, let us consider a beam of particles of type ${}^{A_a}_{Z_a}a$, of energy E_α moving in the direction x against a potential $V(x)$, as shown in figure (3.2).

The incident wave (moving to the right) may be represented as an incident plane wave which is partially reflected (moving to the left) as it strikes the wall of the potential well of depth V_0. So the wave function outside the potential well will consist of two parts as given below;

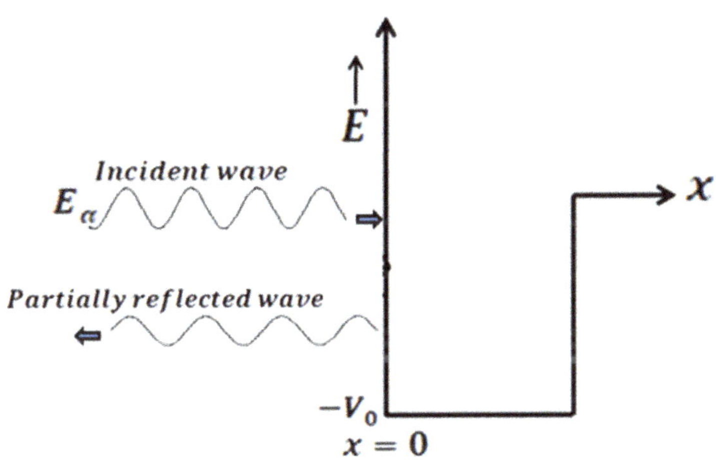

Figure 3.2. Incident projectile beam (wave) is partially reflected at $x = 0$ from the wall of the potential well and is partially transmitted, as described in the text.

$$\Psi_{\text{outside}}(x) = Ae^{ikx} + Be^{-ikx} \text{ for } x < 0, \text{ here } k = \frac{\sqrt{2ME}}{\hbar} \tag{3.11}$$

Inside the potential well there will be a transmitted wave and no reflected wave, as such;

$$\Psi_{\text{inside}}(x) = Ce^{iKx}; \text{ for } x > 0, \text{ here, } K = \frac{\sqrt{2M(E + V_0)}}{\hbar} \tag{3.12}$$

For waves with $x < 0$ the wave number is k and the wave for $x > 0$ has wave number K. The complex exponent denotes oscillations. These wavenumbers are fixed by the Schrödinger equation and;

$$k^2 = \frac{2mE}{\hbar^2}, \text{ and } K^2 = \frac{2m(E - V_0)}{\hbar^2}$$

A part of the incident wave is reflected, and a part is the transmitted wave. Since for the well behaved wave functions not only their values but the derivatives of the wave functions should also match at the boundary ($x = 0$) of the potential well, therefore, by applying these conditions and solving the equations, one obtains values of coefficients A and B in terms of C, k and K as,

$$A = \frac{1}{2}C\left(\frac{K}{k} + 1\right), \text{ and} \tag{3.13a}$$

$$B = \frac{1}{2}C\left(1 - \frac{K}{k}\right) \tag{3.13b}$$

The amplitude of the incident wave is;

$$|\Psi_{\text{incident}}(x)|^2 = \Psi^*_{\text{incident}}\Psi_{\text{incident}}(x)$$

$$|\Psi_{\text{incident}}(x)|^2 = (Ae^{ikx})^*(Ae^{ikx}) = A^*A = |A|^2 \tag{3.14a}$$

Similarly, the amplitude of the reflected wave is;

$$|\Psi_{\text{reflected}}(x)|^2 = |B|^2 \tag{3.14b}$$

It is known that the square of the amplitude of the wave is directly proportional to the intensity of the wave. So, if we are interested in finding how much of the incident wave tunnels through the potential well or what is the transmission coefficient, then by knowing the required amplitudes one can estimate it. The transmission coefficient represents the probability of the transmitted wave flux with respect to that of the incident wave. This coefficient is generally used to describe the probability of a particle tunnelling through the barrier. The transmission coefficient may be obtained as;

$$T_a = \frac{|A|^2 - |B|^2}{|A|^2} \tag{3.15}$$

Substituting the values of coefficients A and B, from equations (3.13) in the above expression; one obtains;

$$T_\alpha = \frac{\left[\left\{\frac{1}{2}C\left(\frac{K}{k}+1\right)\right\}^2 - \left\{\frac{1}{2}C\left(1-\frac{K}{k}\right)\right\}^2\right]}{\left\{\frac{1}{2}C\left(\frac{K}{k}+1\right)\right\}^2}$$

Solving the above expression one gets;

$$T_\alpha = \frac{4kK}{(k+K)^2} \tag{3.16}$$

As can be seen from above equation, the transmission coefficient T_α is small if $k \ll K$, i.e., at low energies and approaches unity at relatively higher energies, or when $k \cong K$. Now, substituting the value of T_α in equation (3.10), we get;

$$\sigma_C(\alpha) = \pi(R+\lambda_\alpha)^2 . \frac{4k_\alpha K_\alpha}{(k_\alpha + K_\alpha)^2} \tag{3.17}$$

The above expression may be used to calculate the cross-section for the formation of the CN through the incident channel α referred to above. It may be remarked that this expression is more accurate than $\sigma_C(\alpha) = \pi R^2$, as it takes into account the quantum mechanical tunnelling effect through the potential barrier as described above.

Now, one needs to obtain the value of $G_C(\beta)$ which is the relative probability for the CN at an excitation energy U_C to break up the emitting particle of type ${}^{A_b}_{Z_b}b$. It may be remarked that the value of $G_C(\beta)$ depends on the excitation energy of the CN. Further, according to Heisenberg's uncertainty principle, the excitation energy U_C cannot be sharply defined. The uncertainty in the excitation energy is defined by a quantity called level width Γ_β (i.e., the spread in energy of the level). As such, if the τ^C is the mean life of the CN at an excitation energy U_C^*, the excited nucleus may decay into a large number of possible reaction channels, say $Y + b$; $Y_1 + b_1$; $Y_2 + b_2$; $Y_3 + b_3$; $Y_4 + b_4$;..., as shown in figure 3.3. Note that number of levels is increasing as the excitation energy increases. Using Heisenberg's uncertainty relation;

$$\Gamma\ (\text{Total level width}) = \frac{\hbar}{\tau^C} = \hbar\lambda \tag{3.18}$$

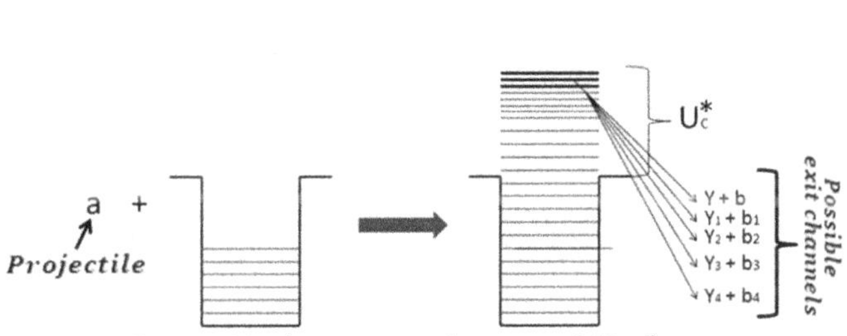

Figure 3.3. Schematic representation of a nuclear reaction.

Here, $\lambda = \frac{1}{\tau^C}$ is the decay constant.

If the excitation energy U_C^* is higher than the separation energy of different particles from the CN, the system may decay by particle emission as well as by γ-emission, as represented by various possible channels in figure 3.3. Further, at relatively higher excitation energies, corresponding to the energy continuum of the CN, the levels become so close that they overlap completely and instead of talking of well separated levels one talks about the density of levels, or rather level width. Thus, total level width for decay into various channels will be the sum of level widths for decay into individual channels. If, Γ_β, Γ_{β_1}, Γ_{β_2}, ... are the partial level widths with corresponding decay constants λ_β, λ_{β_1}, λ_{β_2}, ... then,

$$\Gamma_{\text{Total}} = \Gamma_\beta + \Gamma_{\beta_1} + \Gamma_{\beta_2} + \ldots \tag{3.19}$$

and

$$\lambda_{\text{Total}} = \lambda_\beta + \lambda_{\beta_1} + \lambda_{\beta_2} + \cdots \tag{3.20}$$

If one assumes that the CN is formed in a particular quantum state which can decay through channel β with a decay constant λ_β, then the level width and the associated decay constant are related by the expression;

$$\Gamma_\beta = \hbar\lambda_\beta = \frac{\hbar}{\tau_\beta} \tag{3.21}$$

where, τ_β is the mean lifetime of the level β.

Now the relative probability $G_C(\beta)$ for decay into a particular channel β may be obtained from the ratio of decay constant for channel β to the total decay constant, i.e.,

$$G_C(\beta) = \frac{\lambda_\beta}{\lambda} = \frac{\text{Decay constant for channel } \beta}{\text{Total decay constant}} \tag{3.22}$$

or,

$$G_C(\beta) = \frac{\Gamma_\beta}{\Gamma_{\text{Total}}} \tag{3.23}$$

The decay into individual reaction channels via the decay constants λ_β and the corresponding partial level width Γ_β strongly depends on the barrier penetration factor. Similarly, the quantum mechanical tunnelling as described above for calculating the transmission coefficient T_α also affects the formation of the CN in the entrance channel. In the case where the same CN is formed via the inverse reaction channel, ${}^{A_Y}_{Z_Y}Y + {}^{A_b}_{Z_b}b$, i.e., through channel β, and we assume that the compound system is in thermodynamic equilibrium, all energetically permissible states are occupied with equal probability. Then using the reciprocity theorem, it can be shown that;

$$\frac{\sigma(\alpha, \beta)}{\lambda_\alpha^2} = \frac{\sigma(\beta, \alpha)}{\lambda_\beta^2} \tag{3.24}$$

where, $\sigma(\alpha,\beta)$ is the cross-section for the formation of the CN through channel α and its decay through channel β, while $\sigma(\beta,\alpha)$ is the cross-section for the formation of the same CN (with the same quantum numbers) through inverse reaction channel β that leads to its decay into channel α. This leads to;

$$\frac{\sigma_C(\alpha).\frac{\Gamma_\beta}{\Gamma_{\text{Total}}}}{\lambda_\alpha^2} = \frac{\sigma_C(\beta).\frac{\Gamma_\alpha}{\Gamma_{\text{Total}}}}{\lambda_\beta^2} \tag{3.25}$$

or,

$$\sigma_C(\alpha)k_\alpha^2\Gamma_\beta = \sigma_C(\beta)k_\beta^2\Gamma_\alpha \tag{3.26}$$

or

$$\frac{k_\alpha^2\sigma_C(\alpha)}{\Gamma_\alpha} = \frac{k_\beta^2\sigma_C(\beta)}{\Gamma_\beta} = \ldots = F(U_C^*) \tag{3.27}$$

where k_α and k_β, are the wavenumbers associated with energies E_α and E_β. Here, F is a constant and is a function only of the excitation energy U_C^* of the CN. As such, the level width Γ_β may be written as;

$$\Gamma_\beta = \frac{k_\beta^2\sigma_C(\beta)}{F(U^*)} \tag{3.28}$$

and

$$G_C(\beta) = \frac{\Gamma_\beta}{\Gamma_{\text{Total}}} = \frac{k_\beta^2\sigma_C(\beta)/F(U^*)}{\sum_{\gamma=1}^{\text{All channels}} k_\gamma^2\sigma_C(\gamma)/F(U^*)} \tag{3.29}$$

or

$$G_C(\beta) = \frac{k_\beta^2\sigma_C(\beta)}{\sum k_\gamma^2\sigma_C(\gamma)} \tag{3.30}$$

where the value of $\sigma_C(\beta)$ may be obtained from equation (3.17) as;

$$\sigma_C(\beta) = \pi(R + \lambda_\beta)^2.\ \frac{4k_\beta K_\beta}{(k_\beta + K_\beta)^2}; \tag{3.31}$$

and

$$\sigma_C(\gamma) = \pi(R + \lambda_\gamma)^2.\ \frac{4k_\gamma K_\gamma}{(k_\gamma + K_\gamma)^2} \tag{3.32}$$

Substituting the values from above in equation (3.7) one obtains;

$$\sigma_C(\alpha,\ \beta) = \pi(R + \lambda_\alpha)^2.\ \frac{4k_\alpha K_\alpha}{(k_\alpha + K_\alpha)^2}.\ \frac{k_\beta^2\sigma_C(\beta)}{\sum k_\gamma^2\sigma_C(\gamma)} \tag{3.33}$$

or

$$\sigma_C(\alpha,\ \beta) = \pi(R+\lambda_\alpha)^2.\ \frac{4k_\alpha K_\alpha}{(k_\alpha + K_\alpha)^2} \bullet \left(\frac{k_\beta^2 \pi\,(R+\lambda_\beta)^2\, 4k_\beta K_\beta/(k_\beta + K_\beta)^2}{\sum k_\gamma^2 \pi\,(R+\lambda_\gamma)^2\, 4k_\gamma K_\gamma/(k_\gamma + K_\gamma)^2} \right) \quad (3.34)$$

As may be seen from the above, it is possible to obtain the decay probability through a given channel $\sigma_C(\alpha,\ \beta)$, if the cross-section $\sigma_C(\gamma)$ for the formation of the CN by all possible channels is known and if the Bohr's independence hypothesis is valid.

It may be remarked that the independence hypothesis holds good for isolated levels because of the fact that decay of the level formed depends only on the property of the level and not on how it was formed. At relatively higher energies the levels become so close to each other that the level width becomes comparable to the level spacing, then each level may be considered to be formed due to the superposition of a few states having nearby but different energies and phases. In such a case it is likely that the above hypothesis may not hold. On the other hand, when there is full overlapping or superimposition of levels due to the availability of many exit channels, the phase relationship between overlapping levels may not exist. Under the random phase approximation, the independence hypothesis becomes valid. Such a region of overlapping levels is called continuum region. Here the formation and decay of the CN become independent of each other due to the fact that any effect of the formation of the CN gets averaged over many overlapping states. The above description where it is assumed that overlapping levels have random phases is referred as the statistical model assumption. The validity of statistical assumption in the continuum region was confirmed by the pioneering experiment of Ghoshal [5]. In a statistical model, it is assumed that a CN is formed with excitation energy high enough so that many states may be excited by the interaction of the incident particle with the target nucleus.

3.3 Verification of Bohr's independent hypothesis

Bohr's CN hypothesis was confirmed by the pioneering experiments done by S N Ghoshal [5] in 1950. In his stacked foil experiments Ghoshal [5] studied the nuclear reactions forming the same CN at nearly the same excitation energy via two different channels, viz.,

$$\alpha + {}^{60}_{28}\mathrm{Ni} \rightarrow {}^{64}_{30}\mathrm{Zn}^* \quad (3.35)$$

and

$$p + {}^{63}_{29}\mathrm{Cu} \rightarrow {}^{64}_{30}\mathrm{Zn}^* \quad (3.36)$$

The experiments for measuring excitation functions for alpha induced reactions were carried out using the 40 MeV α-beam from the 60 inch cyclotron, employing activation technique. In the case of the α + nickel experiment, thin Ni foils of enriched ${}^{60}_{28}\mathrm{Ni}$ were prepared by electroplating metallic nickel on copper foil and the copper foil was then dissolved by $AgNO_3$ solution. The abundance of ${}^{60}_{28}\mathrm{Ni}$ in the enriched sample was more than 85%. Further, the proton excitation functions were obtained by the same method using the 32 MeV proton beam obtained from the Berkeley linear accelerator. Ordinary copper foils, consisting of ${}^{63}_{29}\mathrm{Cu}$ (69.1%) and ${}^{65}_{29}\mathrm{Cu}$ (30.9%) were used as the samples. The ${}^{63}_{29}\mathrm{Cu}$ interacting with a proton formed ${}^{64}_{29}\mathrm{Cu}$ giving rise to positron emission which was stopped by a 300 mg cm^{-2} aluminum absorber foil. When the proton is absorbed by ${}^{65}_{29}\mathrm{Cu}$ it forms ${}^{65}_{30}\mathrm{Zn}$ which has a half-life of ≈250 days which is too large as compared to the irradiation time and was, therefore, neglected. The incident energies of α particles and protons were chosen in such a way that the CN ${}^{64}_{30}\mathrm{Zn}^*$ is produced at nearly the same excitation energy. The energy difference between the Q-values of the two reactions is such that a proton beam of energy ε_p in the first irradiation and an alpha particle beam of energy $\varepsilon_\alpha = \varepsilon_p + 7\,\mathrm{MeV}$ in the second experiment produce the CN ${}^{64}_{30}\mathrm{Zn}^*$ in the same state of excitation. Experiments were carried out using the activation technique. In this technique the residual nuclides are identified through their unique radioactive decay properties. In activation technique the samples are bombarded with suitable beams which induce radioactivity in the sample. After the stop of irradiation, the samples are counted by a detector in a fixed geometry. The cross-section for the specific reaction responsible for the induced activity may be determined from the intensity of the induced activity. Using a stack of target samples with appropriate energy degraders in between successive samples, it is possible to irradiate different samples of the stack by incident particles of different energies. Thus, in a single irradiation it is possible to measure reaction cross-sections at different energies and to generate the excitation function for the reaction. In these activation experiments of Ghoshal [5], the cross-sections were measured as a function of incident beam energy over a wide range to obtain the excitation functions for the decay of CN ${}^{64}_{30}\mathrm{Zn}^*$ into the following exit channels:

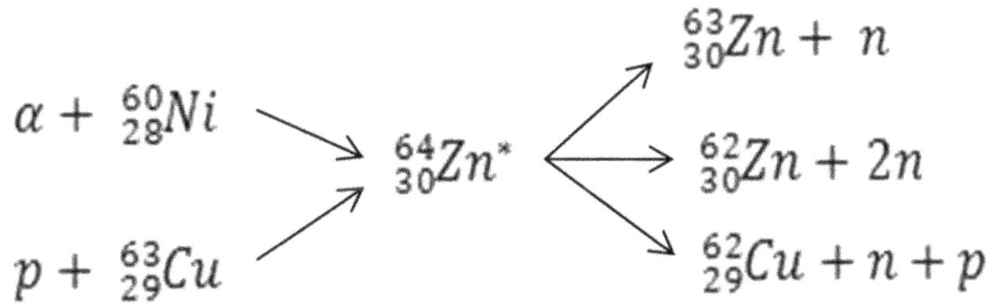

As may be seen from the above, the same CN ${}^{64}_{30}\mathrm{Zn}^*$ is formed via two different entrance channels that decay through three entirely different exit channels. If the CN theory proposed by Bohr is correct then the cross-section for the decay of the CN via respective channels should be independent of the entrance channels that formed the same CN. Ghoshal plotted the excitation functions for each reaction channel as a function of excitation energy as described above, leading to the same exit channel, and found that the ratio of cross-sections for decay into a given channel through different entrance channels was the same within statistical uncertainties. The results [5] of Ghoshal's experiment are illustrated in figure 3.4. As can be seen from the above

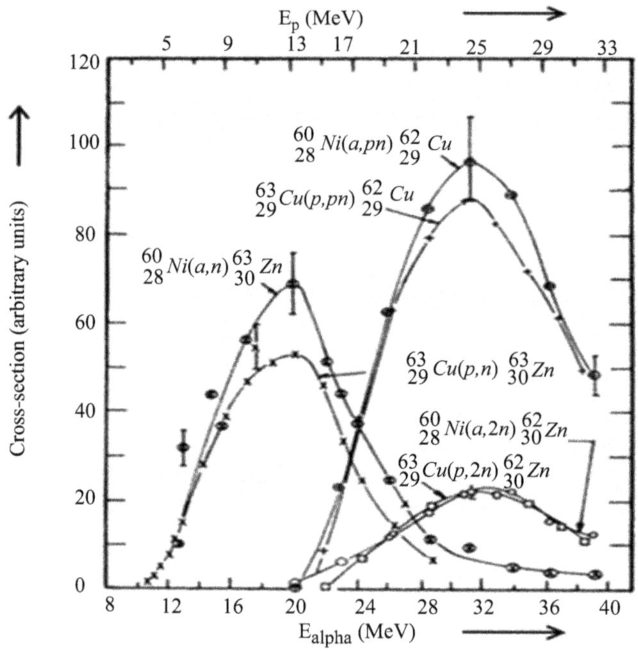

Figure 3.4. Measured cross-sections for $(\boldsymbol{p}, \boldsymbol{n})$, $(\boldsymbol{p2n})$, $(\boldsymbol{p}, \boldsymbol{pn})$ reactions on $^{63}_{29}$Cu and for $(\boldsymbol{\alpha}, \boldsymbol{n})$, $(\boldsymbol{\alpha}, \boldsymbol{2n})$, $(\boldsymbol{\alpha}, \boldsymbol{pn})$ reactions on $^{60}_{28}$Ni (adapted from [5] with permission from American Physical Society).

figure, the cross-sections are of the same order within the uncertainty, over the entire range of energy, for the two projectiles giving rise to the same exit channel i.e.,

$$\frac{\sigma(\alpha,\, n)}{\sigma(p,\, n)} = \frac{\sigma(\alpha,\, 2n)}{\sigma(p,\, 2n)} = \frac{\sigma(\alpha,\, pn)}{\sigma(p,\, pn)}, \tag{3.37a}$$

or

$$\sigma(\alpha,\, n)\colon \sigma(\alpha,\, 2n)\colon\ \sigma(\alpha,\, pn) = \sigma(p,\, n)\colon \sigma(p,\, 2n)\colon\ \sigma(p,\, pn) \tag{3.37b}$$

This clearly indicates that formation and decay of the CN are independent of each other. Ghoshal's experiments for the first time gave a direct proof of the independence hypothesis of compound reaction theory propounded by Bohr. S N Ghoshal born in 1923, did his post-graduation from Calcutta University, Calcutta, India, and completed his PhD under Professor Emilio Segre (co-winner of the Nobel Prize with Owen Chamberlain on anti-particles in 1959) at University of California, Berkeley. Ghoshal's pioneering work was to verify Bohr's theory of CN in 1950, which was perhaps the hottest topic during that time in nuclear physics. Later on, several authors verified Bohr's CN theory. Results of some of them are given in the following.

Walter John [6] measured excitation functions for the formation of the same CN ^{210}Po in excited state through α +^{206}Pb channel and compared it to the excitation functions for p+ ^{209}Bi reaction forming the same CN at nearly the same excitation energy. He found the following ratio to hold true;

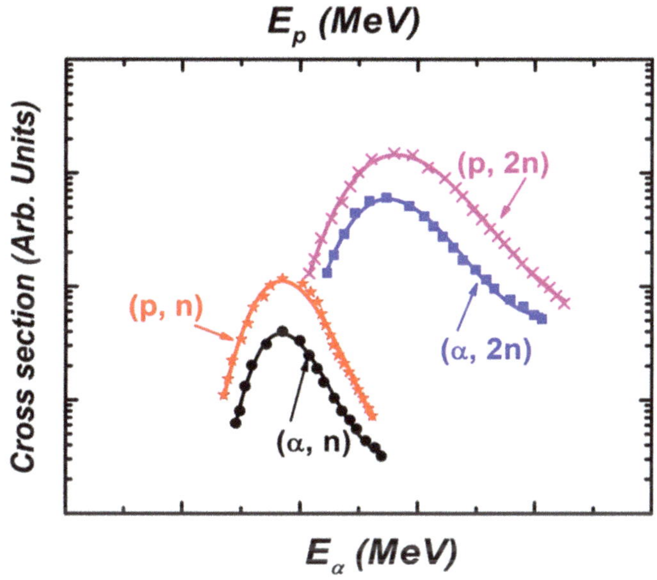

Figure 3.5. A typical representation of comparison of measured excitation functions for alpha and proton induced reactions on different targets forming the same CN, wherein Bohr's CN theory was found to be valid.

$$\sigma(p,\ 2n)\text{: } \sigma(p,\ 3n)\text{: } \sigma(p,\ 4n) = \sigma(\alpha,\ 2n)\text{: } \sigma(\alpha,\ 3n)\text{: } \sigma(\alpha,\ 4n)$$

Since the CN formed is the same in two reactions, the constant values of three cross-section ratios confirms the independence hypothesis of Bohr's CN theory.

Similarly, Rayudu and Yaffe [7] carried out experiments by bombarding Erbium ($^{168}_{68}$Er) with protons and holmium ($^{165}_{65}$Ho) with alpha particles forming the same CN ^{169}Tm*. The measured cross-sections for reactions $^{168}_{68}$Er(p, n) and $^{165}_{65}$Er(α, n); and $^{168}_{68}$Er(p, $2n$) and $^{165}_{65}$Ho(α, $2n$) were compared by plotting them in the same panel, keeping the energy axis in such a way that excitation energy for the two systems matches. As a representative case the typical form of the results was as shown in figure 3.5 where it was found that the ratios $\sigma(p,n)$: $\sigma(p,2n) = (\alpha,n)$: $\sigma(\alpha,2n)$ hold good, indicating the validation of Bohr's CN theory. Several such experiments were carried out by different authors and it was observed that at low energies the CN theory given by Bohr is valid.

3.4 Theoretical formulation

As already mentioned, at higher incident energies a large amount of energy is transferred to the target nucleus and the CN is highly excited. In these conditions the spacing of the levels becomes narrower as compared to their widths. As a result the levels start overlapping with each other. At these energies the energy spread of the incident beam may contain many resonance levels and the measured cross-section is the energy averaged cross-section. The behaviour of energy averaged cross-sections in terms of transmission coefficients was initially given by Hauser-Feshbach (1952) using Bohr's [4] CN assumptions.

Let us consider the following binary reaction,

$$^{A_P}_{Z_P}a + {}^{A_T}_{Z_T}X \rightarrow \left[{}^{(A_P+A_T)}_{(Z_P+Z_T)}\mathrm{C}^*\right] \rightarrow {}^{A_Y}_{Z_Y}Y + {}^{A_b}_{Z_b}b \tag{3.38}$$

The projectile $^{A_P}_{Z_P}a$ fuses with the target nucleus $Z_T A_T$X forming an excited CN [C*] that lives for considerably long till thermodynamic equilibrium gets established in the CN. After the equilibration the CN C* emits the particle $^{A_b}_{Z_b}b$ and it decays leaving behind the residual nucleus $^{A_Y}_{Z_Y}Y$.

3.4.1 Parameters of the entrance channel

In the above description of the binary reaction, the projectile nucleus has atomic number Z_P, and atomic mass number A_P. The linear and angular momentum of the projectile may be represented as p_a and ℓ_a while its spin and parity are, respectively, s_a and π^a. Let the kinetic energy of the projectile be E_a. Similarly, the target nucleus $^{A_T}_{Z_T}X$ has atomic number Z_T, and atomic mass number A_T. The linear and angular momentum of the target nucleus are p_X and ℓ_X, while its spin and parity are, respectively, s_X and π^X. Let the excitation energy of the target be E_A.

3.4.2 Parameters of exit channel

The residual nucleus $^{A_Y}_{Z_Y}Y$ has atomic number $Z_{\mathrm{Y},}$ and atomic mass number A_Y, and excitation energy E_Y. The linear and angular momentum of the residual nucleus are p_Y and ℓ_Y while its spin and parity are, respectively, s_Y and π^Y. In the similar way the emitted particle or the ejectile is $^{A_b}_{Z_b}b$ with atomic number Z_b and atomic mass number A_b. The linear and angular momentum of the emitted particle are p_b and ℓ_b, while its spin and parity are, respectively, s_b and π^b with its kinetic energy is E_b. The CN is in the excited state and is represented by $\left[{}^{(\mathrm{A_P}+A_T)}_{(\mathrm{Z_P}+Z_T)}\mathrm{C}^*\right]$, which has the atomic number $(Z_P + Z_T)$, and atomic mass number $(A_P + A_T)$. Let the CN have excitation energy E_C, spin J_C, linear and angular momentum p_C and ℓ_C, while its spin and parity are, respectively, s_C and π^C.

Now from the conservation of atomic number, atomic mass number, total energy, linear momentum, angular momentum and parity it follows that:

$$Z_a + Z_X = Z_C = Z_Y + Z_b \tag{3.39}$$

$$A_a + A_X = A_C = A_Y + A_b \tag{3.40}$$

$$E_a + E_X = E_C = E_Y + E_b \tag{3.41}$$

$$\vec{p}_a + \vec{p}_X = \vec{p}_C = \vec{p}_Y + \vec{p}_b \tag{3.42}$$

$$\vec{\ell}_a + \vec{s}_a + \vec{J}_X = \vec{J}_C = \vec{\ell}_b + \vec{s}_b + \vec{J}_Y \tag{3.43}$$

$$\pi^a \pi^X (-1)^{\ell_a} = \pi^C = \pi^b \pi^Y (-1)^{\ell_b} \tag{3.44}$$

It may, however, be noted that in equation (3.43) above, the sum of the terms is the quantum mechanical sum of three angular momentum values. As such, the CN angular momentum $\vec{J}_C$ can have a large number of possible values for a given set of $\vec{\ell}_a$, $\vec{s}_a$ and $\vec{J}_X$, each having a single value.In an actual experiment the compound nuclei are formed with a fixed excitation energy, definite charge, and definite mass but with a distribution of angular momentum values.

The quantum mechanical addition or coupling of the three angular moments l, s and j may be done according to two different coupling schemes: the L-S coupling scheme and the J-J coupling scheme. In L-S coupling scheme, first the total angular momentum $\vec{j}_a = \vec{\ell}_a + \vec{s}_a$ is obtained, where $\vec{j}_a$ is called the channel spin. Next, $\vec{j}_a$ is quantum mechanically added to $\vec{J}_X$ to get the spin of the CN $\vec{J}_{CN}$. The L-S coupling scheme, is generally used because of the fact that optical model transmission coefficients required for calculations in a statistical compound reaction mechanism, are available mostly with this scheme of coupling. Therefore, the entrance and the exit channel spins are given as the sum of orbital and spin angular momentum as;

$$\vec{\ell}_a + \vec{s}_a = \vec{j}_a; \vec{\ell}_b + \vec{s}_b = \vec{j}_b \tag{3.45a}$$

and

$$\vec{j}_a + \vec{J}_A = \vec{J}_{CN} = \vec{j}_b + \vec{J}_B \tag{3.45b}$$

Here, $\vec{j}_a$ and $\vec{j}_b$ are, respectively, the entrance and exit channel spins as obtained above.

From the point of view of the calculations it is easy and convenient to represent the reaction in terms of the input or the entrance channel α and the outgoing or the exit channel β with specific properties as discussed above.

3.5 Hauser-Feshbach formulations

In this formalism one considers the interacting particles to be spinless. For such particles, it has been shown [8, 9] using the diffusion theory that for the reaction with incident channel α to the exit channel β, the cross-section $\sigma_{(\alpha,\beta)}$ is given by the expression;

$$\sigma_{(\alpha,\beta)} = \frac{\pi}{k_\alpha^2} \langle | \delta_{(\alpha,\beta)} - S_{(\alpha,\beta)} |^2 \rangle \tag{3.46}$$

where, k_α is the wavenumber of the relative motion in the given entrance channel α while and $S_{(\alpha,\beta)}$ is the scattering matrix element. From the optical model the value of scattering matrix element $S_{(\alpha,\beta)}$ and the transmission coefficient T_α for the incident channel α are related as,

$$T_\alpha = 1 - | S_{\alpha\beta} |^2 \tag{3.47}$$

The probability for the formation of a CN through a particular entrance channel α of a binary reaction may be written in terms of the transmission coefficient T_α as,

$$\sigma_{CN}^{\alpha} = \frac{\pi}{k_\alpha^2} T_\alpha \tag{3.48}$$

It is assumed that the CN decays after the establishment of thermodynamic equilibrium, then the CN decay depends only on its quantum numbers and not on the reaction channel through which it was formed. As such, the cross-section for the formation of the CN by channel α and decay by channel β may be obtained from the product of the cross-section (σ_{CN}^{α}) for the formation of the CN via entrance channel α and the probability for the decay of the CN, $\{G^{CN}(\beta)\}$ through the reaction channel β. Therefore, the cross-section $\sigma_{(\alpha,\,\beta)}$ for the reaction may be written as;

$$\sigma_{(\alpha,\,\beta)} = (\sigma_{CN}^{\alpha})\bullet\ \{G^{CN}(\beta)\} \tag{3.49}$$

Since a given CN may decay through various different exit channels, say, β, γ_1, γ_2, ...γ_n which are allowed by applicable conservation laws, then the decay probability as indicated by the second term of equation (3.49) may be given by;

$$G^{CN}(\beta) = \left(\frac{T_\beta}{\sum_i T_{\gamma i}}\right) \tag{3.50}$$

and

$$\sigma_{(\alpha,\,\beta)} = (\sigma_{CN}^{\alpha})\left(\frac{T_\beta}{\sum_i T_{\gamma i}}\right) = \left(\frac{\pi}{k_\alpha^2}\right)T_\alpha\left(\frac{T_\beta}{\sum_i T_{\gamma i}}\right) = \left(\frac{\pi}{k_\alpha^2}\frac{T_\alpha T_\beta}{\sum_i T_{\gamma i}}\right) \tag{3.51}$$

The above expression for computing the cross-section given by equation (3.51) is referred to as the Hauser-Feshbach formula, and is applicable only for the spinless particles. The above formulations have been derived within the framework of Bohr's independent hypothesis.

In a typical binary reaction when the projectile ${}^{AP}_{ZP}a$ with a definite energy (E_a), spin (s_a) and angular momentum (ℓ_a) fuses with the target nucleus ${}^{AT}_{ZT}X$ which has a given excitation energy (E_A) and spin (J_A) the resulting CN will have a distribution of angular momentum values with certain statistical weight. Let us consider that these compound nuclei with spins J_{CN}^1, J_{CN}^2, J_{CN}^3, ... , J_{CN}^N are populated with the statistical weights w^1, w^2, w^3,...,w^N respectively. The CN spin J_{CN} is the quantum mechanical sum of three angular momentum values, ℓ_a, j_a and J_A. In the case of a given reaction many incident partial waves with different values of orbital angular momentum ℓ_a contribute to the reaction. The maximum value of the projectile orbital angular momentum $\ell_a^{\max}$ may be estimated using the classical expression for angular momentum, i.e., $l = Rxp$. The maximum value of R in this expression may be taken as the sum of the radii of the target and projectile nuclei, i.e., $R = r_0(A_P^{1/3} + A_T^{1/3})$, where $r_0 = 1.2$ fm, while A_P and A_T are, respectively, the projectile and target atomic mass numbers. Further, the linear momentum of relative motion p may be obtained from the centre-of-mass (c.m.) energy of the incident particle using the relation, $E_a^{CM} = \left(\frac{A_T}{A_T + A_P}\right)E_a$. Here, E_a is the kinetic energy of the incident particle in laboratory frame of reference. The possible values of the spin J_{CN} of the CN may vary from $J_{CN} = | J_A - s_a |$ to $(\ell_a^{\max} + j_a + s_a)$ in steps of unity. The statistical weight factor of the CN with a spin J_{CN} is given by expression $\left\{\frac{2J_{CN}+1}{(2J_A+1)(2s_a+1)}\right\}$. Further, all the reaction channels corresponding to the possible allowed values of j_a and ℓ_a will give rise to the formation of the CN with spin J_{CN} and all channels corresponding to the possible allowed values of j_b and ℓ_b will contribute to the decay of the CN with spin J_{CN}. For the cases when the projectile nucleus and the target nucleus have non-zero spins the cross-section for the formation of the CN $\sigma_{(a,\ A,\ E_a)}^{CN}$ will be composed of contributions from various possible values of CN angular momentum values J_{CN}. Therefore, the cross-section for the CN formation at a given energy E is given by;

$$\sigma_E^{CN} = \sum_{J_{CN}^{\min}}^{J_{CN}^{\max}} \sigma_{(E)}^{J_{CN}} \tag{3.52a}$$

while

$$\sigma_{(E)}^{J_{CN}} = \frac{(2J_{CN} + 1)}{(2J_A + 1)(2s_a + 1)} \sum_{\ell_a^{\min}}^{\ell_a^{\max}} T_{(E,\ l_a)}^{J_{CN}} \tag{3.52b}$$

The cross-section for a given value of J_{CN} depends on two factors, the statistical factor $\left\{\frac{2J_{CN}+1}{(2J_A+1)(2s_a+1)}\right\}$ and the transmission coefficient $T_{(E_a,\ \ell_a)}^{J_{CN}}$, which decreases with increasing value of ℓ_a. In view of this, the maximum contribution to the cross-section σ_E^{CN} comes from value(s) of J_{CN} ranging from $J_{CN}^{\min}$ and $J_{Cn}^{\max}$.

The cross-section $\sigma(a,b)$ for the reaction mentioned above, can then be written as,

$$\begin{aligned}\sigma(a,\ b) = \frac{\pi}{k_a^2} & \sum_{J_{CN} = |J_A - s_a|}^{\ell_a^{\max}+J_A+s_a} \sum_{\pi^{CN}=} \\ & \pm \frac{(2J_{CN} + 1)}{(2J_A + 1)(2s_a + 1)} \\ & \sum_{j_a = |J_{CN}-J_A|}^{J_{CN}+J_A} \sum_{\ell_a = |J_A - j_a|}^{J_A+j_a} \sum_{j_b = |J_{CN}-J_B|}^{J_{CN}+J_B} \sum_{\ell_b = |j_b - s_b|}^{j_b+s_b} K\end{aligned} \tag{3.53}$$

where, the term,

$$K = \frac{T_{(E_a,\ J_A,\ j_a,\ \ell_a)} T_{(E_b,\ J_B,\ j_b,\ \ell_b)}}{\sum_\gamma T_\gamma} \delta(a,\ \pi^{CN}) \delta(b,\ \pi^{CN}) \tag{3.54}$$

In the above expression (3.54), the terms $T_{(E_a,\ J_A,\ j_a,\ \ell_a)}$ and $T_{(E_b,\ J_B,\ j_b,\ \ell_b)}$ are the transmission coefficients for the incoming and outgoing channels, respectively. The functions represented by $\delta(a,\ \pi^{CN})$ and $\delta(b,\ \pi^{CN})$ take care of the conservation of parity in the process.

For the scattering angle ϑ, one can give the angular distribution of emitted particle in the c.m. frame using the expression,

$$\sigma^{\vartheta}(a,\ b) = \sum_{\text{even} L} C_L P_L(\cos\vartheta) \tag{3.55}$$

where, C_L is given by;

$$C_L = \frac{\pi}{k_a^2} \sum_{J_{CN},\ \pi^{CN}} \frac{(2J_{CN} + 1)}{(2J_A + 1)(2s_a + 1)} \sum_{(j_a,\ell_a j_b,\ell_b,)} XA^{J^{CN}}_{(J_A,\ \ell_a,\ j_a,\ J_B,\ \ell_b,\ j_b,\ L)} \tag{3.56}$$

Here, the term

$$A^{J^{CN}}_{(J_A,\ \ell_a,\ j_a,\ J_B,\ \ell_b,\ j_b,\ L)} = \frac{(-1)^{(J_B - s_b - J_A + s_a)}}{4\pi}(2J + 1)(Z_1)(Z_2) \tag{3.57}$$

where, Z_1 and Z_2 are given by;

$$Z_1 = (2j_a + 1)(2l_a + 1)C(1100\colon L\ 0)W(j_a j_a \ell_a \ell_a : L s_a) \tag{3.58}$$

and

$$Z_2 = (2j_b + 1)(2\ell_b + 1)C(1100\colon L\ 0)W(j_b j_b \ell_b \ell_b : L s_b) \tag{3.59}$$

Here, W and C are Racah and Clebsch–Gordan coefficients, respectively. Thus, by putting the values of W and C, in expressions (3.58) and (3.59), Z_1 and Z_2 may be obtained. Using the values of Z_1 and Z_2, one may calculate the value of C_L. Putting the value of C_L in equation (3.55) one may get the angular distribution of ejectiles in the c.m. frame. The calculated angular distribution may be compared with the experimental values to understand the validity of theoretical assumptions used above. The Hauser-Feshbach theory has been found to predict the cross-sections of nuclear reactions rather well at relatively lower energies. However, as energy increases theoretical predictions of Hauser-Feshbach formalism starts diverging from experimental data, which may be due to onset of some other reaction mechanism.

3.6 Weisskopf-Ewing formulation

In principle, the Hauser-Feshbach formula gives cross-section from incident channel α to exit channel β as energy averaged over well separated Briet–Wigner resonances. It is strictly true at lower energies where the resonances do not overlap, i.e.

$$\frac{<\Gamma>}{D} \ll 1 \tag{3.60}$$

Here $<\Gamma>$ is the average total width of the resonances of a given J_C value and D is the average spacing of these resonances. Further, the Hauser-Feshbach derivation assumes spinless particles in the entrance channel but rigorously conserves total angular momentum and parities. If width fluctuation corrections are also included in Hauser-Feshbach derivation then it is not possible to split the result into multiplication of two factors; the cross-section for the formation of the CN through channel α and its subsequent decay via channel β. This essentially means the breakdown of independence hypothesis.

In simpler and earlier Weisskopf-Ewing formulation, parity and total angular momentum are not conserved. If conservation of total angular momentum and parity are neglected, then it can be shown that the branching ratios can be expressed entirely in terms of reaction cross-sections, without requiring the explicit use of transmission coefficients. Before proceeding further let us look to situations where conservation of total angular momentum and parity may be overlooked.

(i) If the energy of the CN is sufficiently high so that nearly all channels into which it can decay are dominated by integrals over the level density then explicit inclusion of parity and total angular momentum may not be required. Of course the decay to isolated discrete levels of the residual nucleus must be small.

(ii) In the case where above condition (1) is fulfilled then width fluctuations will become unimportant.

(iii) Transmission coefficients are independent of the spin J_C of the states reached. This condition is generally satisfied as transmission coefficients are very nearly independent of the target spin.

(iv) The spin dependence of level density is proportional to $(2J_C + 1)$ and should not depend on parity. This condition is generally fulfilled if spins are not unusually large as level density is generally written as;

$$\rho_c(E_c, J_c, \Pi_c) = \rho_c(E_c)(2J_1 + 1)\exp\left[\frac{(J_c + 1/2)^2}{2\sigma^2(E_c)}\right] \tag{3.61}$$

Here σ is the spin cut-off parameter. The spin cut-off parameter σ is generally related to the effective moment of inertia. This parameter may depend on the atomic mass number A, the level density parameter a or the nuclear temperature T. In some cases, it has been predicted that the spin cut-off parameter σ will increase with the excitation energy E.

In the statistical theory there are basically three assumptions, which are quite important. These assumptions are:

(i) When the incident particle of suitable energy hits on a target nucleus it forms a CN and the available excitation energy is distributed among the nucleons of the CN randomly, within 10^{-15} s, after the formation of the CN. Further, once the equilibrium has been established, the sufficient amount of excitation energy more than the binding energy of the particle in the nucleus, may get concentrated on any one particle which can then escape from the nucleus.

(ii) For sufficiently higher values of excitation energy, the overlapping levels may get excited and the number of excited levels becomes quite large to be treated individually. This requires Fermi statistics to be taken into consideration.

(iii) Further, considering Bohr's hypothesis, the mode of decay of the CN is independent of the mode of its formation and depends only on its energy (E), angular momentum (J) and the parity (π).

In order to obtain the average reaction cross-section $\sigma(j,k)$, using Weisskopf-Ewing statistical model [10] one gets,

$$\overline{\sigma}_{jk} = \sigma_{\mathrm{Comp}}(j)\left(\frac{G_k}{G}\right) \tag{3.62}$$

where, j is the incident particle, k the outgoing particle and $\sigma_{\mathrm{Comp}}(j)$ is the cross-section of the CN. Further, $\left(\frac{G_k}{G}\right)$ is the branching ratio for the decay.

However, from statistical theory, the total cross-section of the above reaction averaged over many resonance structures may be given as [4],

$$\overline{\sigma}_{jk} = \sum_{J,\pi} \sigma_{jk} J^{\pi} \tag{3.63}$$

Further, with the assumption that the formation and decay of the CN are independent of each other, we have,

$$\overline{\sigma}_{jk} = \left[\sigma_{\mathrm{Comp}} J^{\pi}(j)\right]\left[\frac{G_t J^{\pi}}{\Sigma G_t J^{\pi}}\right] \tag{3.64}$$

The summation over t in equation (3.64) runs over all possible reaction alternatives available to the CN. As equation (3.64) is now valid for each spin and parity, i.e., J, π values separately, the factorization of the cross-section as represented in equation (3.64) above does not imply that the CN formed forgets its total angular momentum and the parity.

The relation between CN cross-section and the branching ratio can be found by applying the reciprocity theorem. As such,

$$\frac{\overline{\sigma}_{jk} J^{\pi}}{\overline{\sigma}_{kj} J^{\pi}} = \frac{K_k^2}{K_j^2} \tag{3.65}$$

where, K_k and K_j are, respectively, the wavenumbers for the relative motion of the given pair k and j.

From equations (3.64)and (3.65), we have;

$$G_j J^{\pi} = K_j^2 \sigma_{\mathrm{Comp}} J^{\pi}(j) \tag{3.66}$$

Substituting the value of $G_t J^\pi = k_t^2 \sigma_{\text{Comp}} J^\pi(t)$ back in equation (3.64), one can obtain the Weisskopf-Ewing formula for the reaction cross-section as,

$$\bar{\sigma}_{jk} = \left[\sigma_{\text{Comp}} J^\pi(j)\right]\left[\frac{k_t^2 \sigma_{\text{Comp}} J^\pi\,(t)}{\sum k_t^2 \sigma_{\text{Comp}} J^\pi\,(t)}\right] \tag{3.67}$$

The formulations obtained above may be used to calculate cross-section for any desired reaction.

3.7 Nuclear level density

When a nucleus is excited, at lower excitation well-separated discrete excited levels are reached. However, with the increase of excitation energy, overlapping levels with energy separations less than the level width are obtained. In this region of excitation it is not possible to talk about a single excited level, instead one talks of energy level density, i.e., the number of levels per unit energy interval at some excitation energy E^*. The nuclear level densities are used in statistical calculation of compound nuclear decay and, therefore, useful for both pure as well as applied research. As such, nuclear level densities are important in many areas of physics including the nuclear reaction model studies as well as in astrophysics. A nuclear model that may correctly reproduce the sequence, energy, spin etc, of nuclear energy levels for a large number of nuclei must have mostly correct elements of physics. This is why considerable efforts have been made towards both; gaining experimental information about nuclear levels and developing models that may predict level densities. Traditionally, for most of the nuclei the level density is obtained on the basis of experimental information from low lying discrete energy levels and from the neutron resonance spacing.

3.7.1 Models for level density

Not only at high energies but even at relatively lower energies where nuclear levels are well separated, the number of levels in a given energy interval or energy bin are sufficiently large. Therefore, instead of talking of individual level, it is more appropriate to talk about level density, i.e., number of levels in a given energy bin or per unit energy interval. Another closely related quantity which is often used in literature is the state density. The difference between the level density and the state density lies in the fact that a state is one of the $(2J + 1)$ spin projections of a level of spin J. Therefore, the state density $\rho_{\text{state}}(U)$ of states corresponding to a level of spin J at excitation energy U may be written as;

$$\rho_{\text{state}}(U) = \sum_J (2J + 1)\rho_{\text{level}}(U,\ J) \tag{3.68}$$

Here, $\rho_{\text{level}}(U,\ J)$ is the density of levels of spin J at energy U. It may be noted that the sum over J in the above expression extends over all the J values for which levels are present. Under the assumption that the distribution in J_z of states is Gaussian, it may be shown that state density may be written in terms of the parameter σ, (not to be confused with cross-section) called the spin cut-off parameter defined as;

$$\sigma = (\langle J_z^2 \rangle)^{1/2} = \left(\frac{1}{3}\langle J(J+1)\rangle\right)^{1/2} \tag{3.69}$$

The expression for state density in terms of spin cut-off parameter σ becomes;

$$\rho_{\text{state}}(U) = (2\pi)^{1/2}\sigma \sum_J \rho_{\text{level}}(U, J) \tag{3.70}$$

The spin cut-off parameter σ, is generally energy dependent.

The formulas for the level density are usually separated into two multiplicative factors; a function $f(J)$ that takes into account the spin part and the total level density $\rho(U)$ that increases exponentially with energy;

$$\rho(U, J) = f(J)\rho(U) \tag{3.71}$$

Bethe [11], perhaps for the first time, derived an analytical expression for nuclear level density in an equidistant non-degenerate single-particle-level approach, which may be treated as the zeroth-order approximation of Fermi gas model. Later, Rosenzweig and Kahn [12] showed that the effect of the periodic shell structure can be approximately accounted for by introducing an energy shift in Bethe formula. This gave rise to what is called the backshifted Fermi gas model [13] in which both the energy backshift U_0 and the level density parameter 'a' of Bethe's formula are treated as free parameters to fit the experimental data. Backshifted level density is given as;

$$\rho(U) = \frac{1}{12\sqrt{2}\sigma} \frac{\exp\left(2\sqrt{a(U - U_0)}\right)}{a^{1/4}(U - U_0)^{5/4}} \tag{3.72}$$

Although, the level density parameter 'a' is often treated as an adjustable parameter to fit the experimental data, however, it is equal to $\left(\frac{\pi^2}{6}\right)g$ where 'g' is the density of single-particle states. In semi-classical approximation the level density parameter 'a' may be written as;

$$a = 2\left(\frac{\pi}{3}\right)^{4/3} \frac{m_0 r_0^2}{\hbar^2} A\left(1 + \beta_s A^{-1/3}\right) \tag{3.73a}$$

and

$$\sigma^2 = \langle m^2 \rangle g t = It, \quad U = at^2 \tag{3.74b}$$

Here, m_0 is the nucleon mass, r_0 the nuclear radius parameter, A the mass number, β_s the surface component of the single-particle-level density, t the nuclear temperature, I the moment of inertia of the heated nucleus. Assuming the nucleus as a rigid rotator, the moment of inertia I_0 may be given as,

$$I_0 = \frac{2}{5}\frac{m_0 r_0^2}{\hbar^2}A^{5/3} \tag{3.74}$$

The expression given by equation (3.72) neglects the possible dependence of the level density on parity of levels. One may define a parameter, parity ratio $\pi(U)$ as;

$$\pi(U) = \frac{\rho^+(U)}{\rho^+(U) + \rho^-(U)} \tag{3.75}$$

Here, $\rho^+(U)$ and $\rho^-(U)$ are, respectively, the densities for even and odd parity levels. It is expected that $\pi(U)$ will approach 0.5 at higher energy where probability of even and odd parity levels is nearly equal, but at lower energy and for even-A nuclide it is expected to approach 1. For odd-A nuclei at lower energy $\pi(U)$ may converge to 1 or zero depending on the parity of the orbital of the odd particle; if it is even then $\pi(U)$ will be one.

Another expression frequently used for level density and referred to as the constant temperature formula is given as,

$$\rho_{CT} = \frac{1}{T}e^{(U-U_0')/T} \tag{3.76}$$

Both T and U'_0 in the above expression are treated as free parameters to reproduce the experimental data.

The main difference between the Fermi gas model and the constant temperature model for calculating the level density lies in the fact that while in the former level density is obtained from the possible ways in which a given excitation energy may be distributed in single-particle states, i.e., through methods of statistical mechanics, while in the constant temperature approach nuclear temperature is defined in terms of the derivative of the level density through the following equation,

$$T \equiv \left[\frac{\mathrm{d}\rho(U)}{\mathrm{d}U}\right]^{-1} \tag{3.77}$$

In the constant temperature model it is assumed that within a given range of excitation energy, any increase of excitation energy does not increase the temperature of the nucleus. This is possible if on increasing the excitation energy the number of degrees of freedom of the nuclear system increases so that the additional excitation energy gets distributed in these additional degrees of freedom. This is in close analogy to the first order phase transitions when matter changes from solid to liquid or from liquid to gas phase. In other words, the nuclei behave as if it is a mixture of ice and water, the temperature of which remains constant on heating as more ice gets converted into water absorbing energy without increase of temperature. This phenomenon is often explained as the breaking of nuclear Cooper pairs with the excitation energy leaving mean energy per excited particle unaltered. Several recent experiments indicated that analysis of light particle spectra based on Fermi gas like temperature approach does not fit the data even at higher energies of the order of 30 MeV or so, while the same data is well reproduced by a constant temperature approach. Very interesting information in support of the constant temperature model comes from the data of low energy neutron induced fission. It is known that in low energy neutron induced fission two fission fragments of different masses are formed from which neutrons are emitted. In an experiment, the average number of neutrons emitted from the heavier and the lighter fragments were measured when ^{237}Np was made to undergo fission by neutrons of 0.8 and 5.55 MeV energies. It was recorded in the experiment that the average number of neutrons emitted only from the heavy fragment increased with the increase of the incident neutron energy, while the average number of neutrons emitted from the lighter fragment remained almost constant. This means that the additional excitation energy pumped into the fashioning system by the higher energy neutrons was almost totally taken by the heavier fragment. According to the Fermi gas model approach, in a normal case the additional excitation energy should have been shared by the two fragments in the ratio of their masses and, therefore, the result of the experiment could not be explained by the Fermi gas model approach. However, according to the constant temperature approach the two fission fragments will be formed at two different temperatures, the temperature of the heavier fragment will be lower and, therefore, almost total additional excitation energy will be taken by it. Thus, it may be said that the constant temperature approach appears to be valid up 30 MeV.

In the past there was a feeling that the backshifted Fermi gas model gives a better description of level density at higher excitation energies and that constant temperature model at lower excitations. Gilbert and Cameron [14] therefore, gave a composite formula where at higher energies the level densities are calculated using Fermi gas model and at lower energies using the constant temperature approach. The smooth matching of the level density and its derivative is demanded at some energy U_x such that,

$$U_x = U_0 + T\ \ln\rho_{\text{Fermi}}(U_x) \tag{3.78}$$

Both in the backshifted Fermi gas model and the constant temperature model, the shell and pairing effects are simply plugged-in without any physical reasoning. In order to evolve these effects from microscopic description it is required to have a model that takes into account [15] non-uniform distribution of single-particle levels on one hand and residual interaction effects on the other. Although attempts have been made to develop such models, microscopic calculations of level densities are very laborious, time consuming and, therefore, they are not useful for analysis of available experimental data on level densities.

The effects of pairing, particularly in even–even nuclei, may be compared to the property of superconductivity in fluids. This analogy is the basis of superfluid and microscopic generalised superfluid models. The superfluid property of the nucleus may be characterised by the value of the correlation function $\Delta_{0\tau}$ that directly determines the even–odd differences in nuclear binding energies and the energy gap $2\ \Delta_{0\tau}$ in the spectrum of quasi-particle excitations of even–even nuclei. The critical temperature t_c of the phase transition from a superfluid state to normal state is related to the correlation function through the following relation,

$$t_c = 0.567\Delta_0 \tag{3.79}$$

The excitation energy of the nucleus at critical temperature may be written as,

$$U_c = \frac{\pi^2}{6}gt_c^2 + \frac{1}{4}g\Delta_0^2 - n\Delta_0 \tag{3.80}$$

In the above expression for excitation energy at critical excitation, $n = 2$ for odd–odd nuclei, $n = 1$ for odd nuclei and $n = 0$ for even–even nuclei. Above the critical temperature, the pairing correlation decreases the excitation energy by the amount of the condensation energy E_{cond}, given as,

$$E_{\text{cond}} = \frac{1}{4}g\Delta_0^2 - n\Delta_0 \tag{3.81}$$

The coherent collective effects in level density may be included by multiplying the level density due to quasi-particle excitation $\rho_{qp}(U)$ by enhancement coefficients $K_{\text{vibr}}(U)$ and $K_{\text{rot}}(U)$ which, respectively, describe the enhancements due to vibration and rotational modes. Therefore,

$$\rho(U) = \rho_{qp}(U)K_{\text{vibr}}(U)K_{\text{rot}}(U) \tag{3.82}$$

In adiabatic approximation, vibrational enhancement coefficient $K_{\text{vibr}}(U)$ has the value I for spherical nuclei and $(I_\perp)$ for deformed nuclei. Here, $I_\perp$ is the moment of inertia of the nucleus about the perpendicular axis.

The value of the vibrational enhancement coefficient in microscopic approach is obtained from the following relation,

$$K_{\text{vibr}} = \prod_i \left[\frac{1 - \exp(-\omega_i^0/t)}{1 - \exp(-\omega_i/t)}\right]^{g_i} \tag{3.83}$$

Here, ω_i^0 and ω_i are, respectively, the excitation energies of quasi-particle excitations and corresponding vibrational excitations and g_i is the degeneracy of these excitations. All microscopic models developed during the last few decades have indicated that the effect of rotational motion diminishes with the increase in the excitation energy, one estimate of this damping of K_{rot} based on SU-3 model may be given as,

$$K_{\text{rot}}(U) = \frac{K_{\text{rot}}^{\text{adia}}(U)}{1 + e^{\frac{(U-U_r)}{d_r}}} \tag{3.84}$$

where, $U_r = 120A^{1/3}\beta^2$ MeV and $d_r = 1400A^{-2/3}\beta^2$ MeV. Here β is the quardrupole deformation parameter. The vibrational enhancement coefficient has been estimated as,

$$K_{\text{vib}} = e^{[\delta S - (\delta U/t)]} \tag{3.85}$$

Here, δS and δU are, respectively, the changes in the entropy and the excitation energy due to vibrational modes and are given as,

$$\delta S = \sum_i (2\lambda_i + 1)[(1 + n_i)\ln(1 + n_i) - n_i \ln n_i] \tag{3.86}$$

$$\delta U = \sum_i (2\lambda_i + 1)\omega_i n_i \tag{3.87}$$

where, ω_i, n_i and λ_i are, respectively, the energies, occupation numbers and multipolarities of vibrational excitations at a given temperature. In order to take into account the damping of collective effect at higher excitations, the occupation number is expressed in terms of their spreading widths γ_i as follows,

$$n_i = \frac{e^{(\gamma_i/2\omega_i)}}{e^{(\omega_i/t)} - 1} \tag{3.88}$$

$$\text{with } \gamma_i = C(\omega_i^2 + 4\pi^2 t^2)$$

The value of C has been estimated from the systematic of the neutron resonance densities of medium weight nuclei and has been found to be $C = 0.007\ A^{1/3}\ \text{MeV}^{-1}$.

In these formulations, the non-uniform distribution of nuclear levels that is the shell effect may be taken into account through the variation in the value of the level density parameter a. Shell effects were found to decrease with the excitation energy and vanish at higher energies where the value of the level density parameter a may be estimated from their semi-classical formulation.

A better, reliable and rigorous description of level densities and other statistical parameters of excited nuclei may be obtained by using the realistic single-particle level schemes. The state equations for thermodynamic functions of an excited nucleus may be given as,

$$S = \sum_i g_i[\beta E_i \overline{n}_i + \ln\{1 + e^{-\beta E_i}\}] \tag{3.89}$$

$$U = \frac{1}{2}\sum_i g_i[((\varepsilon_i - \lambda)^2 + \Delta_0^2)^{1/2} - E_i(1-2\overline{n}_i)] + \frac{\Delta_0^2 - \Delta^2}{G} \tag{3.90}$$

Here, E_i = energy of quasi-particle excitations

$\overline{n}_i$ = the occupational numbers for the corresponding single-particle levels

$\beta = \frac{1}{T}$ = inverse temperature.

Further, N_r, the number of protons or neutrons in a nucleus, λ_r, the corresponding chemical potential, G_r, the pairing force constant and quasi-particle energies are related with each other through the following expressions:

$$N_r = \frac{1}{2}\sum_i g_i\left[1 - \frac{(\varepsilon_i - \lambda_r)}{E_i}(1-2\overline{n}_i)\right] \tag{3.91}$$

$$G_r = \left[\frac{1}{4}\sum_i g_i\left(\frac{1-2\overline{n}_i}{E_i}\right)\right]^{-1} \tag{3.92}$$

If the single-particle level scheme of a nucleus is known, then it is possible to calculate the nuclear level density and other thermodynamic functions for the nucleus.

3.7.2 Sources of information about level density

There are basically four ways to get information about level density.

(i) At lower energies where levels are well separated, detailed spectroscopic studies provide information about level density. Simple counting of levels in a given energy interval may yield level density. Actual counting of levels of known parity and spin also yield parity ratio $\pi(U)$ and spin cut-off parameter σ. This method is limited only for the case where complete spectroscopic data is available and also there is no overlap of levels, which becomes a big problem for heavy nuclei.

(ii) Study of neutron resonances at low energies has been used to infer information about compound nuclear level density around neutron binding energies. S-wave neutron resonances are of special interest as the compound nuclear states are restricted to $J_C = J_t \pm 1/2$, values, where J_C and J_t are, respectively, the spins of the compound and the target nuclei. Since, compound nuclear states are restricted only to two values of spins, the levels remain well resolved even when level density is high. However, the main problem in these experiments is the interference of p-wave and higher angular momentum neutrons, the unavoidable presence of which complicates the observed resonance spectrum. Another problem closely associated with resonance measurements is that levels of only one parity and of restricted spin are studied, and therefore, the two important parameters, the parity ratio and spin cut-off parameter, cannot be obtained from these measurements.

(iii) The shape of evaporation spectra of emitted particles in principle may be analysed to get information about level densities. As a matter of fact, spectra of neutrons, protons and alpha particles emitted in compound reactions have been used to infer level densities. Complications in such analysis arise because of the presence of contributions from non-compound processes, Coulomb suppression of charged particle emission, particularly in heavy nuclei, and multiple particle emission at higher energies.

(iv) It has been observed that compound reaction cross-sections, measured with good energy resolution, fluctuate with energy even in the region of overlapping levels where cross-sections are expected to be smooth function of energy. These fluctuations of cross-section with energy are called Ericson fluctuations and are the outcome of interference effects between overlapping levels. The extent of fluctuations is related to the level density. However, as expected, Ericson fluctuations diminish with the increase of excitation energy and ultimately vanish around the excitation energy $\approx$25 MeV.

3.7.3 Parameterization

Since microscopic models for level density are too complicated, tedious to work and in general do not provide level density values that are in any way better than the ones obtained by either the backshifted Fermi gas model or the constant temperature model, the general approach is to parameterize these models by analysing a large amount of data. Parameterization is done by obtaining the best fit value of the level density parameter 'a' and the backshift energy U_0 (also represented by ξ) from the experimental data on level density in the backshifted Fermi gas model or of the backshift energy U_0' for the constant temperature model. Several analyses based on the above method have been carried out in the past. Huang *et al* [16] studied the level densities obtained by counting of levels for nuclei in the range $20 \leqslant A \leqslant 41$ and obtained a good fit for conventional parameterization of $a = \alpha + \beta A$. Also, they found the value of energy shift U_0 to be more than the pairing energy and argued that the shift may be simulating collective effects. Simulating collective effects in energy shift U_0, results in diminishing the relative importance of collective effects at higher energies, which is physically justified.

In another similar analysis that included nuclei in the range $20 \leqslant A \leqslant 70$, and laid stress on the difference of level densities for nuclei lying in the valley of stability to those off the stability line has been done by Al-Quraishi *et al* [17]. They also based their analysis on the data obtained from spectroscopic information on level schemes and counting of levels, therefore, they were able to obtain the spin ratio $\pi(U)$ and the spin cut-off parameter σ from their analysis. They deduced the following two functional forms for the level density parameter 'a';

$$\text{(i) } a = \frac{\alpha A}{\exp[\gamma(z - z_o)^2]} \text{ and (ii) } a = \frac{\alpha A}{\exp[\beta(N - Z)^2]} \tag{3.93}$$

where, z is the atomic number of the nucleus, and z_0 is the atomic number of the beta stable nucleus with mass number A. The value of constant γ obtained from least square fit has been found to be approximately 0.05. Form (i) gave a better fit to the data, but form (ii) reflects the isospin dependence of level density. They also inferred from their analysis that level density for nuclei in the valley of stability is larger than those at the line of stability, at comparable excitation energy.

In another comprehensive and detailed analysis based on spectroscopic data and counting of levels Egidy and Bucurescu [18] considered the spectroscopic data of 310 nuclei in the range ^{18}F to ^{251}Cf and obtained empirical values for the parameters for both the backshifted Fermi gas model as well as for the constant temperature model. In their analysis Egidy *et al* [18] represented the level density $\rho(U, J)$ by equation (3.47) and discussed the nature of both the spin distribution function $f(J)$ and the energy dependent total level density $\rho(U)$. The Ericson formula [19] for spin distribution function has a single free parameter σ and is given by;

$$f(J, \sigma) = \exp\left(-\frac{J^2}{2\sigma^2}\right) - \exp\left(-\frac{(J + !)^2}{2\sigma^2}\right) \approx \frac{2J + 1}{2\sigma^2} e^{-\frac{J(J+1/2)}{2\sigma^2}} \tag{3.94}$$

The spin cut-off parameter is generally related to an effective moment of inertia of the nucleus and may also be a function of the level density parameter a, or the nuclear temperature T, respectively, for the case of backshifted Fermi gas model and the constant temperature approach. In their analysis of 8116 levels in 1556 spin groups, Egidy and Bucurescu [18] found that in the case of even–even nuclei the spin distribution function $f(J, \sigma)$ staggers or oscillates with the magnitude of spin J and cannot be described by a smooth function. This oscillatory behaviour was found for all even–even nuclei, while no such staggering with spin was observed in cases of odd–odd and odd mass nuclei. Based on their observations they proposed the following formula for the spin distribution function $f_{ee}(J, \sigma)$ for even–even nuclei as;

$$f_{ee}(J, \sigma) = f(J, \sigma).\,(1 + x) \tag{3.95}$$

where,

$$x = \begin{cases} +0.227(14) \text{ for even spin values} \\ -0.227(14) \text{ for odd spin values} \\ +1.02(9) \text{ for zero spin levels} \end{cases}$$

Further, they did not observe any mass dependence of x. It was argued that the staggering of spin distribution function with spin for even–even nuclei is already predicted by most of the theoretical models, including the interacting boson approximation.

Regarding the spin cut-off parameter σ, they concluded that the only important dependence is on mass and in general, it may be reasonably well represented by the expression,

$$\sigma^2 = 2.61A^{0.28} \tag{3.96}$$

In another analysis where individual moments in the (U, J) plane were calculated for each nucleus for which levels were known, the following dependence of spin cut-off parameter on mass A and excitation energy U was observed,

$$\sigma^2 = 0.391.\,A^{0.675}(U - 0.5.\,Pa')^{0.312} \tag{3.97}$$

Here, Pa' is the so called deuteron pairing energy.

Egidy *et al* [18] determined the values of the free parameters of the backshifted Fermi gas model and the constant temperature model for different nuclei from the analysis of their database using equations (3.52) and (3.54), respectively, for the spin distribution function of even–even nuclei and the spin cut-off parameter. They also looked for correlations between the determined values of parameters and other quantities and found that these parameters have strong correlation with shell correction and the deuteron pairing. The value of the shell correction term S was obtained from the difference of mass calculated using liquid drop model and the experimental value of mass for a given nucleus. The deuteron pairing energy has been calculated from the mass excess value $M(A, Z)$ of the mass tables, given as;

$$P'_{ak} = 1/2[M(A + 2,\ Z + 1) - 2.M(A,\ Z) + M(A - 2, Z - 1)] \tag{3.98}$$

The empirical values for parameters of backshifted Fermi gas model obtained by Egidy *et al* [18] are;

$$a = [0.199(7) +\ 0.0096(4)\{S + 0.5\}]A^{0.869} \text{ and } U_0 = -0.381(14) + 0.5P'_{ak} \tag{3.99}$$

And for the constant temperature model, the empirical values are;

$$T = \frac{A^{-\frac{2}{3}}}{[0.0597(2) + 0.00198(10)\{S + 0.5\}]} \text{ and} \qquad (3.100)$$
$$U_0' = [-1.004(21) + 0.5.P_{ak}']$$

Further, they observed strong correlations between T and a; and between U_0 and U_0';

$$T = 5.164 \bullet a^{-0.791} \text{ and } U_0' = U_0 - 0.623 \qquad (3.101)$$

The slope of energy spectra of particles emitted in compound reaction processes has often been used to obtain nuclear temperature and other thermodynamic parameters that lead to the level density. In most of these experiments spectra for evaporated neutrons, protons and alpha particles over a wide energy range have been measured in in-beam experiments. However, in these measurements contribution from pre-equilibrium emission, which is found to be present even at considerably lower excitations, affects the deduced value of level density parameter 'a' and backshift energy U_0. In order to suppress the pre-equilibrium component, heavier systems like $^{19}F + ^{181}Ta$ have been studied in the energy range 90–195 MeV by Fabris *et al* [20] and Caraley *et al* [21]. However, the two groups deduced a contradictory nature of level density parameter 'a'; Plujko *et al* [22] found that level density parameter changes with energy while Fabris *et al* [20] found it to be energy independent. Investigation of Ericson fluctuations [23–25], particularly of total and partial neutron cross-sections, have provided level density parameters in that energy range where data is scare. Therefore, such investigations are very important.

As is known, level densities are influenced by three different interactions, namely pairing interactions, shell effects and collective effects due to residual interactions. Sometimes in formulations of backshifted Fermi gas model only the effect of pairing interaction is included in the backshift energy U_0. As a result, effects due to shell closure and collective interactions get absorbed in level density parameter 'a'. Since level density increases exponentially with (aU), these shell and collective effects become larger at higher energies. This is against the physics of these effects; both of which should diminish with excitation energy. One way of achieving this is to include all the three interactions in backshift energy U_0, which leaves 'a' energy independent. However, Ignatyuk [26] suggested the following form for level density parameter a,

$$a = \bar{a}\left[1 - \frac{W}{U}(1 - e^{-\gamma U})\right] \qquad (3.102)$$

Here, a is the asymptotic value of 'a' reached at higher energies, W is an energy shift while γ is a constant of the order of 0.05.

As $U \rightarrow 0$, $a \rightarrow \bar{a}\ (1-\gamma W)$ and when $U \rightarrow \infty$, $a \rightarrow \bar{a}$. Depending on the choice of W, a may be kept above or below $\bar{a}$ at lower energies.

Two other effects may bring energy dependence in level density parameter a. It has been shown that collective effects may create temperature dependence in effective mass which in turn will make level density parameter temperature dependent.

The level density parameter in the Bethe model is energy independent because of the assumption that the single-particle spectrum has constant density with energy. This assumption is not valid. Firstly, states that have energy more than the total binding energy of the nucleus cannot be treated as compound states coupled to the nucleus and hence nuclear states must truncate at the higher energy side. Secondly, the nuclear potential well is not of infinite depth, and, therefore, low lying nuclear states do not continue up to an arbitrarily low energy. This truncation of nuclear states both towards the higher energy side and the lower energy side constitutes the so called 'roll-over' of nuclear density. At relatively higher excitation energies around 8 MeV and beyond the level density will have a smaller value and change in the value of level density parameter *a* may be substantial at higher energies.

Although considerable enhancement in our understanding of nuclear level density has taken place during the last few decades, the situation regarding the parity ratio and spin cut-off parameter is still not satisfactory on account of the small amount of available experimental data.

3.8 Pre-equilibrium emission

Experimental verification of the compound reaction mechanism put the underlying theory on a firm footing. For a considerable period of time, it was assumed that almost all reactions at low and moderate energies proceed through the compound reaction mechanism, i.e. the CN formed as a result of the fusion of the incident particle and the target nucleus lives long enough to sample a large number of microstates so that a thermodynamic equilibrium is established before the decay to the final state. Reactions at relatively higher energies, say around 50 MeV or so, where it was anticipated that the incident particle might interact only with a few nucleons in the target, may produce localised heating and may eject one or a few nucleons. The timescale for such a process will obviously be much smaller, of the order of 10^{-21} s or so and during this time thermal equilibrium could not be attained. Such processes are termed as direct reactions.

Further, energy redistribution in the CN may also depend on the special properties of the incident particle and interacting target nucleus. As an example, in the case of deuteron–nucleus interaction, redistribution may be weakened; the deuteron has a low binding energy and a large size, which makes the stripping process more amenable, giving rise to the capture of one of the nucleons available in deuteron. One more example that is given for direct reactions is the nucleon scattering from a nucleus which may excite rotational degrees of freedom in the target. Contrary to the CN reactions the direct reactions are considered to be accompanied by the excitation of only a few degrees of freedom.

As the energy increases, both the intuition and some experimental findings suggest that particle emission may take place after the direct stage, but prior to the attainment of the equilibrium in the CN. Such processes are called pre-compound or pre-equilibrium reactions. The timescale involved in such reactions is intermediate between the fast-direct reactions and the slow CN reactions, i.e., typically $\approx 10^{-18} - 10^{-21}$ s. Here, it is important to mention that the verification of such reactions is obtained from the studies of the fluctuations in the measured excitation functions of a large number of reactions to a well-defined residual nucleus state, which show coherence widths that are much higher than those characterising the fluctuations observed in the CN reactions.

With the advancement in the nuclear detectors and pulse processing electronics, attempts were made to study the energy spectra, angular distribution and double differential cross-sections of the particles emitted in the nuclear reactions. A typical energy spectrum of the charged particles emitted during nuclear reactions, at a given angle at moderate excitation energy is shown in figure 3.6. Although a similar figure is given earlier, in chapter 1, however, it is presented again for ready reference. As may be seen from this figure, there lies a broad peak at lower energies followed by the continuum and at relatively higher energies there are several discrete sharp peaks. These isolated peaks at higher energies may be ascribed to the direct reactions where only a few degrees of freedom are involved, while the broad Maxwellian peak at low energies may be attributed to the compound nuclear processes where, because of many interactions and degrees of freedom statistical treatment is applicable. It is mostly referred to as the evaporation region after a model proposed by Weisskopf [27]. In this region Bohr's independence hypothesis which says that decay of the CN is independent of its formation is valid. The energy spectrum for this part of the graph does not depend on the entrance channel but only on the excitation energy and good quantum numbers of the CN. Further, because of the statistical assumption the angular distribution of emitted particles is symmetric about 90° in the c.m. frame [28]. However, there is no satisfactory explanation of the structureless continuum in these two approaches. It may be explained by the mechanism which neither involves statistically large nor very few degrees of freedom, i.e., a process intermediate in nature between the compound and the direct reaction mechanism. This intermediate process is considered as the pre-equilibrium or pre-compound process. The particles emitted during this process are referred to as the pre-equilibrium particles. The component at lower energies has an angular distribution symmetric about 90° that has a cross-section much greater than that given by CN theory. At the higher energy side of the continuum,

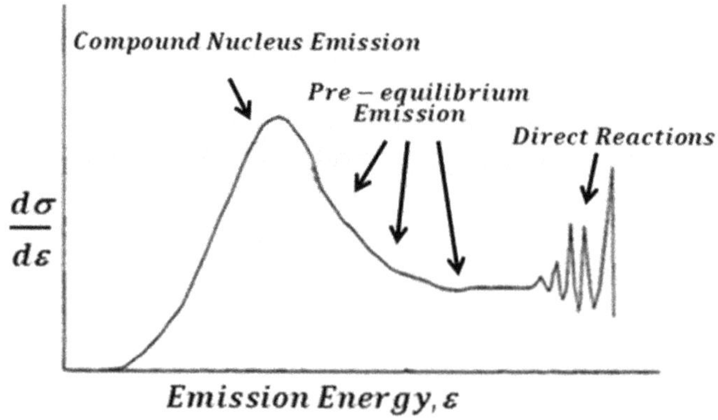

Figure 3.6. Typical energy spectrum of emitted particles in nuclear reactions at moderate energies.

distribution of pre-equilibrium particles is forward peaked but lacks the diffraction-like structure and also the nuclear structure dependence characteristic of the emitted particles in the direct reactions [29].

It was indicated by Weisskopf [30] way back in 1961 that separation of nuclear reaction mechanism into compound and direct reactions is too simplistic. The predicted cross-sections according to the pure compound nuclear process is too small compared to the measured ones. Also, the measured angular distributions in this intermediate region are found to be anisotropic indicating the existence of memory preserving and direct-like reaction process. As such, an intermediate process which embodies both compound and direct reaction like features exists. These reactions are referred to as pre-equilibrium/pre-compound or, usually in a quantum mechanical context, multi-step processes. The pre-equilibrium emission takes place after the first stage of the reaction but quite long before the statistical equilibrium of the CN is established. It is assumed that the incident particle step-by-step creates more complex states of the compound system and gradually loses its memory of the initial energy and direction. Pre-equilibrium processes provide a sizeable part of the reaction cross-section for projectiles of moderate energies. It may also be mentioned that the failure of the CN model and the independence hypothesis were indicated in a few papers in the early 1960s. Scientists started looking into this discrepancy soon after some important papers by Griffin [31] were reported. Nearly at the same time the pre-equilibrium emission was identified experimentally in various reports.

3.8.1 Some early evidence of pre-equilibrium emission

With the advancement in accelerator technology and the development of isochronous cyclotrons which became popular by the 1960s, projectiles of higher energies were available to study nuclear reactions. The measured inexplicable spectral components became ever more obvious in a large range of experimental observations [32]. These results could not be explained with the available compound and direct reaction models.

One of the early evidences regarding the presence of the pre-compound reactions was obtained when comparison of CN formation cross-section and the total cross-section calculated from the optical model was made. In this regard the deduced experimental excitation functions measured by Rao and Yaffe [33] for CN formation for protons on $^{181}_{73}$Ta, were plotted along with that estimated from the optical model. The comparison made in the energy range ≈20–60 MeV showed that the measured CN cross-section falls very fast with increasing energy, but the total cross-section remains almost constant. This observed behaviour between experimental and optical model estimations is shown in figure 3.7. This type of typical difference between the cross-sections is (see figure 3.7) attributed to the direct reactions and quite more due to the pre-compound reactions [29].

From the experiments of R W West [34], and descriptions of Kalbach [35] a representation of deduced angle integrated proton spectrum obtained with 40 MeV α-particles incident on $^{93}_{41}$Nb target is shown in figure 3.8. In this figure, the dashed curve shows the amount of cross-section representation that can be accounted for by

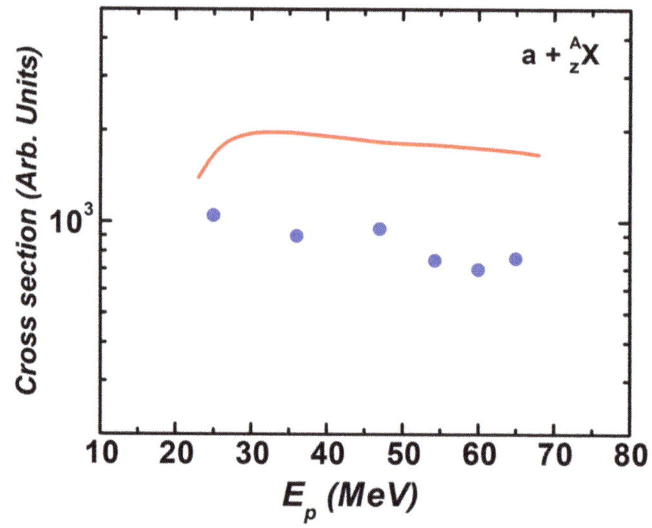

Figure 3.7. The typical observed experimental excitation functions for CN formation cross-sections for $a + {}^A_Z X$ reaction as a function of proton beam energy. The solid dark (red) line is the representation of corresponding reaction cross-section estimated with the optical model.

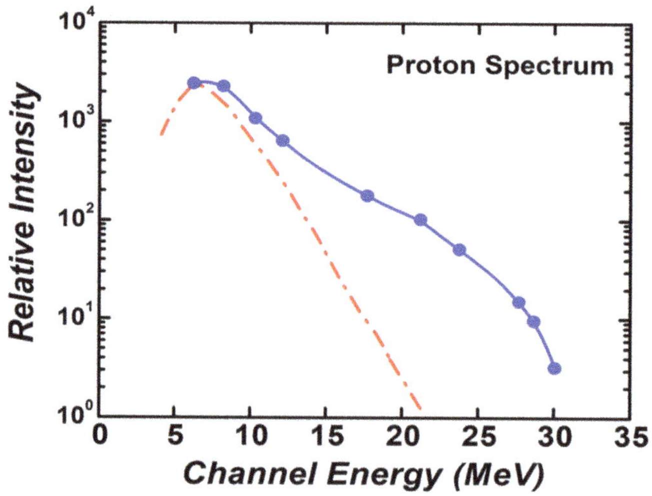

Figure 3.8. Heavy line with dark blue solid circles represents the observed trend of experimental proton spectrum data, while the dashed line is a typical representation of the CN calculations. The difference of the two curves (i.e., blue dark line with solid circles and the dashed red line) may be accounted for by a suitable theoretical model used for simulating pre-equilibrium emission.

the CN model calculations or CN contribution; however, the heavy line shows the typical variation of observed data shown in reference. The observed results, if averaged in energy may satisfactorily account for the upper ≈8 MeV or above part of the spectrum but how much of the intermediate region represents the cross-section due to direct reactions or due to intermediate or pre-equilibrium processes can be explained only by the modelling of such reactions. Such initial experimental evidences indicated that intuition of particles emitted before equilibration of the CN are likely to be true as found in many other experiments.

Proton emission in 42 MeV alpha particle bombardments of several elements has also been studied by West [34]. They measured energy spectra, angular distributions and absolute cross-sections of proton emitted from bombardments of ^{59}Co, ^{60}Ni, ^{62}Ni, ^{63}Cu, ^{65}Cu, Nb, Rh, Pd, and Pt targets by 42 MeV α-particles. The measurements for almost all the reactions indicated that the low energy particles have a symmetric angular distribution, however, the high energy particles were found to have forward peaked angular distribution, a characteristic of the pre-equilibrium emission. It gave one of the initial evidences of the pre-equilibrium emission.

Bhardwaj and Prasad [36] measured the excitation functions for alpha induced reactions in the $^{197}_{79}$Au target nuclei. The results of the excitation functions were compared with the compound and pre-compound nuclear reaction model calculations, as shown in figure 3.9. Some other literature values where the same reaction was studied are also shown in this figure. Excitation functions have also been calculated simulating the pre-equilibrium emission by a hybrid model [37]. The experimental excitation functions are best reproduced by taking a mixture of both the equilibrium and pre-equilibrium emissions. As can be seen from figure 3.9, the

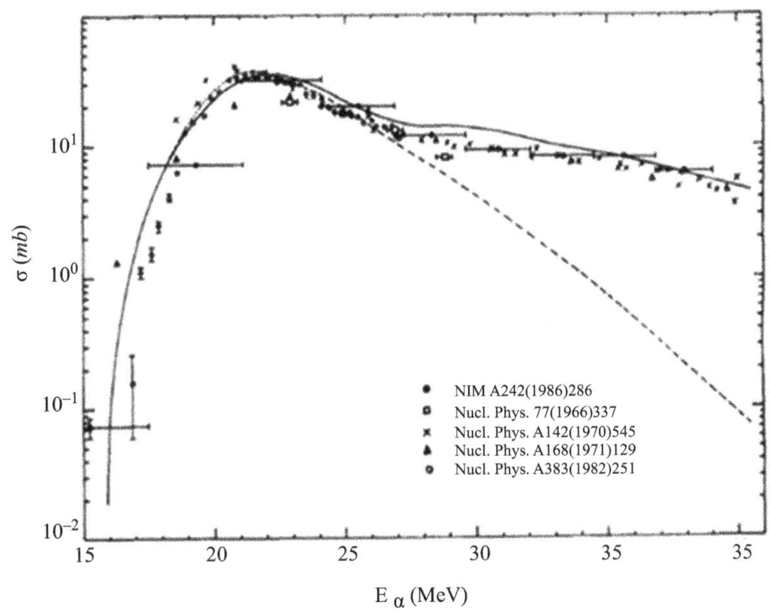

Figure 3.9. Experimental and theoretically calculated excitation function for the reaction ^{197}Au(α,n). The broken line shows calculations done assuming pure compound emission. The full line represents the inclusion of pre-equilibrium emission along with compound process, as discussed in the text (reprinted from [36] with permission from Elsevier).

measured excitation function for the reaction $^{197}_{79}Au(\alpha,n)$ is reproduced nicely, particularly, in the tail portion when the pre-compound emission is considered along with the CN calculations. These and other similar studies conclusively established that pre-equilibrium emission plays an important role in nuclear reactions at moderate and low energies.

Lux and Porile [38] have measured energy spectra of α-particles emitted in reactions of ^{64}Zn, ^{66}Zn, ^{67}Zn, and ^{70}Zn with 12.5 MeV protons and 18 MeV α-particles between 30° and 150° with respect to the to the beam direction. They also measured the angular distribution of α-particles emitted in these reactions. The measured angular distributions are shown in figure 3.10. As can be seen from this figure that (p, α) angular distributions indicate a gradual progression from almost symmetry for ^{64}Zn and ^{67}Zn to moderate forward peaking of angular distribution for ^{70}Zn target. On the other hand, the ascendant features of all the (α, α') angular distribution show the strong degree of forward peaking. It was concluded that the emission of the incident α-particle takes place after undergoing a finite energy loss but well before the establishment of equilibrium in the CN.

Thus, the contribution of the pre-compound process was determined on the basis of the asymmetry in the angular distribution and by comparing the spectra with calculations based on the compound and pre-compound models. In their another work [39] it was shown that the asymmetry in the angular distribution obtained in (p, p') reactions in ^{64}Zn, ^{66}Zn, ^{67}Zn and ^{70}Zn gives a quite good measure of the pre-equilibrium reaction cross-section at the proton energies around 12 MeV. It is

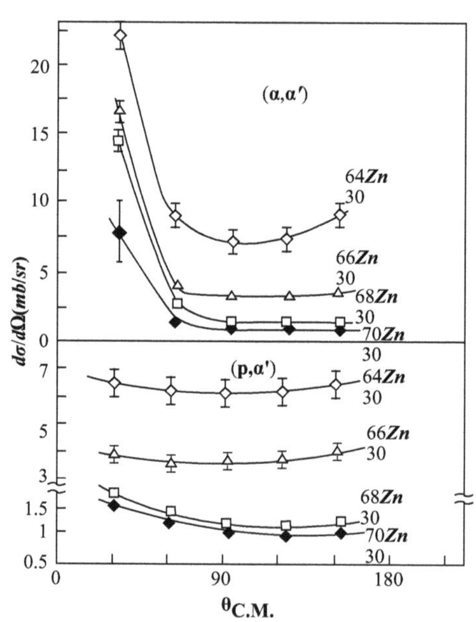

Figure 3.10. Angular distribution of α-particles emitted in (α,α') and (p,α) reactions of ^{64}Zn, ^{66}Zn, ^{63}Zn and ^{70}Zn (adapted from [38] with permission from American Physical Society).

indicated that the measured angular distributions give a lower limit to the contribution of pre-equilibrium reaction. Further, the measured angular distributions were found to be in good agreement with that calculated by the quasi-free scattering model [40]. Both these approaches are found to show that pre-equilibrium emission is of greater importance for the (α, α') reaction as compared to (p, α) reactions. It is also observed that importance of the pre-equilibrium process increases with respect to compound emission with increasing target neutron number and that it becomes the dominant reaction mechanism for the (α, α') reactions of $^{68}_{30}$Zn and $^{70}_{30}$Zn. Some Monte Carlo cascade calculations have also been done for the reactions initiated by low energy protons. From these calculations it has been predicted that nearly about 95% protons emitted before the establishment of statistical equilibrium are found to be produced in the forward direction.

On the basis of measured angular distribution of emitted particles, it is possible to distinguish the compound and the pre-compound emission process. As already mentioned, as the CN lives long enough, the angular distribution of evaporated particles is expected to be symmetric about 90° in the c.m. frame, with respect to the direction of incident beam. On the other hand, the pre-equilibrium emission takes place well before the establishment of the equilibrium of the CN and hence is predominantly at the forward angles. In figure 3.11, the angular distributions of protons emitted in reactions $^{68}_{30}$Zn(α, p) and $^{68}_{30}$Zn(p, p') are shown. The numbers indicated to the right of each curve represent the energy interval in MeV, of the emitted protons. The curves are arbitrarily displaced from each other. It may be observed [39] in this figure that higher energy particles emitted in pre-equilibrium emission exhibit relatively higher degree of forward peaking in angular distributions. As such, the amount of forward–backward asymmetry in angular distribution of emitted particles produced in nuclear reactions may be considered as a measure of the pre-equilibrium contribution. However, this should be considered as a lower limit of pre-compound contribution because a fraction of such emission undoubtedly may take place at backward angles as well.

There is ample evidence based on experimental measurements of energy spectrum, angular distribution, and excitation functions where it was conclusively shown that without the inclusion of pre-compound contributions in the theoretical calculations the experimental data could not be explained. These studies conclusively established that pre-compound emission contribution is significant even at these low energies. In other words, the pre-compound emission is a competing mode of reactions to that of CN emission. These studies initiated efforts for developing models for the pre-equilibrium process. Initially semi-classical models were put forward and later some quantum mechanical models were also proposed. In recent years with the availability of heavy ion beams the experimental investigations on pre-equilibrium emission have been extended to reactions initiated by heavy ions. It has been observed that the pre-compound emission has a significant contribution even at energies ≈5–7 MeV A^{-1}. The details of some of the important models to describe the experimental data within the framework of compound and pre-compound emission processes are given in the next chapter.

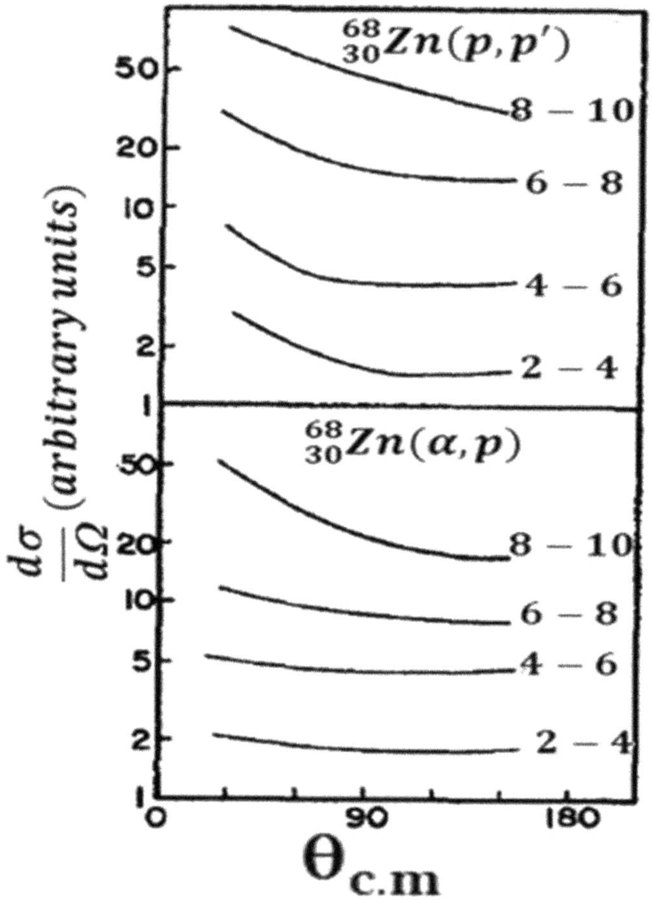

Figure 3.11. Angular distribution of protons emitted in reactions $^{68}_{30}$Zn(α, p) and $^{68}_{30}$Zn(p, p') (adapted from [39] with permission from Elsevier).

Bibliography

[1] Geiger H, Harling J and Marsden E 1909 On a diffuse reflection of the α-particles *Proc. Roy. Soc.* A **82** 495

[2] Rutherford E 1911 The scattering of alpha and beta particles by matter and the structure of the atom *Phil. Mag. Ser.* **6** 669

[3] Fermi E, Amaldi E, D'Agostino O, Rasetti F and Segre E 1934 Artificial radioactivity produced by neutron bombardment *Proc. Roy. Soc. Ser.* A **146** 483

[4] Bohr N 1936 Neutron capture and nuclear constitution *Nature* **137** 344

[5] Ghoshal S N 1950 An experimental verification of the theory of compound nucleus *Phys. Rev.* **80** 839

[6] John W Jr 1956 Excitation functions for (α,xn) reactions on lead *Phys. Rev.* **103** 704

[7] Rayudu G V S and Yaffe L 1963 Reactions produced on erbium by protons of energies between 6 and 87 MeV *Can. J. Chem.* **41** 10

[8] Hodgson P E 1971 *Nuclear Reactions and Nuclear Structure* (Oxford: Clarendon)

[9] Hilaire S 2000 Workshop on nuclear data and nuclear reactors *ICTP, Trieste, Italy*
[10] Weisskopf V F and Ewing D H 1940 On the yield of nuclear reactions with heavy elements *Phys. Rev.* **57** 472
[11] Bethe H A 1937 Nuclear dynamics: theoretical *Rev. Mod. Phys.* **9** 69
[12] Rosenzweig N 1957 Influence of shell structure on the level density of a highly excited nucleus *Phys. Rec.* **108** 817
Kahn P B and Rosenzweig N 1966 Density of states of degenerate Fermi systems with periodic level schemes *Phys. Lett.* **22** 307–9
[13] Dilg W, Schantl W, Vonach H and Uhl M 1973 Level density parameters for the back-shifted Fermi gas model in the mass range $40 < A < 250$ *Nucl. Phys.* A **217** 269
Von Egidy T, Schmidt H H and Behkami A N 1986 Nuclear level densities and level spacing distributions from ^{20}F to ^{244} *Am Nucl. Phys.* A **454** 109–27
Von Egidy T, Schmidt H H and Behkami A N 1988 Nuclear level densities and level spacing distributions: Part II *Nucl. Phys.* A **481** 189–206
[14] Gilbert A and Cameron A G W 1965 A composite nuclear level density formula with shell corrections *Can. J. Phys.* **43** 65–139
[15] Guoxiong L, Liaoyuan D, Zhoongfu H and Guochun Q 1999 Research on GSM level density formula and its parameters *Nucl. Phys. Rev.* **16** 246
[16] Po-lin Huang S M, Grimes and Massey T N 2000 Level densities for nuclei with $20 \leqslant A \leqslant 41$ *Phys. Rev* C **62** 024002
[17] Al-Quraishi S I, Grimes S M, Massey T N and Resler D A 2001 Are the level densities for *r*- and rp-process nuclei different from nearby nuclei in the valley of stability? *Phys. Rev.* C **63** 065803
[18] Von T, Egidy D and Bucurescu 2012 Statistical nuclear properties (level densities, spin distributions) *J. Phys. Conf. Ser.* **338** 0123028
[19] Ericson T 1960 The statistical model and nuclear level densities *Adv. Phys.* **9** 425
[20] Fabris D *et al* 1994 Alpha emission as a probe of the level density in highly excited $A \approx 200$ nuclei *Phys. Rev.* C **50** R1261
[21] Caraley A L, Haenry B P, Lestone J P and Vandenbosch R 2000 Investigation of level density parameter using evaporative α-particle spectra from ^{19}F + ^{181}Ta reaction *Phys. Rev.* C **62** 054612
[22] Plujko V, Gorbachenko O, Bondar B and Rovenskykh E 2013 Nuclear level density within extended superfluid model with collective state enhancement *Nucl. Data Sheets* **118** 240–2
[23] Abfalterer W P, Finlay R W and Grimes S M 2000 Level widths and level densities of nuclei in $32 \leqslant A \leqslant 60$ mass region inferred from fluctuation analysis of total neutron cross sections *Phys. Rev.* C **62** 064312
[24] Mishra V, Boukharouba N, Grimes S M, Doctor K, Pedroni R S and Haight R C 1991 Determination of the level density of ^{29}Si from ericson fluctuations *Phys. Rev.* C **44** 2419
[25] Abfalterer W P, Finlay R W, Grimes S M and Mishra V 1993 Level densities of ^{28}Al, ^{29}Si, and ^{41}Ca inferred from fluctuation measurements *Phys. Rev.* C **47** 1053
[26] Ignatyuk A V, Istekov K K and Smirenkin G N 1979 Phenomenological description of the energy dependence of the level density parameter *Sov. J. Nucl. Phys.* **29** 450
[27] Weisskopf V F 1937 Statistics and nuclear reactions *Phys. Rev.* **52** 295
[28] Hauser W and Feshbach H 1952 The inelastic scattering of neutrons *Phys. Rev.* **87** 366
[29] Hodgson P E, Gladioli E and GadioliErba E 2003 *Introductory Nuclear Physics* (Oxford: Oxford University Press)

[30] Weisskopf V F 1961 Problems of nuclear structure *Phys. Today* **14** 18
[31] Griffin J J 1966 Statistical model of nuclear structure *Phys. Rev. Lett.* **17** 478
Griffin J J 1967 Energy dependence of average direct reaction cross sections and partial nuclear level densities *Phys. Lett.* B **24** 5
[32] Bertrand F E and Peelle R W 1973 Complete hydrogen and helium particle spectra from 30- to 60-MeV proton bombardment of nuclei with A = 12 to 209 and comparison with the intranuclear cascade model *Phys. Rev.* C **8** 1045
[33] Rao C L and Yaffe L 1963 Nuclear reactions induced in tantalum by protons of energy up to 84 MEV *Can. J. Chem.* **41** 2516
[34] West R W 1966 Proton emission in 42-MeV alpha-particle bombardments of several elements *Phys. Rev.* **141** 1033
[35] Kalbach C 1975 Pre-equilibrium models in general: the Griffin model in particular *Acta Phys. Slov* **25** 100
[36] Bhardwaj H D and Prasad R 1986 Excitation functions for ^{197}Au(α, xn) reactions in the 10–40 MeV energy range *Nucl. Instrum. Methods Phys. Res.* A **242** 286
[37] Blann M 1972 Importance of the nuclear density distribution on pre-equilibrium decay *Phys. Rev. Lett.* **28** 757
[38] Lux C R and Porile N T 1978 Compound and pre-compound emission of α particles in reactions of even Zn isotopes with low-energy protons and ^{4}He ions; *Phys. Rev.* C **18** 148
[39] Lux C R and Porile N T 1977 Compound and pre-compound processes in (p, p′) and (α, p) reactions of even Zn isotopes *Nucl. Phys.* A **277** 413
[40] Blann M, Mignerey A and Scobel W 1976 Equilibration processes in nuclear reactions: nucleons to heavy ions *Nukleonika* **21** 335

IOP Publishing

Pre-equilibrium Emission in Nuclear Reactions
Fundamentals, measurements and analysis
B P Singh, Manoj K Sharma and R Prasad

Chapter 4

Models for pre-equilibrium emission

Nuclear particles or light clusters emitted from an excited composite system during the course of its equilibration are termed as pre-equilibrium emission. Few semi-classical and at least one quantum mechanical model have been developed to explain the process of pre-equilibrium emission. Several computer codes based exclusively on the compound reaction mechanism and some including both the compound and the pre-equilibrium processes are available in the literature. Salient features of some important pre-equilibrium models and details of computer codes frequently used to analyse the experimental data on pre-equilibrium emission are presented in this chapter.

4.1 Introduction

Analysis of experimental data on nuclear reactions at moderate and low energies indicates that these reactions occur at two distinct timescales. The fast component that occurs in time $\approx 10^{-21}$ s, the time taken by a moderate energy nucleon to cross an average nucleus, corresponds to 'direct reactions' in which the projectile remains in the continuum. The slower component that is estimated to have interaction times of the order of 10^{-16} s is referred to as the compound reactions. In compound processes the incident ion or a part of it is trapped in a quasi-bound state and, therefore, these processes are relatively slow. The differential cross-section data for direct processes is found to be strongly forward peaked, while those of pure compound process are symmetric around 90°; that is consistent with the theory. However, the component of the reaction that occurs in the timescale intermediate to the above mentioned two times, are referred to as the pre-compound (PC) or pre-equilibrium (PE) component. The compound reaction process assumes that the projectile trapped in quasi-bound state surfs through all possible states permitted by conservation laws and remains bound till a thermodynamic equilibrium is established. As a result, differential cross-section associated with the compound process retains symmetry around 90°. The PE process, however, takes into account the emission that may occur during the process of equilibration and, therefore, particles emitted during this process carry some

doi:10.1088/978-0-7503-5077-8ch4

information about the incident channel; the differential cross-section is forward peaked but not so strongly as that for direct processes (see figure 4.1). As expected, PE-processes become more important as the projectile energy increases. At moderate excitation energies, the reaction dynamics assumes that two-body interactions between the target and projectile nucleons build up the energy distribution process. The compound nuclear model assumes that two-body interactions drive the target–projectile composite system from states of lower complexity to states of higher complexity, that have higher state densities and therefore, the system essentially moves in the forward direction, i.e., to more complex states as per the Golden rule. Ultimately, after a sufficient number of two-body interactions, the system attains thermal equilibrium; where the probability of going to the next complex state is the same as that of going back to the previous state of lesser complexity. The compound reaction mechanism (compound nucleus, CN) assumes that particle and light ion emission from the excited composite system formed as the result of absorption of the projectile by the target does not take place till the thermodynamic equilibrium is established. However, both the intuition and experimental observations unambiguously indicate that particles and light ions may be emitted from the composite excited system during the process of equilibration. The light and heavy ions emitted during the process of equilibration are termed as PE or PC particles and are characterised by energies higher than what is expected from compound emission after the establishment of thermal equilibrium, forward peaked angular distributions and stretched distribution in angular momentum space. Some classical and at least one quantum mechanical model have been proposed to describe PE-emission. A brief description of these models is presented in the following.

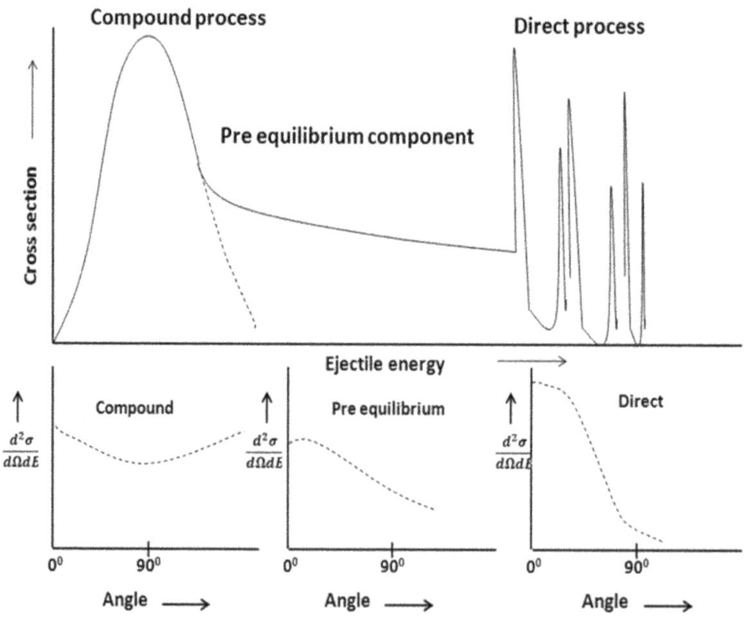

Figure 4.1. Ejectile energy spectrum and double differential cross-sections for different components.

4.2 Intra-nuclear cascade model (INC)

The intra-nuclear cascade model is perhaps the first model, proposed by Serber [1] in 1947, essentially to generate the energy spectra of particles emitted in direct reactions at high energies. However, the concept of the model may well be used for calculating the cross-section and energy spectra of light ions emitted in low and moderate energy nuclear interactions. The INC model is based on the fact that in nuclear reactions at moderate energy the de Broglie wavelength of the incident ion is comparable or shorter than the average nucleon–nucleon separation in the target nucleus and, therefore, the interactions may be described in terms of particle–particle collisions. A schematic representation of the INC model is shown in figure 4.2, where an energetic incident ion enters the target and interacts with a nucleon at location A ejecting an ion out of the nucleus and being scattered to location B where it undergoes another interaction. The resulting two scattered particles may undergo a different number of collisions before ions with different energies may be emitted. As shown in figure 4.2 the energy of the emitted particle depends on the number of two-body collisions that have taken place before the emission. For example, a particle designated 1 is emitted after a single collision and has maximum energy, while particle number 2 is emitted after four collisions and, therefore, has smaller energy as compared to particles 1 and 3, that are ejected after one and three collisions, respectively. Goldberger [2] in 1947, used this model for the first time and carried out hand calculations of nucleon energy spectra at intermediate energies. They used Monte Carlo technique for these simulations. In 1958, Metropolis *et al* [3] carried out Monte Carlo simulations based on the INC model using a computer. However, the standard methods in implementing the INC model were formalised with the publication of results by Bertini [4] in 1968–1969.

The inelastic ion–nucleus collision may be divided into two phases: a fast phase that takes place within (10^{-23}–10^{-22}) s of INC results in a highly excited nucleus and is followed by fission and pre-equilibrium emission; a slower phase, that follows the

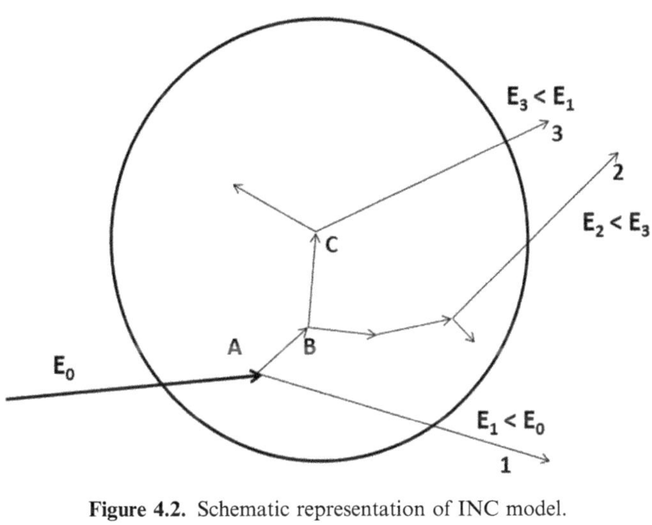

Figure 4.2. Schematic representation of INC model.

faster phase, results in the formation of the equilibrated CN (10^{-18}–10^{-16} s) and particle evaporation. In general, a Boltzmann equation must be solved to treat the physical process of the collision in detail. In most computer simulations, including that of Bertini, the Boltzmann equation is solved on the average. The nuclear model used in these calculations assumes a three-region approximation to the continuously changing nuclear density distribution within the nucleus. Relativistic kinematics is applied throughout the cascade which stops when all the particles which can escape, have done so. Then conformity with the energy conservation law is checked. The INC is implemented in the following sequence of steps:

(a) The spatial location where the projectile hits the target is selected uniformly over the entire area projected by the target.
(b) Total free particle scattering cross-sections and region dependent nucleon densities are used to determine the path length for the projectile.
(c) The momentum of struck nucleon, the type of reaction and the four momentum of the reaction products are determined.
(d) If Pauli's exclusion principle allows and the energy of the particle is larger than the assumed cut-off energy, step (b) is performed to transport the products.

In most INC model computer simulations the target nucleus is represented by a Fermi gas and the Fermi energy is calculated in a local density approximation with Fermi momentum $p_F(r) = \left(\frac{3\pi^2\rho(r)}{2}\right)^{1/3}$. The nucleon binding energies are either calculated from specific mass formula or experimental values are used, wherever available. It is also possible to use an asymptotic high temperature mass formula, if required.

For light targets with $1 < A < 4$, the nucleus is assumed to have a single layer of radius 0.8 fm. However, for heavier targets if $1 < A < 11$, the nucleus is composed to be of three concentric spheres $i = (1, 2, 3)$ with radii

$$r_i(\alpha_i) = \sqrt{C_1^2\left(1 - \frac{1}{A}\right) + 6.4\sqrt{-\log(\alpha_i)}}$$

where,

$$\alpha_i = (0.01, 0.3, 0.7) \text{ and } C_1 = 3.3836\, A^{1/3} \text{ fm}.$$

Nuclei with $A > 11$ are also assumed to have three concentric spheres with their radii defined as,

$$r_i(\alpha_i) = C_2 \log\left(\frac{1 + e^{-\frac{C_1}{C_2}}}{\alpha_i} - 1\right) + C_1$$

where, $C_2 = 1.7234$ fm.

The potential energy of nucleon N is given by

$$V_N = \frac{p_F{}^2}{2m_N} + BE_N(A, Z)$$

Impulse distribution in each region follows Fermi distribution with zero temperature,

$$f(p) = cp^2$$

and

$$\int_0^{p_F} f(p)dp = n_p \text{ or } n_n.$$

Here, n_p and n_n are the number of protons and neutrons in the region.

The momentum p_F corresponds to Fermi energy

$$E_F = \frac{{p_F}^2}{2m_N} = \frac{\hbar^2}{2m_N}\left(\frac{3\pi^2}{v}\right)^{2/3}$$

It may be noted that Fermi energy depends on n/v of the particles, which is different for each particle and in each region.

Since the target nucleus is assumed to be a completely degenerate Fermi gas, all levels from the bottom till the Fermi energy are filled, and, therefore, the minimum energy allowed to a collision product is greater than Fermi energy. Thus, accepting only secondary nucleons that have $E_N > E_F$, the Pauli's exclusion principle is taken into account.

Path lengths of nucleons in the target nucleus are sampled according to the local nucleon density and free nucleon–nucleon cross-sections. Angles of the products after collision are sampled from the experimental data of differential cross-sections. Total cross-sections are generated from Letaw's formulation [7].

Computer code GEANT4 [5] which is a Monte Carlo particle detector simulation toolkit incorporates Bertini's INC model. The code provides a rather flexible framework for modular implementation of various kinds of hadronic interactions. The hadronic interactions used in the code are based on physics processes that generate the final state; as such several different models may be used to generate a given final state, for example in the case of inelastic scattering, different models and formulations may be used for different energy ranges. The fast component of the particle–nucleus inelastic interaction in code GEANT4 uses Bertini's INC model and also Griffin's [6] exciton model. Results of theoretical calculations using the INC model based on GEANT4 code [5] shows that the code represents experimental data satisfactorily only at high energies larger than 100 MeV.

4.3 The exciton model

The exciton model was first proposed by Griffin [6] in 1966. Although the model was initially proposed for nucleon induced reactions, it has since been extended to include light and heavy ion projectiles. The essential features of the model may be summarised as follows:

1. The target nucleus is generally assumed to be in ground state have single-particle states completely filled up to Fermi level.

2. The incident particle (nucleon) initiates two-body interactions with nucleons of the target creating particles and holes. Particles are nucleons above the Fermi energy and holes are unoccupied states below it.
3. The hole at Fermi energy due to the incident nucleon is not taken into account.
4. Both particles (above Fermi energy) and holes (below Fermi energy) are called excitons. Particles are often represented by letter p and holes by letter h. Total number of particles and of holes, i.e., $(p + h) = n$ gives the total number of excitons in a configuration. A configuration is identified by its exciton number n. There may be several different sub-configurations within a given exciton number n, having different values of p and h such that $(p + h) = n$. Often the set of configurations with the same number of excitons n is called an exciton class.
5. It is assumed that the reaction proceeds through two-nucleon collisions in which energy is conserved. In more detailed computations momentum conservation may also be taken into account. A particular configuration of a given exciton class may pass through one of the following three steps during the evolution of the reaction: (a) pre-equilibrium particles are emitted; (b) the state transforms to some other sub-configuration of the same class; and (c) the state proceeds to a state with $(n \pm 2)$ excitons.
6. In exciton model framework the initial state of the system may be represented as a one exciton state with one particle and no hole ($n = 1$, $1p + 0\,h$) as shown in figure 4.3, where E^* is the excitation energy.

Two-body interactions may take the initial state to one of the several possible sub-configurations of $(n \pm 2) = 3$-exciton class. Negative values of exciton number n are not allowed. Pre-equilibrium particle emission may take place from all those sub-states where there are unbound particles. For example, the sub-state of 3-exciton class shown in figure 4.4(a) may emit a pre-equilibrium particle with energy $\varepsilon_p = (E^* - B_e - \varepsilon_h)$. Here, B_e and ε_h are, respectively, the binding energy of the particle and the energy of the hole. The excitation energy E^* is given by,

$$E^* = E - B_e - \varepsilon_p$$

There may be several other sub-states of 3-exciton class which may have one or more unbound particles and also those which have no unbound particles, as shown in figure 4.4(b).

The next two-body interaction may take the system back to one exciton state ($n = 1$) or to the $n = 5$ exciton state. However, the probability of going to $n = 5$ is much larger than that of reverting back to a 3-exciton state. This is because the level density of the 5-exciton state is much larger than that of the 3-exciton state, and according to the Golden Rule of transitions, the system predominantly moves in the direction of higher level density. For small values of n the level density for $(n + 2)$ is considerably larger than that for the $(n-2)$ states or for the n-states. As such, in the initial two-body interactions when n is not very large the system moves predominantly in the direction $n \rightarrow (n + 2) \rightarrow (n + 4) \rightarrow (n + 6)$.... However, for sufficiently large values of n, the level densities for the n, $(n + 2)$ and $(n-2)$ states are almost

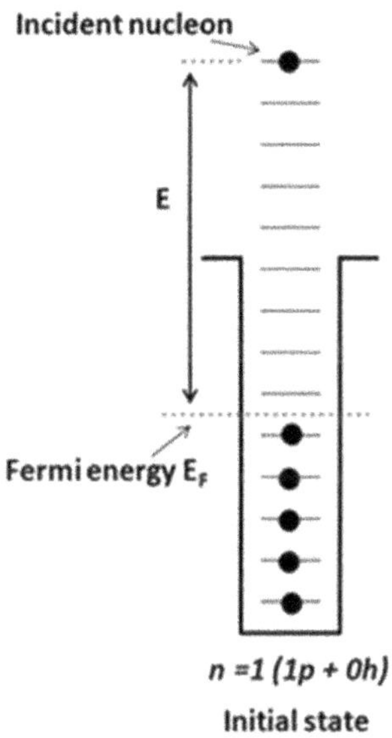

Figure 4.3. Initial state in exciton model.

same, and, therefore, the system may go in the forward or the backward direction with equal probability or it may simply pass from one sub-configuration to another of the same exciton class. This characterises the state of thermal equilibrium. Figure 4.4(c) shows how two-body interactions drive the initial state to states of higher exciton complexity. The length of filled arrows in the figure show the relative probability of going forward while the unfilled arrows indicate the probability for the system to move back. Since for lower values of exciton number n the state density increases rapidly with the value of n, the lengths of filled arrows increases successively.

If $\varepsilon_i^{\text{particle}}$ and $\varepsilon_k^{\text{hole}}$, respectively, denote the energy of the ith particle and the energy of the kth hole, then the total excitation energy E^* may be represented as;

$$E^* = \sum_1^{n^p} \varepsilon_i^{\text{particle}} + \sum_1^{n^h} \varepsilon_k^{\text{hole}} \tag{4.1}$$

Here n^p and n^h are, respectively, the numbers of particles and of holes. Further, the exciton number 'n' of the class is given by $n = n^p + n^h$.

Equation (4.1) holds both for transitions within a given exciton class ($\Delta n = 0$) and also for any transition to ($n\pm2$) states ($\Delta n = 2$). This happens because the excitation energy deposited by the projectile is fixed and does not depend on the complexity of the states. Equation (4.1) further assumes that no energy is associated with the creation and annihilation of particle–hole pairs. This is true only in classical limits. On the other hand, treating target nucleus with single-particle levels gives a quantum mechanical flavour to the treatment. It is for this reason that these models for pre-equilibrium emission are termed as semi-classical models.

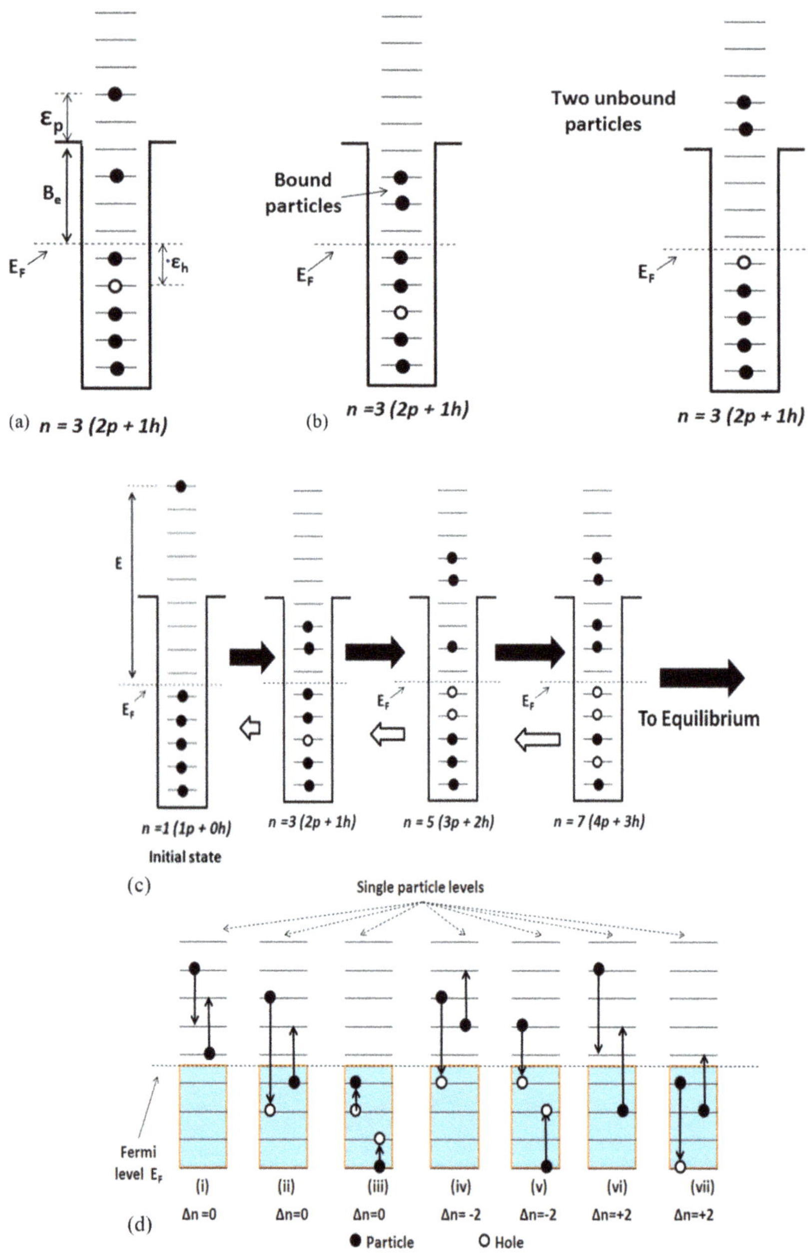

Figure 4.4 (a) A sub-state of exciton number $n = 3$ class with one unbound particle. (b) Sub-states of 3-exciton class with two bound and two unbound particles. (c) Time evolution of thermal equilibrium via two-body interactions. (d) Seven types of two-body interactions.

For calculations on equilibrating system the internal transition rates are needed. It is generally assumed that the transition probability per unit time for each interaction is given by the Golden Rule,

$$\lambda(n^p, n^h \rightarrow n^{p'}, n^{h'}) = \frac{2\pi}{\hbar} \overline{|M^2|} \omega_{\text{final}}$$

Here $\overline{|M^2|}$ is the averaged matrix element and ω_{final} the density of the accessible final states. According to Williams [21] following seven types of two-body interactions may take place.

(i) **Particle–particle scattering:** Two excitons above the Fermi level scatter to unoccupied states above the Fermi level, as shown in figure 4.4(d) (i).

(ii) **Particle–hole scattering:** A particle above the Fermi level scatters with a particle below the Fermi level. One of these particles is scattered to an unoccupied state above the Fermi level and the other to an unoccupied state below the Fermi level (figure 4.4(d) (ii)).

(iii) **Hole–hole scattering:** Two particles below the Fermi level scatter to unoccupied states below the Fermi level (figure 4.4(d) (iii)).

(iv) **(p,h) annihilation with particle scattering:** Two particles above the Fermi level scatter to states of which one is an unoccupied state below the Fermi level (hole annihilation); the other state which must have higher energy, therefore, the other particle is scattered to an unoccupied state above the Fermi level, as shown in figure 4.4(d) (iv).

(v) **(p,h) annihilation with hole scattering:** A particle above the Fermi level scatters with a particle below the Fermi level. The particle above the Fermi level gets scattered to an unoccupied state below the Fermi level and the other particle is also scattered to an unoccupied state below the Fermi level (figure 4.4(d) (v)).

(vi) **(p,h) creation with particle scattering:** A particle above the Fermi level scatters with a particle below the Fermi level and both particles are scattered to unoccupied states above the Fermi level (figure 4.4(d) (vi)).

(vii) **(p, h) creation with hole scattering:** Two particles below the Fermi level are scattered to an unoccupied state below the Fermi level and the other to an unoccupied state above the Fermi level.

There is a finite non-zero probability of the emission of a particle or a light cluster from any intermediate state as the system is driven from low n values towards high n values with time. If particle or cluster emission takes place from the initial states of low n, they will have higher energies and will be emitted in the forward direction because of the input linear momentum which has not been completely redistributed as yet. These pre-equilibrium particles or clusters are emitted in the first few steps of two-body interactions. There after the remaining system moves towards equilibrium and eventually evaporates particles and/or light clusters giving rise to compound emission. Exciton model, therefore, describes both the pre-equilibrium and compound emission phases of the reaction process in a single unified model.

Cline and Blann, cast the excitation model in the form of an energy conserving, time dependent master equation. They argued that the fractional probability $q(n, t)$ that the system is in an exciton state n at time t, then the rate of change of the probability with time $\left(\frac{dq(n,t)}{dt}\right)$ may be given by;

$$\left(\frac{dq(n, t)}{dt}\right) = \text{(rate of increase in the population of state n at time } t \text{ –, rate of depletion of state } n \text{ at time, } t) \tag{4.2}$$

The following states will increase the population of exciton state n at time t:

(i) Decay of exciton $(n + 2)$ state (probability $q(n + 2,t)$) via transition $(\Delta n = -2)$;
(ii) Decay of exciton $(n-2)$ state(probability $q(n-2,t)$) via transition $(\Delta n = +2)$;
(iii) Decay of exciton n sub-state (probability $q(n,t)$) via transition $(\Delta n = 0)$.
The following decays from exciton state n will deplete the state:
(iv) Decay to exciton state $(n + 2)$ via transition $(\Delta n = +2)$;
(v) Decay to exciton state $(n-2)$ via transition $(\Delta n = -2)$;
(vi) Decay to another sub-state of exciton n via transition $(\Delta n = 0)$;
(vii) Depletion due to the emission of PE-particles/clusters.

Now if the interstate transition rates $\lambda^+(n)$, $\lambda^-(n)$, $\lambda^0(n)$ are associated, respectively, to the transitions $(\Delta n = +2)$, $(\Delta n = -2)$ and $(\Delta n = 0)$, then equation (4.2) becomes;

$$\left(\frac{dq(n, t)}{dt}\right) = W(n)q(n, t) =$$

$$\{\lambda^+(n-2)q(n-2, t) + \lambda^-(n+2)q(n+2, t) + \lambda^0(n)q(n, t)\}$$

$$-[\{\lambda^+(n)q(n, t) + \lambda^-(n)q(n, t) + \lambda^0(n)q(n, t)\} - W(n)q(n, t)] \tag{4.3}$$

Here $W(n)q(n, t)$ represents the depletion of exciton n state by the emission of PE-ejectiles. However, it may be observed in equation (4.3) that term $\lambda^0(n)q(n, t)$ cancels out. Equation (4.3) is often referred to as the master equation.

The theory of detailed balance has been used to obtain rate of pre-equilibrium emission. If a particle b of reduced mass μ_b, energy $\in_b$, and spin s_b, is emitted from a state of excitation energy E^*, of n^p particles and n^h holes leaving a residual nucleus of excitation energy U with $(n^p - n^p{}_b)$ particles and n^h holes, where $n_b{}^p$ is the number of particles in b, then the emission rate from n-exciton class may be written as,

$$\frac{d\lambda_{\text{emission}}}{d\varepsilon}(n, \varepsilon)d\varepsilon = W(n, \varepsilon_b) = (2s_b + 1)\frac{\mu_b\varepsilon_b\sigma_{\text{inv}}(\varepsilon_b)}{\pi^2\hbar^3}\frac{\omega\left(\left(n^p - n_b^p\right), n^h, U\right)}{\omega(n^p, n^h, E^*)}Q_b(n)\phi_b \tag{4.4}$$

In equation (4.4) $\sigma_{\rm inv}(\varepsilon_b)$ is the inverse cross-section for the formation of the same nucleus by particle b of energy ε_b and may be calculated using the optical model. $Q_b(n)$ and ϕ_b are, respectively, the factors that differentiate between neutrons and protons and the factor that gives the probability of finding a preformed b, if b is a complex particle or cluster. ϕ_b is 1 if b is a neutron, proton or gamma ray. $\omega(n^p, n^h, E^*)$ is the density of states of an n-exciton class with $n^{\rm p}$ particles and $n^{\rm h}$ hole of excitation energy E^*. The $W(n)q(n, t)$ represented by equation (4.3) may be obtained from integrating equation (4.4) over all particles b and over all energies ε_b.

4.3.1 The state densities

The simplest estimate of density of state of an n-exciton class may be obtained assuming constant single-particle and single-hole densities with the constraint of energy conservation, along with the assumption that both holes and particles within themselves are indistinguishable. For example, in the case of 3-exciton class with 2-particles and 1-hole, the density $\omega(p = 2, h = 1, E)$ may be written as;

$$\omega(p = 2, h = 1, E) = \frac{g_p^2 g_h}{2!1!}\int d\varepsilon_{p1}d\varepsilon_{p2}d\varepsilon_{h1}\delta\left(E - \varepsilon_{p1} - \varepsilon_{p2} - \varepsilon_{h1}\right) = \frac{g_p^2 g_h E^2}{2!1!2!} \quad (4.5)$$

Equation (4.5) may be easily generalized to,

$$\begin{aligned}\omega(p, h, E) &= \frac{g_p^p g_h^h}{p!h!}\int\prod_{j=1}^{p} d\varepsilon_{pj}\prod_{k=1}^{h} d\varepsilon_{hk}\delta\left(E - \sum_{j=1}^{p}\varepsilon_{pj} - \sum_{k=1}^{h}\varepsilon_{hk}\right)\\ &= \frac{g_p^p g_h^h E^{n-1}}{p!h!(n-1)!}\end{aligned} \quad (4.6)$$

Here g stands for the single-particle level density at the top of the Fermi level. Generally one takes $g_p = g_h = g$.

Expression (4.5) for density of states does not take into account Pauli exclusion principle. Williams [9] obtained the corrected densities as,

$$\omega(p, h, E) = \frac{g^n\left(E - A_{p,\,h}\right)^{n-21}}{p!h!(n-1)!} \quad (4.7)$$

where the Pauli blocking correction is

$$A_{p,\,h} = \frac{1}{4g}[p(p-1) + h(h-1)] \quad (4.8)$$

When the Fermi energy is placed symmetrically between the last occupied and the first unoccupied levels as shown in figure 4.5. In expressions for state densities derived so far no consideration has been given to the fact that single-particle and hole states are limited to the depth of the potential well. The effect becomes significant at higher energies. Correction for this effect was obtained by Betak and Dobes [10] and corrected densities are given as,

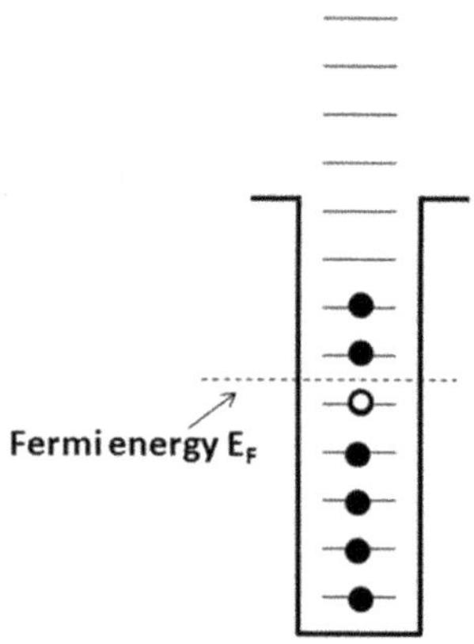

Figure 4.5. Fermi level placed symmetrically between the last occupied and the first unoccupied levels.

$$\omega(p, h, E) = \frac{g^n}{p!h!(n-1)!}\sum_{j=0}^{h}(-1)\binom{h}{j}\left(E - jV - A_{p,h,j}\right)^{n-1} x\theta\left(E - jV - A_{p,h,j}\right) \quad (4.9)$$

Kalbach [11] approximated it as

$$\omega(p, h, E) = \frac{g^n\left(E - A_{p,h,j}\right)^{n-1}}{p!h!(n-1)!} f(p, h, E) \quad (4.10)$$

where

$$f(p, h, E) = \sum_{j=0}^{h}(-1^j)\binom{h}{j}\left(\frac{E - jV}{E}\right)^{n-1} x\theta(E - jV) \quad (4.11)$$

Coming back to equation (4.4), with the above prescription for the state densities, the density of state $\omega((n^p - n_b^p), n^h, U)$ may be written as,

$$\omega((n^p - n_b^p), n^h, U) = \frac{g^n}{n^p!\, n^h!(n-1)!}\sum_{j=0}^{n^h}(-1)^j\binom{n^h}{j} \left(E^* - A_{n^p, n^h} - jE_F\right)^{(n-1)}\theta\left(E^* - E_{pp} - jE_F\right) \quad (4.12)$$

In equation (4.12) the variables have the following meaning:

$n^p$and n^h respectively the number of particles and of holes;

n = number ofexcitons = $n^p + n^h$;

E_F the Fermi energy;

E_{pp} the minimum energy required to excite n^p particles and n^h holes satisfying Pauli's blocking.

It is given as,

$$E_{pp} = \frac{(n^p)^2 + (n^h)^2 + n^p - n^h}{2g} \tag{4.13}$$

θ is the function that takes care of the finite depth of holes, g the single-particle state density at the top of Fermi level and $g = 6a/\pi^2$ a the level density parameter

$$A_{n^p,\, n^h} = \frac{n^p(n^p - 1) + n^h(n^h - 1)}{4g} \text{ the Pauli correction factor}$$

The following expressions for interstate transition rates have been derived by Williams [8] and Betak *et al* [9]

$$\lambda^0(n) = \frac{2\pi}{\hbar} < M^2 > \frac{g}{2n}\left[gE^* - \frac{1}{2}\left\{(n^p)^2 + (n^h)^2\right\}\right] \\ [n^p(n^p - 1) + n^h(n^h - 1) + 4n^p n^h] \tag{4.14}$$

$$\lambda^+(n) = \frac{2\pi}{\hbar} < M^2 > \frac{g}{2(n + 1)} \\ \left[\frac{\left\{gE^* - A_{(n^p+1),\,(n^h+1)}\right\}^{(n+1)} - \left\{C((n^p + 1),\, (n^h + 1))\right\}^{(n+1)}}{(gE^* - A_{n^p,\, n^h})^{(n+1)}}\right] \tag{4.15}$$

and

$$\lambda^-(n) = \frac{2\pi}{\hbar} < M^2 > \frac{gn^p n^h}{2}(n - 2)\left[1 - \left\{\frac{C((n^p - 1),\, (n^h - 1))}{gE^* - A((n^p - 1),\, (n^h - 1))}\right\}^{(n-3)}\right] \tag{4.16}$$

Here, $((n^p), (n^h))$, the finite depth correction is given by

$$C((n^p), (n^h)) = gn^p(E^* - E_F) - n^h A(n^p, n^h) \tag{4.17}$$

The averaged squared matrix element for two-body interaction $<M^2>$ is often treated as a free parameter to match the experimental data. The following expression is also used for the averaged matrix element, $<M^2> = \frac{c}{A^3 E^*}$, here, A is the atomic mass number of the CN, E^* the excitation energy and c a free parameter. An early parameterization obtained by Kalbach-Cline [12] gives $c = 230$ MeV^3. The dependence of $<M^2>$ on mass number A may be derived from shell model matrix elements. However, it is obvious that the matrix element for the two-body interaction of coupling states is not likely to be so simple.

4.3.2 Equilibrium solution of master equation

If it is assumed that no pre-equilibrium emission takes place, that is, the term $W(n)q(n, t)$ in master equation (4.3) is zero, the master equation reduces to,

$$\left(\frac{dq(n,\,t)}{dt}\right) = \{\lambda^+(n-2)q(n-2,\,t) + \lambda^-(n+2)q(n+2,\,t)\} \\ - [\{\lambda^+(n)q(n,\,t) + \lambda^-(n)q(n,\,t)\}] \tag{4.18}$$

Recalling the identity,

$$\omega(n,\,E)\lambda^+(n) = \omega(n+2,\,E)\lambda^-(n+2) \tag{4.19}$$

It may be observed that when $q(n,\,t)\propto\omega(n,\,E)$

The equilibrium is obtained when,

$$\frac{dq(n,\,t)}{dt} = 0 \tag{4.20}$$

When particle emission rates are smaller than the transition rates, the system will evolve to the normalized equilibrium solution before decaying;

$$q(n,\,t) = \frac{\omega(n,\,E)}{\sum\limits_{n=3,5,7..} \omega(n,\,E)} \tag{4.21}$$

4.3.3 Integrated emission rates

The solution to the master equation in vector form may be written as,

$$\overrightarrow{q(n,\,t)} = \exp[-\Lambda^{-1}t]q(n,\,0) \text{ with } q(n,\,0) = \sigma_{abs}\delta_{p2}\delta_{h1} \tag{4.22}$$

where the matrix Λ^{-1} is given by,

$$(\Lambda^{-1})_{nm} = \lambda(n)\delta_{m,\,n} - \lambda^-(n+2)\delta_{m,\,n+2} - \lambda^+(n-2)\delta_{m,\,n-2} \tag{4.23}$$

The time-integrated emission spectrum is,

$$\frac{d\sigma}{d\varepsilon} = \sum_n \int_0^\infty \frac{\mathrm{d}\lambda_{\text{emission}}}{\mathrm{d}\varepsilon}(n,\,\varepsilon)q(n,\,t)dt = \sum_{n,m} \frac{\mathrm{d}\lambda_{\text{emission}}}{\mathrm{d}\varepsilon}(n,\,\varepsilon)\Lambda_{n,\,m}q(m,\,0) \tag{4.24}$$

Here,

$$\Lambda_{n,\,m}q(m,\,0) = \sigma_{abs}\tau_n \tag{4.25}$$

The mean lifetime of an exciton state is defined as

$$\tau_n = \int_0^\tau q(n,\,t)\mathrm{d}t$$

Once mean lifetime is known, the average differential cross-section for the emission of particle b with energy ε_b may be given as,

$$\frac{\mathrm{d}\sigma^{\text{pre}}}{\mathrm{d}\varepsilon_b}(a,\,b) = \sigma(a)\sum_n \omega_b(n,\,\varepsilon_b)\tau_n \tag{4.26}$$

4.3.4 Never-come-back approximation

There are several different prescriptions for calculating the mean lifetime of an exciton state. However, in the simplest method it is assumed that in the beginning of the two-body interaction cascade the system moves only in the forward direction, that is, towards the more complex states and does not move backwards towards the state of less complexity. This is called the never-come-back assumption and in this case $\lambda^{-}(n)$ is zero. In never-come-back assumption the average differential pre-equilibrium emission cross-section is given as,

$$\frac{d\sigma^{PE}(a, b)}{d\varepsilon_b} = \left[\frac{\omega_b(n, \varepsilon_b)}{\omega(n) + \lambda^{+}(n)} + \sum_{n=n_0+2, \Delta n=2}^{\bar{n}} \left\{\prod_{j=n_j, \Delta j=2}^{n-2} \frac{\lambda^{+}(j+2)}{\omega(j) + \lambda^{+}(j+2)}\right\} \frac{\omega_b(n, \varepsilon_b)}{\omega(n) + \lambda^{+}(n)}\right] \tag{4.27}$$

Akkermans [13] worked out an exact solution for equation (4.27), he first obtained the mean lifetime $\tau(N)$ for the state that has maximum number of excitons N.

$$\tau(N) = T_N H_N \sum_{j=n_0, \Delta j=2} q_0(j) \left\{\prod_{i=j, \Delta i=2}^{N-2} \lambda^{+}(i) T_i H_i\right\} \tag{4.28}$$

In equation (4.28) $T_i = \frac{1}{\{\lambda^{+}(i) + \lambda^{-}(i) + \omega(i)\}}$

$$H_{j+1} = \frac{1}{\{1 - H_j \lambda^{+}(i) T_i \lambda^{-}(i+1) T_{i+1}\}}$$

and

$$H_1 = 1$$

Further, in equation (4.28) for $j = N$, the term in the curly bracket is replaced by 1 and the remaining mean lives may be calculated using the recursion formula;

$$\lambda^{+}(n)\tau(n) = \frac{1}{2}\tau(n+2) - \lambda^{-}(n+4)\tau(n+4) - q_0(n+2) \tag{4.29}$$

where,

$$n_o \leqslant n \leqslant N - 2 \text{ and } \tau(N+2) = 0$$

4.3.5 Gamma emission

It is also possible in the initial stages of the interaction pre-equilibrium photons are emitted from the intermediate states. An n-exciton state may decay to an n or to an $n-2$ exciton state by the emission of a dipole gamma (giant dipole resonance). Pre-equilibrium photon emission may be taken into account using Brink–Axel [14] formalism, according to which the differential dipole gamma emission rate may be written as,

$$\frac{d\lambda^{e\gamma}}{d\varepsilon}(n, \varepsilon) = \frac{e^2\sigma_\gamma(\varepsilon)}{\pi^2\hbar^3c^2}\sum b(k \to n, \varepsilon)\frac{\omega(k, E - \varepsilon)}{\omega(n, E)} \tag{4.30}$$

Here b represents the gamma strength function for the designated transition. In order to be consistent with equilibrium emission, the following relation must hold

$$b(n \to n + 2, \varepsilon) + b(n \to n, \varepsilon) = 1$$

4.3.6 Emission of protons and neutrons

In the discussion so far, no difference is made between a proton and the neutron. However, there are important differences: the emission of protons is inhibited because of the Coulomb barrier and also the binding energies of neutrons and protons are significantly different.

These differences may be taken into account by appropriate correction factors in their emission rates. One takes for protons,

$$\frac{d\lambda^{e\pi}}{d\varepsilon}(n, \varepsilon) = (2s + 1)\frac{\mu\varepsilon\sigma_\pi(\varepsilon)}{\pi^2\hbar^3}R_\pi(n)\frac{\omega(n - 1, E - \varepsilon - B_\pi)}{\omega(n, E)} \tag{4.31}$$

and for neutrons

$$\frac{d\lambda^{e\nu}}{d\varepsilon}(n, \varepsilon) = (2s + 1)\frac{\mu\varepsilon\sigma_\nu(\varepsilon)}{\pi^2\hbar^3}R_\nu(n)\frac{\omega(n - 1, E - \varepsilon - B_\nu)}{\omega(n, E)} \tag{4.32}$$

In the above expressions subscripts π and ν stand, respectively, for proton and neutron. $R_x(n)$ is the fraction of the particles of type x (proton π, neutron ν) in exciton class n. Final PE-emission rate that enters the master equation is,

$$\lambda^{\text{emission}}(n) = \lambda^{e\pi}(n) + \lambda^{e\nu}(n) = \int\left(\frac{d\lambda^{e\pi}}{d\varepsilon}(n, \varepsilon) + \frac{d\lambda^{e\nu}}{d\varepsilon}(n, \varepsilon)\right)d\varepsilon \tag{4.33}$$

4.3.7 Two-component exciton states

The correction factors $R_\nu(n)$ and $R_\pi(n)$ take care of the difference in binding energy and Coulomb suppression of proton over neutron in exit channels only, i.e., only during pre-equilibrium emission of neutrons and protons. However, so far as transitions from one class of exciton state to another class are considered no distinction is made between neutrons and protons. Physically, the transition rates for neutron and protons must also be different. Further, assuming that single-particle densities for both neutrons and protons are the same is also not physically justified. Similarly, no attention is paid to the difference in the properties of the holes created by the deficiency of neutron or proton. The two-component model corrects for all these deficiencies.

Kalbach [15] and Koning *et al* [16] applied corrections for the two-component nucleus and corrected the master equation as,

$$\frac{dq(n_\pi, n_\nu, t)}{dt} = \lambda_\pi^-(n_\pi + 2, n_\nu)q(n_\pi + 2, n_\nu, t)$$
$$+ \lambda_\nu^-(n_\pi, n_\nu + 2)q(n_\pi, n_\nu + 2, t) + \lambda_\pi^{\nu 0}(n_\pi + 2, n_\nu - 2)q(n_\pi + 2, n_\nu - 2, t)$$
$$+ \lambda_\nu^{\pi 0}(n_\pi - 2, n_\nu + 2)q(n_\pi - 2, n_\nu + 2, t) \tag{4.34}$$
$$+ \lambda_\pi^+(n_\pi - 2, n_\nu)q(n_\pi - 2, n_\nu, t) + \lambda_\nu^+(n_\pi, n_\nu - 2)q(n_\pi, n_\nu - 2, t)$$
$$- \lambda(n_\pi, n_\nu)q(n_\pi, n_\nu, t)$$

Here,

$$n_\pi = (p_\pi + h_\pi) \text{ and } n_\nu = (p_\nu + h_\nu)$$

It may be noted that in the above expressions n_π, n_ν, p_π, h_π, p_ν and h_ν, respectively, means total number of excitons due to protons, total number of excitons due to holes, number of excited protons, number of holes due to protons, number of excited neutrons and number of holes due to neutrons.

Further, in the two-component model the exciton densities become;

$$\omega(p_\pi, h_\pi, p_\nu, h_\nu, E) = \frac{g_\pi^{n_\eta} g_\nu^{n_\nu}\left(gE - A_{p_\pi, h_\pi, p_\nu, h_\nu}\right)^{(n-1)}}{p_\pi!\, h_\pi!\, p_\nu!\, h_\nu!(n-1)!} f(p, h, E) \tag{4.35}$$

with

$$A_{p_\pi, h_\pi, p_\nu, h_\nu} = \frac{p_\pi(p_\pi - 1) + h_\pi(h_\pi - 1)}{4g_\pi} + \frac{p_\nu(p_\nu - 1) + h_\nu(h_\nu - 1)}{4g_\nu} \tag{4.36}$$

It is obvious that new transition rates will have more collision probabilities;

$$\lambda_\pi^+(n)\omega(n, E) = \lambda_\pi^-(n + 2)\omega(n + 2, E)$$

$$= \frac{2\pi}{\hbar}\int_0^E d\varepsilon\Big(|M_{\pi\pi}|^2\, \omega(2,1,0,0, \varepsilon)\omega(1,0,0,0, \varepsilon)\omega(p_\pi - 1, h_\pi, p_\nu, h_v, E - \varepsilon)\Big)$$
$$+ |M_{\pi\pi}|^2\, \omega(1,2,0,0, \varepsilon)\omega(0,1,0,0, \varepsilon)\omega(p_\pi, h_\pi - 1, p_\nu, h_v, E - \varepsilon)$$
$$+ |M_{\pi\nu}|^2\, \omega(1,1,1,0, \varepsilon)\omega(0,0,1,0, \varepsilon)\omega(p_\pi, h_\pi, p_\nu - 1, h_v, E - \varepsilon)$$
$$+ |M_{\pi\nu}|^2\, \omega(1,1,0,1, \varepsilon)\omega(0,0,0,1, \varepsilon)\omega(p_\pi, h_\pi, p_\nu, h_{v-1}, E - \varepsilon) \tag{4.37}$$

4.3.8 Conservation of angular momentum

In its initial version the exciton model considered particles and holes and energy conserving two-body interactions. As already discussed, the exciton model was extended to include two components: neutron and proton both for emission and transitions from one to the other exciton state. The model can be further generalized to include the conservation of angular momentum at each collision. For example, for the case where no distinction is made between neutron and proton and dropping the symbol for time t, one may write,

$$\frac{\mathrm{d}q(n, J)}{\mathrm{d}t} = \lambda^{-}(n+2, J)q(n+2, J) + \lambda^{+}(n-2, J)q(n-2, J) + \lambda^{0}(n, J)q(n, J) \tag{4.38}$$

Here,

$$\lambda(n, J) = \lambda^{+}(n, J) + \lambda^{-}(n, J) + \lambda^{\mathrm{emission}}(n, J)$$

Further, the exciton level densities are given in terms of the densities of states as follows,

$$\rho(p, h, E, J) = \omega(p, h, E)R_n(J) \tag{4.39}$$

where the spin distribution $R_n(J)$ is given by,

$$R_n(J) = \frac{2J+1}{\pi^{1/2}(2\sigma_n^2)^{3/2}} e^{\left(-\frac{(J+1/2)^2}{2\sigma_n^2}\right)} \tag{4.40}$$

and the spin cut-off parameter is often taken as $\sigma_n^2 = (0.24{+}0.0038E)A^{2/3}$.

Inter-exciton class transition rates may be written in terms of the average reduced matrix elements as,

$$\begin{aligned}\lambda^{+}(n, E, J)\rho(n, E, J) &= \lambda^{-}(n+2, E, J)\rho(n+2, E, J)\\ &= \frac{2\pi}{\hbar}\sum_{\vec{j}+\vec{j'}=\vec{J}} |\widetilde{M_j}|^2 \int_0^E \mathrm{d}\varepsilon\\ &\quad (\rho(2,1, \varepsilon, j)\rho(1,0, \varepsilon, j)\rho(p-1, h, E-\varepsilon, J'))\\ &\quad + \rho(1,2, \varepsilon, j)\rho(0,1, \varepsilon, j)\rho(p, h-1, E-\varepsilon, J')\end{aligned} \tag{4.41}$$

And emission widths are given in terms of transmission coefficients

$$\frac{\mathrm{d}\lambda_x^{\mathrm{emiss}}}{\mathrm{d}\varepsilon}(n, \varepsilon, J) = \frac{2\pi R_x(n)}{\hbar} \sum_{\left(\vec{l}+\vec{s}+\vec{J'}=\vec{J}\right)} T_{xl}(\varepsilon)\frac{\rho(n-1, E-\varepsilon-B_x, J')}{\rho(n, E, J)} \tag{4.42}$$

In the above expression x stands for π or ν.

4.3.9 The generalised exciton model

In generalised exciton model [17, 18] evolution of the reaction is described in terms of two factors: (i) the exciton number n; and (ii) the direction of the projectile Ω. It is further assumed that the transition rates may be factorized into two multiplicative parts: (a) the usual transition rates; and (b) an angular part that is proportional the angular distribution of free nucleon–nucleon scattering. Mathematically, it may be written as;

$$\lambda_{n\leftarrow m}(\Omega \leftarrow \Omega') = \sigma_{NN} G(\Omega, \Omega') = \lambda_{n\leftarrow m} \frac{1}{\sigma_{NN}} \frac{\mathrm{d}\sigma_{NN}}{\mathrm{d}\Omega} \tag{4.43}$$

Under these assumptions the master equation becomes,

$$\frac{\mathrm{d}q(n, \Omega)}{\mathrm{d}t} = \sum_{m=n-2}^{n+2} \lambda_{n\leftarrow m} \int \mathrm{d}\Omega' \, G(\Omega, \Omega') q(\mathrm{m}, \Omega') - \lambda(n) q(n, \Omega) \tag{4.44}$$

Here,

$$\lambda(n) = \lambda_{n\leftarrow m} + \lambda^{\mathrm{emission}}(n)$$

It is assumed that emission rates have their usual form but particle is assumed to be emitted in direction Ω. Further, the scattering kernel $G(\Omega, \Omega')$ obviously depends on the angle between directions Ω and Ω'. As such,

$$G(\Omega, \Omega') = \sum_{l=0} \mu_l Y_{lm}(\Omega) Y^*_{lm}(\Omega') \tag{4.45}$$

Since the system has axial symmetry, the master equation may be expanded into Legendre polynomials as,

$$\frac{\mathrm{d}q(n, l)}{\mathrm{d}t} = \mu_l(\lambda^+(n-2)q(n-2, l) + \lambda^0(n)q(n, l) + \lambda^-(n+2)q(n+2, l) - \lambda(n)q(n, l))$$

with

$$q(n, \Omega) = \sum_{1} q(n, l) P_1(\cos\theta).$$

The emission spectra of PE-component may be obtained as,

$$\frac{\mathrm{d}^2\sigma}{\mathrm{d}\varepsilon \mathrm{d}\Omega} = \sigma_{\mathrm{abs}} \sum_{n=3,5,7\ldots} \lambda^{\mathrm{emission}}(n) \tau(n, \Omega) \tag{4.46}$$

where

$$\tau(n, \Omega) = \int_0^\infty q(n, \Omega, t) \mathrm{d}t$$

4.3.10 Question mark about the assumption of exciton model

While calculating internal transition rates between sub-states of a given exciton class, it is assumed in the exciton model that each sub-state is equally probable. Bisplinghoff [22] has shown that this is not automatically valid for exciton classes $n > 3(2p + 1h)$. For example, figure 4.6 shows four different sub-states of exciton class $n = 5$ and the probability of the occurrence of each of the sub-configurations is likely to be different. However, if it is assumed that a local thermal equilibrium exists between all sub-configurations of a given exciton class, then it can be shown that each sub-state will be equally probable. Though assumption of *a priori* equal

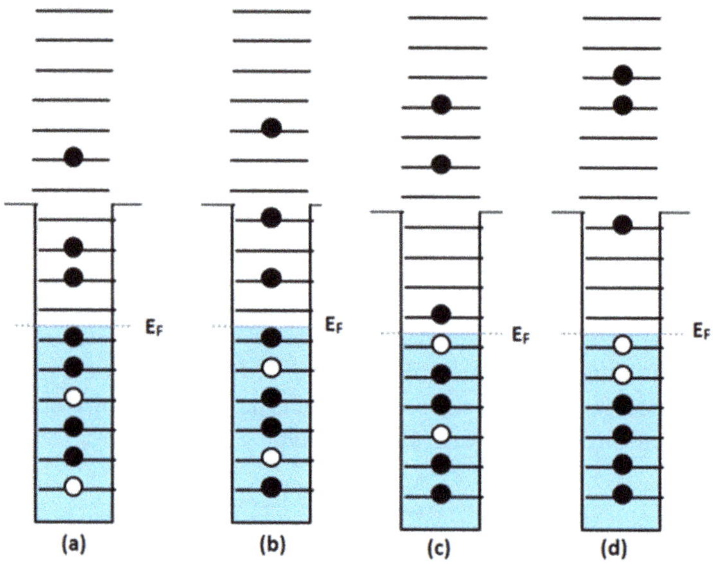

Figure 4.6. Four sub-states of $n = 5$ exciton class.

probabilities of sub-states is strictly not valid according to the Golden Rule, it is used to make calculations simple as it permits the use of exciton densities of states and transition densities.

4.4 Harp–Miller–Berne (HMB) model

As the name suggests, the HMB model was proposed by Harp, Miller and Berne [19] in 1968, earlier than the exciton model. Figure 4.7 shows the underlying concept of the HMB model where the energy scale is divided into bins of some suitable size, generally, of 1 MeV. One then computes the number of single-particle energy levels in each bin, mostly assuming a free Fermi gas model, though it is possible to use a more realistic model for these calculations. The number of single-particle levels in each bin is designated as occupation number of the bin. In the initial stage at time τ_0 all bins up to Fermi level are completely filled and all other bins, except the one corresponding to the incident particle are empty. Two-body interactions are assumed to take place between nucleons which change the bin population (occupation number) with time.

Probabilities for transition of a nucleon from one bin into another bin, and probabilities for emission of nucleons of energy larger than the binding energy from each bin are calculated. The occupation number distribution which is known initially changes with time by the scattering of nucleons into and out of the bins. The free nucleon scattering data is often used to calculate the desired probabilities. The time evolution of occupation number distribution is described by a set of coupled differential equations, the master equation, of the following type,

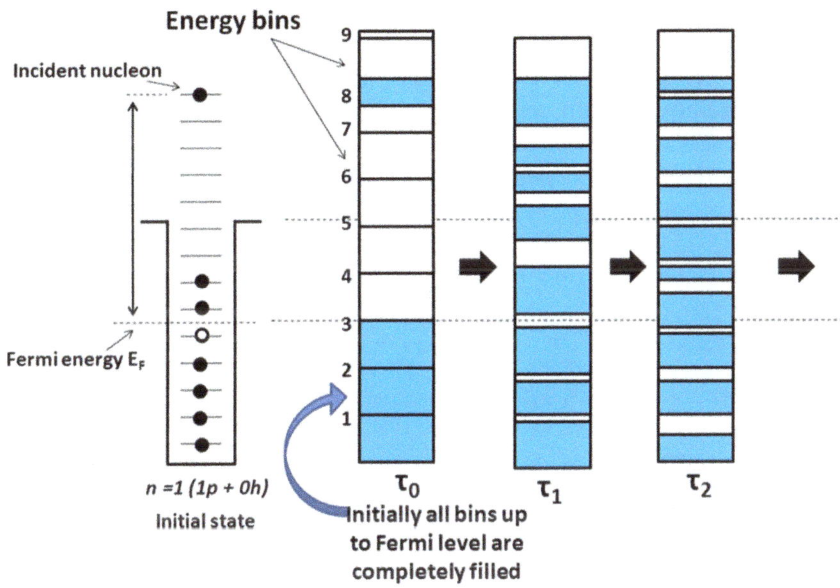

Figure 4.7. Schematic representation of Harp–Miller–Berne (HMB) model.

$$\begin{aligned}\frac{\mathrm{d}(n_i,\, g_i)}{\mathrm{d}t} = &\sum_{j,k,l} \omega_{(k_l,\, i_j)} g_k n_k g_l n_l (1 - n_i)\left(1 - n_j\right) g_k g_j \\ &- \sum_{j,k,l} \omega_{(k_l,\, i_j)} g_i n_i g_j n_j (1 - n_k)(1 - n_l) g_i g_{l - n_i g_i \lambda_i (i')}\end{aligned} \tag{4.47}$$

The above equation gives the net rate of change of the number of nucleons in the ith bin with g_i number of single-particle states per MeV and n_i number of nucleons. The first term in the equation represents the gain in the number of nucleons in ith bin due to scattering from other bins, the second the net loss of nucleons from the ith bin due to scattering to other bins and the third term gives the reduction of nucleons due to emission to the continuum.

It is possible to include nuclear structure effects, shell effects, spin and parity conservations etc in the model. However, the model is not capable of predicting the angular distribution of emitted pre-equilibrium particles.

4.5 Hybrid model

A limitation of exciton model is that it predicts only the relative spectral yields. The reason is that the square of the two-body interaction matrix element $|M^2|$ which enters into the expression of transition rates is considered as an adjustable parameter. To overcome this deficiency Blann and Mignerey [20] derived $\lambda^{+}(n, E)$ from nucleon–nucleon scattering cross-sections.

The mean free path of a nucleon inside the nucleus divided by the average nucleon velocity inside the nucleus gives the collision frequency. From this collision frequency Blann made an approximation to obtain internal transition rates as,

$$\lambda^{\mathrm{Bl}}(n, E) = (1.4 \times 10^{21}E - 6 \times 10^{18}E^{2})s^{-1} \tag{4.48}$$

Here E is the excitation energy in MeV. Further, Blann gave the following expression for pre-equilibrium emission probability as,

$$W_{\upsilon}^{\mathrm{Bl}}(\varepsilon) = \sum_{n=n_0,\Delta n=2}^{\bar{n}} P_{\upsilon}^{n}\frac{\rho(n, U, \varepsilon)g}{\rho(n, E)}\frac{\lambda_C(\varepsilon)}{\lambda^{Bl}(n, E) + \lambda_C(\varepsilon)}D_n^{\mathrm{Bl}} \tag{4.49}$$

where,

$$D_n^{\mathrm{Bl}} = \prod_{k=n_0,\Delta k=2}^{n} (1 - \overline{P_{k-2}}) \tag{4.50}$$

and

$$\overline{P_k} = \sum_{\nu=\nu_1,\nu_2} \int_0^{(E-B_\nu)} P_\nu(n, \varepsilon)\mathrm{d}\varepsilon \tag{4.51}$$

Also,

$$P_\nu(n, \varepsilon)\mathrm{d}\varepsilon = P_{\upsilon}^{n}\frac{\rho(n, U, \varepsilon)g}{\rho(n, E)}\frac{\lambda_C(\varepsilon)}{\lambda^{Bl}(n, E) + \lambda_C(\varepsilon)}D_n^{\mathrm{Bl}} \tag{4.52}$$

In the above expressions P_{υ}^{n} stands for the number of particles of type ν in the n-exciton state, while $\rho(n,U,\varepsilon)$ is the density of states of n-excitons such that if a particle is emitted from this state it leaves with energy ε and the residual state has the excitation energy U. The variable ν may have two values ν_1 and ν_2, respectively, the neutrons and the protons. $\lambda_C(\varepsilon)$ represents the transition rate (number of transitions per unit time) of a particle of energy ε into the continuum. The superscript Bl specifies that the term was introduced by Blann. G stands for the single-particle-level density of the composite nucleus. It is simple to realise that the average dwell time for which the system remains in the state of n-excitons at energy E is,

$$\tau_n = \frac{D_n}{\lambda^{Bl}(n, e) + \lambda_c(E)} \tag{4.53}$$

Since there is no adjustable parameter in equation (4.49), the expression predicts the absolute yield.

It may be observed that the hybrid model is similar to the exciton model with the difference that the internal transition rates in hybrid model are obtained from free nucleon scattering data instead of from the two-body interaction matrix element $|M^2|$. The emission spectrum of PE-particles in the hybrid model is given by:

$$\frac{\mathrm{d}\sigma_\nu^{\mathrm{pre}}}{\mathrm{d}\varepsilon} = p_\nu(\varepsilon)\sigma_{\mathrm{abs}} \tag{4.54}$$

Here σ_{abs} is the absorption cross-section and $p_\nu(\varepsilon)$ the probability for the emission of particle ν with energy ε. Also,

$$P_{\nu}(\varepsilon) = \sum_{n=n_0,\Delta n=2}^{\bar{n}} R_{\nu}(n) g \frac{\omega(n-1, E-\varepsilon-B_{\nu})}{\omega(n, E)} \frac{\lambda^{\text{emission}}(\varepsilon)}{\lambda^{\text{emission}}(\varepsilon) + \lambda^{+}(\varepsilon)} \tag{4.55}$$

where $R_{\nu}(n)$ is the number of particles of type ν in the n-exciton state with binding energy B_{ν}. The transition rate

$$\lambda^{+}(\varepsilon) = \frac{\text{average velocity of a nucleon inside the nucleus } v}{\text{mean free path at energy } \varepsilon} = \rho_0 < \sigma(\varepsilon)$$

$$> \sqrt{\frac{2(\varepsilon + v)}{m}} \tag{4.56}$$

4.5.1 Geometry dependent hybrid model

To improve the agreement between the experimental data on pre-equilibrium emission and theoretical calculations using a hybrid model, Blann *et al* introduced the geometry dependent hybrid model. This model essentially takes into account the variation of nuclear density with the impact parameter.

As the first step, one first calculates the angular momentum l associated with the impact parameter b of the projectile nucleon. This is done by using the semi-classical approach where b is related to l with the relation,

$$kb = \ell + \frac{1}{2}$$

Next, the average nuclear density $\overline{\rho_0(\ell)}$ is calculated using the Woods–Saxon density formulation as follows,

$$\overline{\rho_0(\ell)} = < \frac{\rho_0}{1 + \exp\left[\left(\sqrt{b^2 + z^2} - R_0\right)/a\right]} > \tag{4.57}$$

The average density is used to determine the Fermi energy as well as the mean free path MFP(ε), given as;

$$MFP(\varepsilon) = \frac{1}{\rho_0 < \sigma(\varepsilon)>}$$

The pre-equilibrium spectrum may be calculated as

$$\frac{d\sigma_{\nu}}{d\varepsilon} = \frac{\pi}{k^2} \sum_{\ell} (2\ell + 1) T_{\ell} q_{\nu}(\ell, \mathcal{E}) \tag{4.58}$$

Here T_{ℓ} is the transmission coefficient for particle ν of angular momentum l and $q_{\nu}(\ell, \mathcal{E})$ is the probability of particle ν with angular momentum l and energy $\mathcal{E}$. Particle ν may be a neutron or a proton.

It is possible to relate the mean free path MFP with the imaginary part of the optical potential. A propagating wave decays exponentially as $e^{-2k_i r}$ in a region of complex potential $(V + iW)$. The propagation vector is given as,

$$k_i = \mathrm{Im}\left\{\frac{\sqrt{2m(\varepsilon + V + iW)}}{\hbar}\right\} \approx \frac{W}{\hbar v} \tag{4.59}$$

Or

$$MFP(\varepsilon) = \frac{\hbar v}{2W} \text{ and } \lambda^{+}(\varepsilon) = \frac{v}{MFP(\varepsilon)} = \frac{2W}{\hbar}$$

4.6 Quantum mechanical model for pre-equilibrium process

The first quantum mechanical model for pre-equilibrium processes was put forward by Feshbach [23] which was further refined later by Feshbach, Kerman and Koonin [24] which is frequently referred to as the FKK model. The FKK model assumes that the projectile initiates two-body interactions which build-up the successive states of the system. It is further assumed that the reaction Hilbert space is made up of two parts: a sub-space where residual two-body interactions always leave the intermediate states in which all particles are in bound states. Successive states having only bound particles form the Q-chain. Similarly, the other sub part of reaction Hilbert space corresponds to those two-body interactions where at least one particle is unbound in continuum. States that have at least one unbound particle form the P-chain. Each event of the P-chain is called a statistical multi-step direct (MSD) event while each step of the Q-chain is termed as statistical multi-step compound (MSC) event.

The complexity of the successive step increases in both P- and Q-chains as one move ahead. It is also assumed that interactions both in P- and in Q-chains may change the complexity of the wave function only by one unit, which means that for example one may reach the P_4 or P_2 boxes from P_3 box only. This is called chaining hypothesis and is a characteristic of two-body interactions. Since steps of P-chain contain at least one particle in continuum, emission of particles from steps of P-chain may take place directly. Further, particles emitted from the first few steps of the P-chain are likely to retain some memory of their direction of motion and are likely to be emitted in forward direction. A member of the Q-chain may decay into the near y element of the P-chain which may emit a particle. Formally, these states of P- and Q-chains may be described by the projection operators P and Q operating on the total wave function ψ, with $P + Q = 1$. The MSD reactions may be described by the propagation through the states $P\psi$ and MSC reactions by propagation through the states $Q\psi$. This division of reaction into two parts is shown in figure 4.8. Propagation through the Q-chain ultimately leads to the equilibrated CN which decays via particle evaporation. Hauser-Feshbach or Weisskopf–Ewing theories for CN decay may be used to obtain the compound reaction cross-sections and particle spectra.

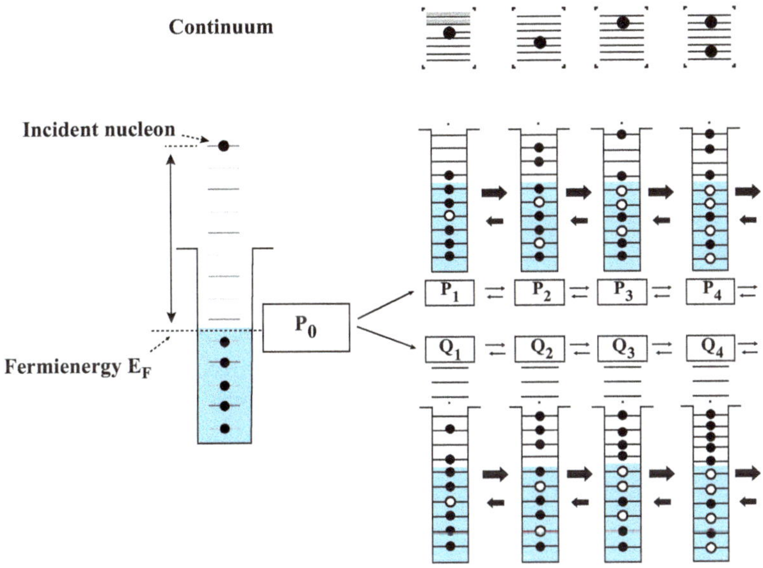

Figure 4.8. Schematic representation of P- and Q-chains of the FKK model.

Many different prescriptions have been proposed to treat the MSD process. Some important theories of MSD are those developed by: (i) Tamura *et al* [25]; (ii) Lenske *et al* [26]; (iii) Udagawa *et al* [27]; and Nishioka *et al* [28].

4.6.1 MSD theory of FKK

Feshbach *et al* used the division of a reaction into two components P and Q, along with the assumption that there is statistical distribution of nuclear energy levels, obtaining the following expression for cross-sections, which is an incoherent sum of a one-step and multi-step terms. Further, each multi-step term may be written as a convolution of a one-step term

$$\frac{\mathrm{d}^2\sigma(\boldsymbol{k}, \boldsymbol{k}_0)}{\mathrm{d}\Omega \mathrm{d}U} = \sum_{n=1}^{n} \frac{\mathrm{d}^2\sigma^n(\boldsymbol{k}, \boldsymbol{k}_0)}{\mathrm{d}\Omega \mathrm{d}U} \tag{4.60}$$

And the one-step cross-section is given as,

$$\frac{\mathrm{d}^2\sigma^{(1)}(\boldsymbol{k}, \boldsymbol{k}_0)}{\mathrm{d}\Omega_1 \mathrm{d}U_1} = \frac{2\pi m}{\hbar^2 k_0}\rho(k)\rho_{(1p,\,1h)}(U_1)(|\ M_{1,0}(\boldsymbol{k}_1, \boldsymbol{k}_2)|)^2 \tag{4.61}$$

The n-step cross-section is given as,

$$\frac{\mathrm{d}^2\sigma^n(\boldsymbol{k}, \boldsymbol{k}_0)}{\mathrm{d}\Omega \mathrm{d}U} = \int \frac{\mathrm{d}k_1}{(2\pi)^3}\cdot \cdots \frac{\mathrm{d}k_n}{(2\pi)^3} \frac{\mathrm{d}^2\omega_{i,\,n}(\boldsymbol{k}, \boldsymbol{k}_0)}{\mathrm{d}\Omega \mathrm{d}U}$$

$$x\frac{d^2\omega_{(n,\,n-1)}(\boldsymbol{k_n},\,\boldsymbol{k_{n-1}})}{d\Omega_n dU_n}\cdots\cdots\frac{d^2\omega_{(2,1)}(\boldsymbol{k_2},\,\boldsymbol{k_1})}{d\Omega_2 dU_2}\frac{d^2\sigma^{(1)}(\boldsymbol{k_2},\,\boldsymbol{k_1})}{d\Omega_2 dU_2} \tag{4.62}$$

In the above expression ω stands for the transition probability from $(n-1)$th to nth stage whereby the particle momentum changes from $\boldsymbol{k_{n-1}}$ to $\boldsymbol{k_n}$.

Also,

$$\frac{d^2\omega_{(n,\,n-1)}(\boldsymbol{k_n},\,\boldsymbol{k_{n-1}})}{d\Omega_n dU_n} = 2\pi^2\rho(k_n)\rho_{1p,\,1h}(U_n)(|\,M_{n,\,n-1}(k_n,\,k_{n-1})\,|)^2 \tag{4.63}$$

where, $\rho_{1p,\,1h}$ is the level density of the residual nucleus at excitation energy U_n.

$M_{n,\,n-1}$ represents the distorted wave Born approximation (DWBA) matrix element for the transition from the $(n-1)$th stage to nth stage. The state density of the leading particle in continuum is given by

$$\rho(k_n) = \frac{mk_n}{(2\pi)^3\hbar^2} \tag{4.64}$$

Koning and Akkermans [28], however, showed that these equations from (4.60)–(4.64) may be written in an equivalent but more compact form expressing them in terms of energy and solid angle in place of the momentum k. The modified forms of these equations with energy and solid angle as variables are as follows.

The total FKK cross-section for MSD processes may be written as,

$$\frac{d^2\sigma(E_k,\,\Omega\leftarrow E_{k0},\,\Omega_0)}{d\Omega dE_k} = \sum_{n=1}^{\infty}\frac{d^2\sigma^{(n)}(E_k,\,\Omega\leftarrow E_{k0},\,\Omega_0)}{d\Omega dE_k} \tag{4.65}$$

The first step cross-section is given as,

$$\frac{d^2\sigma^{(1)}(E_k,\,\Omega\leftarrow E_{k0},\,\Omega_0)}{d\Omega dE_k} = \sum_J \rho_{1p,\,1h}(E_{k0}-E_k)\left[\frac{d\sigma(E_k,\,\Omega\varphi\leftarrow E_{ko},\,\Omega_0)}{d\Omega}\right]^{DWBA} \tag{4.66}$$

The n-step cross-section ($n>1$) may be written in the following recursive form,

$$\begin{aligned}\frac{d^2\sigma^{(n)}(E_k,\,\Omega\leftarrow E_{k0},\,\Omega_0)}{d\Omega dE_k} &= \frac{m}{4\pi^2\hbar^2}\int d\Omega_{n-1}\int dE_{k-1}\\ &\times\frac{d^2\sigma^{(1)}(E_k,\,\Omega\leftarrow E_{k-1},\,\Omega_{n-1})}{d\Omega dE_k}\frac{d^2\sigma^{(n-1)}((E_{kn-1},\,\Omega_{n-1}\leftarrow E_{k0}\Omega_0))}{d\Omega_{n-1}dE_{n-1}}\end{aligned} \tag{4.67}$$

The MSD theory of Feshbach may be considered as an extension of the direction reaction theory for the discrete states to continuum. If one starts from the DWBA it is required to use energy averaging and statistics to obtain convolution-type FKK cross-sections. Initially, FKK formula was derived using statistical approximations only, however, there was criticism of the way distorted waves were treated in the original derivation (this is known as the χ-$\chi\chi$ problem). Feshbach, later justified the formula arguing that an extra energy averaging is required before the application of statistical approximations, otherwise interference effects, that should normally cancel out, are erroneously carried over to the statistical multi-step expression and enhances the strength of the multi-step term that is not physically justified. It has, however, been shown that the extra energy averaging is mathematically equivalent to the conjugate outgoing distorted wave being replaced by an energy averaged incoming distorted wave.

As regards the application of statistical laws to the treatment of MSD processes, generally two types of statistics are employed: statistics associated with the leading particle and the statistics of residual systems. Leading particle statistics involves the creation of many states in a random manner by the leading particle from a given state. Residual system statistics refers to random configuration mixing which also means that the distribution amplitudes are random. However, Feshbach held the view that in statistical reaction where level densities are substantially high everything is random and, therefore, no distinction can be made between the leading particle statistics and the residual system statistics.

Authors like Tamura *et al* and Nishioka *et al* make a physical distinction between the leading particle statistics and the residual system statistics; so much so that these two authors invoked only the residual system statistics in their models, explicitly forbidding leading particle statistics on account of the very small timescale at which this interaction takes place. The interesting result is that the first-step expressions obtained by FKK, Tamura and Nishioka are identical but multi-step results in the case of FKK are much simpler, while those of Tamura and Nishioka are very complicated.

4.6.2 Multi-step compound processes (MSCs)

Apart from the chaining hypothesis, another assumption of multi-step processes is the random phase approximation, which assumes that the phases of components are randomly distributed. This approximation has an immediate consequence for the Q-chains. Since all the states involved in Q-chain are bound, the appropriate quantum numbers are angular momentum and parity. Any sum over n-stages becomes sum over these quantum numbers and, therefore, terms with different parities do not interfere.

In their original work FKK assumed a constant wave function within a nucleus since it makes it easy to calculate a transitional matrix element. However, the Milan group [30] adopted more realistic wave functions for bound and unbound states and calculated transitional matrix elements microscopically. As shown by FKK, the transitional matrix T may be written as the sum of two terms, a direct term that corresponds to the MSD processes and another fluctuating term with vanishing average value. The fluctuating term has its origin in Q-space-chain and is given as,

$$\mathcal{T}_{fi}^{(msc)} = \langle \phi_f^{(-)} V_{PQ} \frac{1}{\left(E - h_{QQ}\right)} V_{QP} \phi_i^{(+)} \rangle \tag{4.68}$$

Here,

$$h_{qq} = H_{QQ} + V_{QP} \frac{1}{\left(E^{(+)} - H_{\text{opt}}\right)} V_{PQ} \tag{4.69}$$

In expression (4.69), $H_{QQ} = QHQ$, H_{opt} is the multi-channel optical model Hamiltonian and other symbols have their usual meaning. The matrix element $\mathcal{T}_{fi}^{(msc)}$ may be written as a linear combination of amplitudes coming from each stage;

$$\mathcal{T}_{fi}^{(msc)} = \sum_{1}^{r} \mathcal{T}_{fi}^{(m)} \tag{4.70}$$

and,

$$\mathcal{T}_{fi}^{(m)} = \langle \phi_f^{(-)} V_{PQ_m} \frac{1}{(E - h_{QQ})} V_{Q_1 P} \phi_i^{(+)} \rangle \tag{4.71}$$

where,

$$V_{PQ_m} \equiv PVQ_m \tag{4.72}$$

Further, Q_m is the projection operator for the sub-space Q_m.

In analogy with the case of MSD process, one may write;

$$Q_k \frac{1}{(E - h_{QQ})} Q_1 = G_k V_{k,\,k-1} G_{k-1} \frac{1}{(E - h_{QQ})} Q_1 \tag{4.73}$$

with,

$$V_{k,\,k-1} \equiv Q_k V Q_{k-1} \text{ and } G_k = \frac{1}{(E - h_{QQ} - V_{k,\,k+1} G_{k+1} V_{k+1,\,k})} \tag{4.74}$$

so the sum over m in equation (4.70) terminates at $m = r$, so that $G = \frac{1}{(E - h_{rr})}$.

With the above substitutions in equation (4.71) one gets,

$$\mathcal{T}_{fi}^{(m)} = \langle \phi_f^{(-)} V_{Pm} G_m V_{m,\,m-1} G_{m-1} \ldots . V_{2\,1} G_1 V_{1P} \phi_i^{(+)} \rangle \tag{4.75}$$

since

$$Q_k \psi_i^{(+)} = Q_k \frac{1}{E - h_{QQ}} V_{1\,P} \phi_i^{(+)} \tag{4.76}$$

In equation (4.76) the left-hand side gives the components in space Q_k of the exact solution of the Schrödinger equation for the system. Using equation (4.73) one gets;

$$Q_k \psi_i^{(+)} = G_k V_{k,\,k} Q_{k-1} \psi_i^{(+)} \tag{4.77}$$

Hence equation (4.76) may be written as

$$\mathcal{T}_{fi}^{(m)} = \langle \phi_f^{(-)} V_{Pm} G_m V_{m,\,m-1} Q_{m-1} \psi_i^{(+)} \rangle \tag{4.78}$$

It is now required to calculate the energy average of $\left(\mid \mathcal{T}_{fi}^{(m)} \mid\right)^2$. Equation (4.74) tells us that the implicit source of energy dependence of $\mathcal{T}_{fi}^{(m)}$ is given by the term $V_{k,\,k+1} G_{k+1} V_{k+1,\,k}$. In order to invoke random phase approximation and for energy averaging it is necessary that many states $\psi_{k,\beta}$ be included in the interval. That leads to the condition,

$$\rho_k \Gamma_{k+1} \gg 1 \text{ or } \Gamma_{k+1} \gg D_k \tag{4.79}$$

Here ρ_k is the density of levels in the kth sub-space, D_k the energy spacing and Γ_{k+1} is the average value of level widths. With these substitutions and with little simplifications it is easy to obtain the following expression for desired average as,

$$\left\langle \left(\left| \mathscr{T}_{fi}^{(m)} \right| \right)^2 \right\rangle = \frac{\Gamma_m^{(f)}}{\Gamma_m} \frac{\langle | \overline{\langle \widetilde{\psi}_{m,\alpha} V_{m,m-1} Q_{m-1} \psi_i^{(+)} \rangle |^2} \rangle}{D_m} \tag{4.80}$$

In the above expression $\Gamma_m^{(f)} = 2\pi \langle | \phi_f^{(-)} V_{Pm} \psi_{m\alpha} |^2 \rangle$ is called the escape width.

Further, if one defines $\Gamma_{m-1}^{\downarrow} = \frac{2\pi}{D_m} \langle | \left\langle \widetilde{\psi}_{m,\alpha} V_{m,m-1} \psi_{m-1,\beta} \right\rangle |^2 \rangle$ as spreading width, one may obtain,

$$\left\langle (| \mathscr{T}_i^{(msc)} |)^2 \right\rangle = \frac{1}{(2\pi)^2} \sum_{m=1}^{r} \frac{\Gamma_m^{(f)}}{\Gamma_m} \left\{ \prod_{k=1}^{m-1} \frac{\Gamma_k^{\downarrow}}{\Gamma_k} \right\} \frac{2\pi \Gamma_1^{(i)}}{D_1} \tag{4.81}$$

With proper normalization of wave functions so that the widths have dimensions of energy, the cross-section for a given channel with quantum numbers designated by γ is given as,

$$\sigma_{f,i}^{(msc)} = \frac{\pi}{k^2} \sum_{m=1}^{r} \frac{\Gamma_m^{(f)}}{\Gamma_m} \left\{ \prod_{k=1}^{m-1} \frac{\Gamma_k^{\downarrow}}{\Gamma_k} \right\} \frac{2\pi \Gamma_1^{(i)}}{D_1} \tag{4.82}$$

The angular distribution may be given by the expression,

$$\frac{d\langle \sigma_\gamma^{(msc)} \rangle}{dU} = \frac{\pi}{k^2} \sum_{m=1}^{r} \sum_{\mu=m-1}^{m+1} \frac{\langle \rho_\mu^{(\gamma)}(U) \Gamma_{m\mu}^{(\gamma)}(U)}{\Gamma_m} \left[\sum_{1}^{m-1} \frac{\Gamma_k^{\downarrow}}{\Gamma_k} \right] \frac{2\pi \Gamma_1^{(i)}}{D_1} \tag{4.83}$$

Equation (4.82) may be considered as the product of three terms; starting from the right the term $\frac{2\pi \Gamma_1^{(i)}}{D_1}$ is the probability of the system for transferring from the initial state to the first sub-state Q_1. The term in curly brackets is the measure of the attenuation because of pre-equilibrium emissions up to the mth channel. The third term $\frac{\Gamma_m^{(f)}}{\Gamma_m}$ is the branching ratio for emission from the mth sub state. Regarding angular distribution, equation (4.83), it may be remarked that it is symmetric around 90° because of random phase approximation and that it already contains a contribution from the pure compound process through the rth term.

According to multi-step compound reaction mechanism, the system passes through several intermediate states before reaching the statistical compound state.

4.7 Computer codes

Several computer codes are available that perform pre-equilibrium calculations using semi-classical models. Most of these codes use exciton model. A brief description of some of the frequently used codes is given below.

(i) **EMPIRE code:**

EMPIRE is a modular nuclear reaction code. The code calculates the emission probability of particles like photons, neutrons, protons, deuterons, tritons, alpha particles etc from direct, compound and pre-equilibrium processes in the energy range up to several hundred MeV. The direct reaction component in the code is simulated using a default spherical optical model that uses the ECIS06 code which is used to calculate the particle transmission coefficients. This section of the code uses optical model parameters proposed by Koning and Delaroche [30]. The statistical Hauser-Feshbach model has been used to calculate the CN decay while the pre-equilibrium decay has been modelled according to the two-component exciton model (PCROSS code) with default mean free path multiplier PCROSS 1.5. Contributions from all the three mechanisms (direct+ compound + pre-equilibrium) are added to get the total cross-section. Both the compound and the pre-equilibrium cross-sections have been found to be very sensitive to the level density parameters used in calculations. EMPIRE has four different options for level densities.

(a) LEVELDEN 0: It is the default option used in EMPIRE, adjusted to RIPL-3 experimental D_{obs} and to discrete levels. This is based on the enhanced generalised superfluid model (EGSM). The model uses superfluid model below critical excitation energy and Fermi gas model above.
(b) LEVELDN 1: This is based on the generalized superfluid model of Ignatyuk *et al* adjusted to RIPL-2 experimental D_{obs} and to discrete levels based on constant temperature model below critical energy and Fermi gas model above it.
(c) LEVELDN 2: Uses Gilbert–Cameron level densities adjusted to RIPL-2 experimental D_{obs} and to discrete levels recommended for compound nuclei excited up to 20 MeV.
(d) LEVELDN 3: Uses RIPL-3 microscopic HFB level densities and neutron resonances spacing for observed level separations.

Apart from exciton model it is also possible to simulate the pre-equilibrium component using MSD and MSC processes. Normally, MSC or MSD calculations are done only for emitted neutrons while exciton model calculations are performed if the emitted particle is a photon or a charged particle.

(ii) **TALYS code:**

Code TALYS is written by A J Koning and S Hilaire [31] as a system for the prediction and analysis of nuclear reactions. The code TALYS can simulate reactions involving neutrons, protons, deuterons, tritons, ^{3}He and alpha particles interacting with target nuclei of mass 12 and heavier in the energy range 1 keV to 200 MeV. Several nuclear models describing reaction mechanisms and nuclear level structure have been incorporated in a single code. Code TALYS may be used to evaluate nuclear reactions from the unresolved resonance region up to intermediate energies.

General features of code TALYS are that it is capable of implementing several modern nuclear reaction mechanisms in a non-approximate way over a wide energy range with targets in the range $12<A<339$. The code may generate partial cross-sections, energy spectra, angular distributions, double differential spectra, excitation functions and isomeric cross-section ratios. The code TALYS automatically refers to nuclear structural parameters like nuclear masses, discrete levels, level density parameters, resonance energies, fission barriers, deformation parameters etc directly from the IAEA RIP database. Pre-equilibrium process in this code may be simulated using semi-classical exciton model or the totally quantum mechanical FKK MSC and/or MSD model. In the latest version of the code the two-component exciton model with any order in multiple pre-equilibrium emission has been incorporated.

Figure 4.9 shows how contributions of compound, pre-equilibrium and direct reaction components give rise to different regions of the ejectile energy spectra. At lower energies less than 4 MeV or so, elastic scattering and capture followed by fission are the only important processes. At higher energies than 4 MeV, compound reaction process dominates, while at still higher energies compound and multiple compound processes become important. Further increase of energy gives rise to pre-equilibrium and multiple pre-equilibrium processes at the cost of compound processes. At

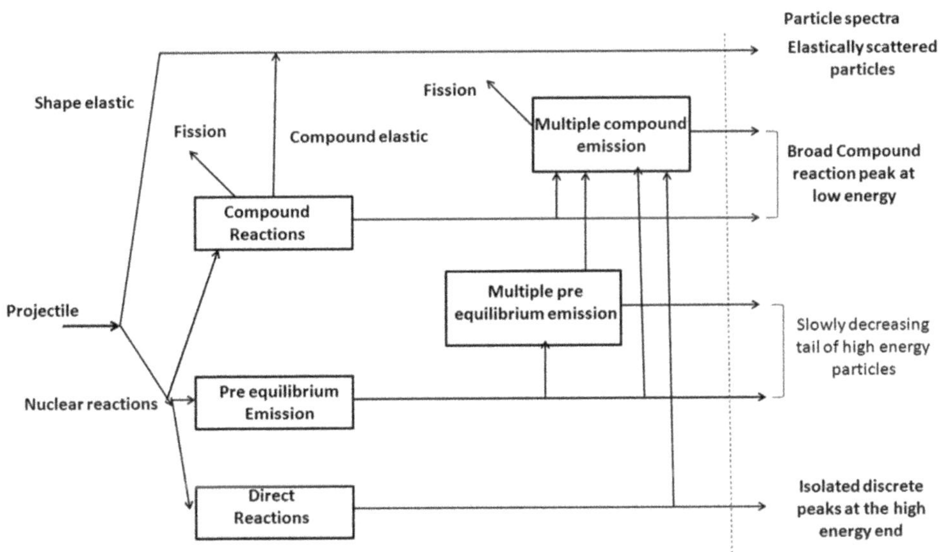

Figure 4.9. Description of a typical nuclear reaction in terms of compound, pre-equilibrium and direct reaction components along with the corresponding parts of the ejectile spectra.

still higher energies multiple pre-equilibrium processes and direct reaction processes replace compound and multiple compound components.

In code TALYS the total cross-section σ_{total} is taken as the sum of the cross-section $\sigma_{\text{shape elastic}}$ for shape elastic scattering and the cross-section for reaction σ_{reaction}, i.e.,

$$\sigma_{\text{total}} = \sigma_{\text{shapeelastic}} + \sigma_{\text{reaction}}$$

It may, however, be said that σ_{reaction} in the last expression does not contain the cross-section for compound elastic scattering.

TALYS code is a versatile code that uses spherical optical model, DWBA, rotational and vibrational coupled channel methods for calculating direct reaction components and the two-component exciton model with any multiple order along with Hauser-Feshbach formalism for compound process for pre-equilibrium calculations. Fission channels are also taken into account at appropriate stages.

(iii) **ALICE-IPPE code:**

ALICE-IPPE is an improved version of code ALICE-91. The code has been primarily developed to create the Medium Energy Nuclear Data Library, MENDL-2. The code calculates pre-equilibrium nucleon spectra using the following expression of geometry dependent hybrid model (GDH),

$$\frac{d\sigma^{\text{pre}}}{d\varepsilon_x} = \pi\Delta^2 \sum_{l=0}^{\infty}(2l+1)T_l \sum_{n=n_0}^{\bar{n}} R_x(n) \frac{\omega(p-1, h, E-Q_x-\varepsilon_x)}{\omega(p, h, E)} \cdot \frac{\lambda_c^x}{\lambda_c^x + \lambda_+^x} \cdot g \cdot D_n \tag{4.84}$$

where,

Δ = reduced wavelength of the incident particle;

T_l = optical model transmission coefficient;

ε_x = energy of the emitted nucleon;

λ_+^x = intra nuclear transition rate;

Q_x = binding energy of nucleon in composite nucleus;

λ_c^x = emission rate of nucleon;

$\omega(p, h, E)$ = density of n − exciton states at energy E having p particles and h holes;

$R_x(n)$ = number of x particles in n − exciton state;

n_0 = initial exciton no. ;

D_n = depletion factor for n − exciton state;

g = single particle k level density.

Pre-equilibrium deuteron emission has been carried out in the framework of the exciton coalescence pick-up model combined with hybrid exciton model. Direct emission of deuteron has also been taken into account using the phenomenological approach.

In this code the spectrum of alpha particles is represented as the sum of three terms corresponding to pick-up, knock-out and evaporation processes.

$$\frac{d\sigma}{d\varepsilon_\alpha} = \frac{d\sigma^{\text{pick-up}}}{d\varepsilon_\alpha} + \frac{d\sigma^{\text{knock-out}}}{d\varepsilon_\alpha} + \frac{d\sigma^{\text{evaporation}}}{d\varepsilon_\alpha} \tag{4.85}$$

Iwamoto–Harada [32] model has been used to describe the alpha pick-up process. Modified approach of Milazzo-Colli and Braga-Marcazzan [33] has been used for alpha knock-out.

Nuclear level density in the code for all nuclides formed in the evaporation chain is calculated using the phenomenological approach [34] based on the generalised superfluid model.

Consideration of gamma emission assumes importance in some cases like reactions induced by medium energy protons in neutron deficient targets. In code ALICE the probability for photon emission has been calculated from the following expression,

$$\Gamma_\gamma(E) = \frac{1}{\pi\eta^2\rho(E)}\cdot \int \varepsilon_\gamma^2\sigma_\gamma(E - \varepsilon_\gamma)\rho(E - \varepsilon_\gamma)d\varepsilon_\gamma \tag{4.86}$$

The photon absorption cross-section σ_γ has been obtained using Lorentz formula

$$\sigma_\gamma(\varepsilon_\gamma) = \sum_{k=1}^{2} \frac{\sigma_k\varepsilon_\gamma^2\Gamma_{k\gamma}^2}{\left(\varepsilon_\gamma^2 - E_{k\gamma}^2\right) + \varepsilon_\gamma^2\Gamma_{k\gamma}^2} \tag{4.87}$$

The nuclear level density for gamma channel has been calculated on the basis of the superfluid model consistently with other reaction channels.

(iv) **GNASAH-FKK code:**

This is a Los Alamos–Livermore code which may be run on PC Pentium 200 and provides a flexible method by which reaction and level cross-sections and emission spectra for neutrons, charged particles and gamma rays resulting from particle and photon induced reactions can be calculated. It is also possible to obtain isomeric cross-section ratios and gamma strength functions using the code. The code uses Kopecky–Uhl prescription for calculating gamma strength function and transmission coefficients. The code GNASH-FKK can read externally calculated pre-equilibrium spectra calculated according to FKK model, can do multi-step pre-equilibrium calculations. GNASH uses Hauser-Feshbach theory to calculate complicated sequences of reactions and includes a pre-equilibrium correction for binary tertiary channels. Gamma competition is considered in detail at each step of de-excitation. Code uses the multi-humped fission barrier model for simulating fission competition.

In code GNASH-FKK, decay sequences consisting of up to 38 compound nuclei with each CN emitting up to five ejectiles, including neutrons, protons, gamma rays, alpha particles etc can be considered. Conservation of angular momentum and of parity are explicitly taken into account at each step of calculation. The code is capable of considering up to 78 discrete levels for each decaying nucleus and energy continuum up to 204 energy bins. The incident particle may be neutrons, protons, deuterons, tritons ^{3}He and alphas. Code uses the following data libraries: STRUCTURE.DAT; for nuclear level energies, spins, parities and gamma branching ratios; TRANS.COEF, for particle transmission coefficients based on spherical or deformed optical model and MASS.DAT; for tables of ground state nuclear masses etc.

(v) **MINGUS code:**

MINGUS is a code developed by Netherlands Energy Research Foundation ECN, Petten, for the quantum mechanical prediction of the whole outgoing energy spectrum and the associated angular distributions for nucleon induced reactions up to 200 MeV. The code, besides using a quantum physical reaction dynamics also uses consistent optical and level density models. Code MINGUS is an extension of the earlier code KAPSIES [36] which was built only for multi-step direct reactions. Code MINGUS is of modular configuration and includes all other reaction mechanisms like multi-step compound and compound besides the multi-step direct processes. The transition amplitudes and DWBA cross-sections required for quantum mechanical calculations of the code are provided by an auxiliary coupled channel code ECIS88 [37].

MINGUS employs the dispersive optical model potential with parameterization given by Johnson *et al* [38] and for protons parameterisation given by Delaroche and Liegeois [39]. For continuum reactions, calculations are done using particle–hole state densities given by Oblozinzky [40]. The total level density is estimated by the backshifted Fermi gas model. Shell effects in calculations have been incorporated using energy dependence of Ignatyuk *et al* [41]. The pre-equilibrium contribution is described by the MSD/MSC mechanism. For MSC model calculations the code follows closely the formalism of Chadwick and Young [42] and for MSD calculations the method given by [36]. Crossover effects between the bound and the unbound configurations, i.e., between P- and Q-chains have been included as given by Marcinkowski and Mariański [43]. In the case of compound reactions, competition of all light particles and gamma rays is taken into account and multiple emission is included until all possible particle channels are closed.

(vi) **Code CCONE:**

Code CCONE was initially developed for nuclear reaction data evaluation by Japan Atomic Energy for JENDL/AC-2008 and JENDL-4 was upgraded to improve its accuracy of prediction of nuclear reaction cross-sections. The CCONE code uses modern nuclear reaction modelling and parameterizations integrating direct, pre-equilibrium and compound reactions with large flexibility such as unrestricted number of nuclei produced by sequential evaporation from the excited composite nucleus. The code uses C++ language and the latest version of the code incorporates multiple pre-equilibrium particle emission. Pre-equilibrium process is treated in the framework of the two-component exciton model with the conservation of the flux of occupation probability in the master equation. The code assumes both for compound and pre-equilibrium emission that the velocity of the recoiling residual nuclei is not changed by particle emission.

Figure 4.10 shows the schematic view of the multiparticle emission exciton model (MPEEM). Pre-equilibrium emission starts from the (1p,0h) state for the first (*Z*, *A*) composite nucleus. As usual, particle and hole annihilation is ignored in the model. The global nucleon optical model potential (OMP) given by Koning and Delaroche [44] has been used for all calculations in the code. The exciton model parameters have been taken from the work of Koning and Duijvestijn [45]. Global level density parameters of Mengoni and Nakajima [46] have been used for statistical calculations. Inelastic scattering cross-sections for giant resonance states, monopole, quadrupole and low and high energy octupole have been calculated using DWBA with parameterization of Kalbach [47]. The shape

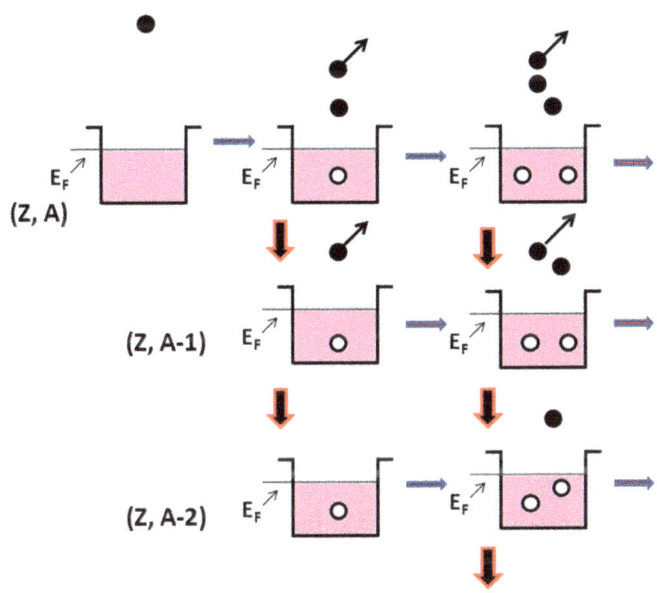

Figure 4.10. Schematic view of multiparticle emission exciton model (MPEEM).

elastic scattering cross-sections, reaction cross-sections for the creation of first composite nucleus, inverse reaction cross-sections required for exciton model calculations, and transmission coefficients for statistical model have been calculated using the global optical model. However, a folding OMP has been used for complex particles like alpha particle.

(vii) **Code EXIFON2.0:**

Several computer programs/codes have been developed to calculate the contribution of either the MSC or the MSD processes in pre-equilibrium emission under different boundary conditions. Salient features of some of these codes/programs are discussed here.

EXIFON2.0 is an easy-to-handle and fast code which predicts cross-sections from one global parameter set in which the only adjustable quantity is the pairing shift. The code is based on an analytical model for statistical multi-step compound theory. It is written in FORTRAN-77. The code was developed by H Kalka at the Institutfuer Kern-und Atomphysik, Technische Universitaet Dresden, Germany. The code EXIFON [48–50] predicts the cross-section within a pure statistical multi-step reaction model, which is based on a Green's function formalism [51, 52] and random matrix physics [53, 54] approach. In this code, the step from simple CN model and single-step direct reaction model towards statistical multi-step theory was mainly influenced by three basic ideas, i.e., consisting of classification of nuclear states by their complexity or the exciton number [55], distinction between bound and unbound configurations as per FKK theory [24] and the treatment of the nuclear Hamiltonian as a random matrix [53]. This code predicts cross-sections including compound, pre-compound, as well as direct (collective and non-collective) processes. The standard parameter set [48] used for calculations in the code EXIFON includes the strength of surface delta interaction $F = 27.5$ MeV, radius parameter $r_0 = (1.21{+}4.0A^{-\frac{2}{3}} - 15A^{-\frac{4}{3}})$, potential depth $V_0 = (52 - 0.3\ E_a$ MeV), Fermi energy $E_F = 33$ MeV, and pairing shift $\Delta = 12.8\ A^{-1/2}$ MeV. The optical model potentials of Wilmore–Hodgson [56] for neutron, those of Perey [57] for protons and for alpha particles that of Huizenga–Igo [58] are used.

The calculations can be done employing the above mentioned global parameter set which describes the emission data up about 100 MeV or so. The subroutines are used to get binding energies and phonon parameters from the nuclear data tables. The formation and the decay of the compound system as well as the direct processes may be calculated by the same interaction strength F, taken from the nuclear structure consideration. The pairing effects in the code EXIFON have been taken into account by using the effective binding energies $B_e^{\text{effective}}$. The effective neutron binding energy $B_n^{\text{effective}}$ and the effective proton binding energy $B_p^{\text{effective}}$ that give a better description of data are defined in this code as,

$$B_n^{\text{effective}} = B_n \pm \Delta_n \text{ for } \begin{pmatrix} \text{odd} \\ \text{even} \end{pmatrix} N \tag{4.88}$$

And,

$$B_p^{\text{effective}} = B_p \pm \Delta_p \text{ for } \begin{pmatrix} \text{odd} \\ \text{even} \end{pmatrix} Z \tag{4.89}$$

where B_n and B_p are the binding energies of neutrons and protons, respectively.

As already mentioned, the standard pairing shift obtained by Dilg *et al* [59] is represented as $\Delta_n = \Delta_p = 12.8\ A^{-1/2}$ MeV. Using the above formalism employed in the code EXIFON within pure statistical multi-step approach it gives a unique and simple description of emission spectra, angular distributions and activation cross-sections in nuclear reactions.

(viii) **Code ACT/STAPRE:**

The nuclear reaction model code ACT [60] is developed on the lines of code STAPRE [61] of Uhl and Strohmaier. The Hauser-Feshbach model [62] is used in this code for the calculation of CN cross-sections, while the exciton model of Griffin [63] is used for simulating the pre-equilibrium emission. Sequential emission of up to six nuclear particles may be considered along with the internal γ-ray cascade in this code. The pre-compound emission of nuclear particles is assumed only in the first step of de-excitation process where the excitation energy of the nucleus is expected to be sufficiently high.

A simple representation of the nuclear reaction $X(a_0, a_1, a_2\gamma)Y$ presented in figure 4.11. The target nucleus $X(E_0, I_0, \prod_0)$ with energy E_0, spin I_0, and parity $\prod_0$ is basically in ground state in a reaction. As indicated in this figure, before reaching the final residual nucleus Y, the compound system passes through many intermediate steps. The system formed by the combination of a projectile a_0, and target nucleus X is termed as the first CN. The residual nuclei formed in the sequential emission of particles $a_1, a_2, \ldots, a_i$ are identified as $(i + 1)$th CN. The combination of particles considered in this code are: neutron; neutron and proton; neutron, proton and alpha particle; neutron, proton, deuteron and alpha particle. It may be mentioned that any other combination of emitted particles is not considered in this code [60]. The incident particle may either be a nucleon or any other complex particle as well, like α-particle. Similar to the code STAPRE [61], the gamma competition may also be considered with the particle emission at any step of de-excitation process in the code ACT [60].

The level densities are calculated within the framework of backshifted Fermi gas model [64]. The formulation of Lang [65], used in the code ACT is given below;

$$\rho(E, I, \prod) = \frac{1}{2}\rho(E, I) \tag{4.90}$$

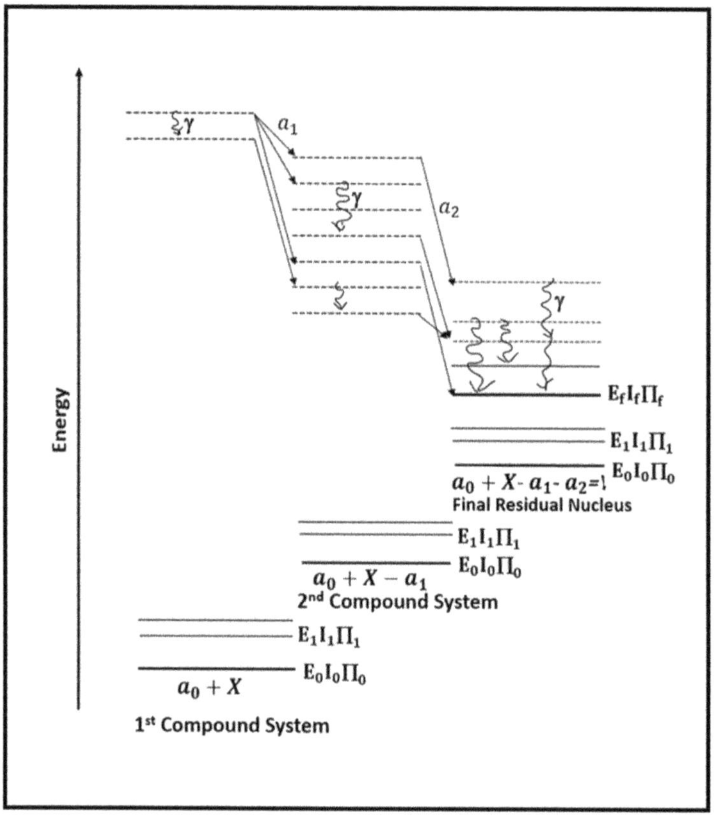

Figure 4.11. Various ways of populating final states ($E_f I_f \prod_f$) for reaction $X(a_0, a_1, a_{2\gamma})Y$.

$$\rho(E, \prod) = \omega(E, M = I) - \omega(E, M = I + 1) \tag{4.91}$$

$$\omega(E, \mathrm{M}) = \frac{1}{12}\left[\frac{2\Theta_{\mathrm{effective}}}{\hbar^2}\right]^2\left[E - \Delta - \frac{\hbar^2 M^2}{2\Theta_{\mathrm{effective}}} + \frac{3}{2}t\right]^{-3/2} \cdot \exp\left[2\left[\mathrm{a}\left(E - \Delta - \frac{\hbar^2 M^2}{2\Theta_{\mathrm{effective}}}\right)\right]^{1/2}\right] \tag{4.92}$$

$$E - \Delta = \mathrm{a}t^2 - \frac{3}{2}t + \frac{\hbar^2 \mathrm{M}^2}{2\Theta_{\mathrm{effective}}} \tag{4.93}$$

The backshifted energy Δ, the level density parameter a and the effective moment of inertia $\Theta_{\mathrm{effective}}$ are required to be given as the input data for all the residual nuclei in the evaporation chain under consideration. The initial exciton number, which is the sum of initially excited number of particles p and holes h is given in the input data file. The expression $\overline{|M|}^2 = F_M A^{-3} E^{-1}$ is used for estimating the square of absolute value of matrix element for two-body interactions. The constant F_M in the above expression of matrix element is used as a free parameter and its value is required to be given in the input data file. The density of single-particle states is chosen as either $g = 6\pi^2(A/8)$ or $g = 6\pi^2(a)$. To enhance the emission of complex particles like α-particles etc, it is assumed in this code that there are preformed alpha clusters in the CN, and hence the probability of pre-formation needs to be given as input parameter in this code. The pre-compound emission is considered at each stage of the decay of the first CN. The configuration of the CN during equilibration is defined by its exciton number which either increases by two units, decreases by two units or remains unchanged.

The internal decay rates are taken from the William's formulae [21] corrected for Pauli principle by Cline [66]. The exciton number dependent/independent expression may be used to obtain the value of matrix element for two-body interaction $|\overline{M}|^2$. The fraction of the population of the compound system that survives pre-compound decay and for any subsequent de-excitation steps, are calculated using Hauser-Feshbach formulations [62]. The initial population of the CN at any arbitrary ith stage is given by $W_i(e, i, \Pi)\Delta E$, which after gamma de-excitation gets modified to $\overline{W}_i(E, I, \Pi)\Delta E$.

In the evaporation chain for gamma emission the spin, parity, excitation energy of each discrete level for all the nuclei are required as the input data. A maximum of 20 levels may be taken. The information regarding discrete levels may not be required if only emission of particles is considered. The Weisskopf–Ewing estimates [67] are used for generating the gamma ray transmission coefficient. For dominant $E1$ radiations the Brink–Axel model [68] is used. The average s-wave neutron radiation width Γ_γ is also required for final normalization. In the input data, optical model transmission coefficient and separation energies for incident as well as outgoing particles are required for the calculations.

The pre-equilibrium emission of nuclear particles give rise to decreases of the population of states that goes to equilibration stage. This decrease or loss is accounted for by the factor q^{pre} such that population of levels

$$W_1(E, I, \Pi)\Delta E = q^{\text{pre}} \frac{\partial \sigma^{\text{HF}}_{a_0\gamma}(E, I, \Pi)}{\partial E} \Delta E \tag{4.94}$$

The contribution of pre-compound emission is shared among different levels in accordance with the compound contribution and a pre-compound part is added to the reduced equilibrium part. The final expression thus becomes;

$$W_2(E, I, \Pi)\Delta E = \left[q^{\text{pre}} \frac{\partial \sigma^{\text{HF}}_{a_0 a_1}(E, I, \Pi)}{\partial E} + \frac{\partial \sigma^{\text{pre}}_{a_0 a_1}}{\partial E} \cdot \frac{\partial \sigma^{\text{HF}}_{a_0 a_1}(E', I', \Pi')/\partial E}{\sum_{a',n'} \partial \sigma^{\text{HF}}_{a_0 a_1}(E', I', \Pi')/\partial E} \right] \Delta E \tag{4.95}$$

For the first chance de-excitation, the differential cross-section is obtained from the population of states of the second CN using the expression

$$\frac{\partial\sigma_1}{\partial E}(E,\, I,\, \prod)\Delta E = W_2(E,\, I,\, \prod)\Delta E \tag{4.96}$$

The terms used above have their usual meaning.

Bibliography

[1] Serber R 1948 Nuclear reactions at high energies *Phys. Rev.* **72** 1114

[2] Goldberger M 1948 The interaction of high energy neutrons and heavy nuclei *Phys. Rev.* **74** 1269

[3] Metropolis N, Bivins R, Storm M, Turkevich A, Miller J M and Friedlander G 1958 Monte Carlo calculations on intranuclear cascades. I. low-energy studies *Phys. Rev.* **110** 185

[4] Bertini H W *et al* 1969 Nucl. Instr. and Meth, 66, 1968; Intranuclear-cascade calculation of the secondary nucleon spectra from nucleon-nucleus interactions in the energy range 340 to 2900 MeV and comparisons with experiment *Phys. Rev.* **188** 1711
Guthrie M P, Alsmiller R G and Bertini H W 1968 Calculation of the capture of negative pions in light elements and comparison with experiments pertaining to cancer radiotherapy *Nucl. Instrum. Meth.* **66** 29–36

[5] The GEANT4 Collaboration 1994 CERN/DRDC/94-29, DRDC/P58

[6] Griffin J J 1966 Statistical model of intermediate structure *Phys. Rev. Lett.* **17** 478

[7] Letaw J R *et al* 1993 Comparison of distributed reacceleration and leaky-box models of cosmic-ray abundances ($3\leqslant Z\leqslant 28$) *Astrophys. J.* **414** 51

[8] Cline C K and Blann M 1971 The pre-equilibrium statistical model: description of the nuclear equilibration process and parameterization of the model *Nucl. Phys.* A **172** 225

[9] Williams Jr F C 1971 Particle–hole state density in the uniform spacing model *Nucl. Phys.* A **166** 231

[10] Betak E and Dobes J 1976 The finite depth of the nuclear potential well in the exciton model of preequilibrium decay *Z. Phys.* A **279** 319

[11] Kalbach C 1983 *Proceedings of the IAEA Advisory Group Meeting on Basic and Applied Problems of Nuclear Level Densities* ed E M R Bhat (Upton, NY: Brookhaven National Lab) Report No. BNL-NSC-51694, 113

[12] Kalbach-Cline C 1973 Residual two-body matrix elements for pre-equilibrium calculations *Nucl. Phys.* A **210** 590

[13] Akkermans J M 1979 Angular distribution in exciton and hybrid models for preequilibrium reactions *Phys. Lett.* B **82** 20

[14] Brink D M 1962 Some aspects of the interaction of light with matter *Doctoral Thesis* Oxford University
Axel P 1955 Electric dipole ground-state transition width strength function and 7-Mev photon interactions *Phys. Rev.* **126** 671

[15] Kalbach C 1986 Two-component exciton model: basic formalism away from shell closures *Phys. Rev.* C **33** 818

[16] Koning A J and Duijvestijn M 2004 A global pre-equilibrium analysis from 7 to 200 MeV based on the optical model potential *Nucl. Phys.* A **744** 15

[17] Akkermans J M, Gruppelaar H and Reffo G 1980 Angular distributions in a unified model of preequilibrium and equilibrium neutron emission *Phys. Rev.* C **22** 73

[18] Mantzouranis G, Weidenmullere H A and Agassi D 1976 Generalized exciton model for the description of preequilibrium angular distributions *Z. Phys.* A **276** 145
[19] Harp G D, Miller J M and Berne B J 1968 Attainment of statistical equilibrium in excited nuclei *Phys. Rev.* **165** 1166
[20] Blann M and Mignery A 1972 Pre-equilibrium decay at moderate excitations and the hybrid model *Nucl. Phys.* A **186** 245
[21] Williams Jr F C 1971 Particle–hole state density in the uniform spacing model *Nucl. Phys.* A **166** 231
Williams Jr F C 1970 Intermediate state transition rates in Griffin model *Phys. Lett.* B **31** 184
[22] Bisplinghoff J 1986 Configuration mixing in preequilibrium reactions: a new look at the hybrid-exciton controversy *Phys. Rev.* C **33** 1569
[23] Feshbach H 1973 *Proc. Int. Conf. Nuclear Physics (Munich, 27 August –1 September, 1973)* ed J de Boer and H J Mang (New York, American Elsevier Pub. Co.: Amsterdam, North Holland Pub. Co.)
Feshbsach H 1958 *Proc. Int. Conf. on Nuclear Reaction Mechanism*, Varenna (CLUED) p 1
[24] Feshbach H, Kerman A and Koonin S 1980 The statistical theory of multi-step compound and direct reactions *Ann. Phys.* **125** 429
[25] Tamura T, Udagawa T, Feng D H and Kan K K 1982 Deep inelastic reactions treated as multi-step direct reaction processes application to (p, p′) reaction *Phys. Lett.* B **66** 109
Tamura T, Lenske H and Udagawa T 1981 Multistep direct reaction analysis of analyzing powers of continuum spectra in (p,α) reactions *Phys. Rev.* C **23** 2769
Tamura T, Udagawa T and Lenske H Multistep direct reaction analysis of continuum spectra in reactions induced by light ions *Phys. Rev.* C **26** 379
[26] Lenske H, Landowne S, Wolter H H, Tamura T and Udagawa T 1983 Reaction mechanism of heavy-ion charge-exchange scattering at intermediate energies *Phys. Rev. Lett.* B **123** 333
[27] Udagawa T, Tamura T and Kim B T 1979 Analysis of continuous spectra of the reactions (^{20}Ne, ^{16}O) and (^{20}Ne, ^{12}C) in terms of multi-step direct reaction theory *Phys. Rev. Lett.* B **82** 349
[28] Nishioka H, Weidenmullerand H A and Yoshida S 1988 Statistical theory of precompound reactions: the multistep direct process *Ann. Phys. (N. Y.)* **183** 166
[29] Koning A J and Akkermans J M 1993 Computational comparison of quantum-mechanical models for multistep direct reactions *Phys. Rev.* C **47** 724
[30] Koning A J and Delaroche J P 2003 Local and global nucleon optical models from 1 keV to 200 MeV *Nucl. Phys.* A **713** 231–310
[31] Koning A J and Hilaire S 2004 TALYS: a nuclear reaction program *NRG-report* 21297/04.62741/P (available at 2004 http://www.talys.eu)
[32] Iwamoto A and Harada K 1982 Mechanism of cluster emission in nucleon induced pre-equilibrium reactions *Phys. Rev.* C **26** 2821
[33] Milazzo-Colli I and Braga G M 1973 α-Emission by pre-equilibrium processes in (*n*,α) reactions *Nucl. Phys.* A **210** 297
[34] Ignatyuk A V 1985 Contribution of collective modes to levelk density of excited nuclei *YadernayaFizika* **21** 20
Blokhin A I, Ignatyuk A V, Pashchenko Yu V and Shubin Yu V 1985 Analysis of threshold reaction excitation function in frame of unified superfluid model *Commun. Acad. Sci. USSR* **49** 962

[35] Kopecky J and Uhl M 1990 Test of gamma-ray strength functions in nuclear model calculations *Phys. Rev.* C **41** 1941
[36] Koning A J and Akkermans J M 1993 Computational comparison of quantum-mechanical models for multistep direct reactions *Phys. Rev.* C **47** 724
[37] Raynal J 1990 Which potentials have to be surface peaked to reproduce large angle proton scattering at high energy? *Austral. J. Phys.* **43** 9–17
[38] Johnson C H, Horen D J and Mahaux C 1987 Unified description of the neutron-^{208}Pb mean field between −20 and +165 MeV from the dispersion relation constraint *Phys. Rev.* C **36** 2252
[39] Delaroche J P and Liegeois C CEA Bruyeres-le-Chatel, Internal Report
[40] Oblozinzky P 1986 Particle–hole state densities for statistical multi-step compound reactions *Nucl. Phys.* A **453** 127
[41] Ignatyuk A V *et al* 1975 Fission of pre-actinide nuclei. Excitation functions for the (α, f) reaction *Sov. J. Nucl. Phys.* **21** 255
[42] Chadwick M B and Young P G 1993 Feshbach–Kerman–Koonin analysis of ^{93}Nb reactions: $P \rightarrow Q$ transitions and reduced importance of multistep compound emission *Phys. Rev.* C **47** 2255
[43] Marcinkowski A and Mariański B 1993 Analysis of low-energy direct reactions and the novel one-step cross section *Nucl. Phys.* A **561** 387
[44] Koning A J and Delaroche J P 2003 Local and global nucleon optical models from 1 keV to 200 MeV *Nucl. Phys.* A **713** 231
[45] Koning A J and Duijvestijn M C 2004 A global pre-equilibrium analysis from 7 to 200 MeV based on the optical model potential *Nucl. Phys.* A **744** 15
[46] Mengoni A and Nakajima Y 1994 Fermi-gas model parameterization of nuclear level density *J. Nucl. Sci. Technol.* **31** 152
[47] Kalbach C 2000 Surface and collective effects in pre-equilibrium reactions *Phys. Rev.* C **62** 044608
[48] Kalka H 1991 *EXIFON—A Statistical Multistep Reaction Code* (Cedex, France: NEA Data Bank)
[49] Kalka H, Tojman M and Seeliger D 1989 Statistical multistep reactions: application *Phys. Rev.* C **40** 1619
[50] Kalka H, Torjman M, Lien H N, Lopez R and Seeliger D 1990 Description of (n, p) and (n, 2n)-activation cross sections for medium-mass nuclei within statistical multistep theory *Z. Phys.* A **335** 163
[51] Ring P and Schuck P 1980 *The Nuclear Many-Body Problem* (New York: Springer)
[52] Migdal A B 1970 *Theory of Finite Fermi Systems and Applications to Nuclei* (New York: Wiley Interscience)
[53] Agassi D, Weidenmuller H A and Mantzouranis G 1975 The statistical theory of nuclear reactions for strongly overlapping resonances as a theory of transport phenomena *Phys. Rep.* **22** 145
[54] Brody T A, Flores J, French J B, Mello P A, Pandey A and Wong S S M 1981 Random-matrix physics: spectrum and strength fluctuations *Rev. Mod. Phys.* **53** 385
[55] Griffin J J 1967 Statistical model of intermediate structure *Phys. Lett.* B **24** 5
[56] Wilmore D and Hodgson P E 1964 The calculation of neutron cross-sections from optical potentials *Nucl. Phys.* **55** 673–94

[57] Perey F G 1963 Optical-model analysis of proton elastic scattering in the range of 9 to 22 MeV *Phys. Rev.* **131** 745

[58] Huizenga J R and Igo G 1962 Theoretical reaction cross sections for alpha particles with an optical model *Nucl. Phys.* **29** 462–73

[59] Dilg W, Schantl W and Vonach H 1973 Level density parameters for the back-shifted fermi gas model in the mass range $40 < A < 250$ *Nucl. Phys.* A **217** 269

[60] Gupta J P 1983 Compound and pre-comp. emission in some reactions induced by neutrons of 10–20MeV *PhD Thesis* Aligarh Muslim University unpublished

[61] Uhl M and Strohmaier B 1981 *Report IRK 76/01* (Cedex, France: NEA Data Bank)

[62] Hauser W and Feshbach H 1952 The inelastic scattering of neutrons *Phys. Rev.* **87** 336

[63] Griffin J J P 1966 Statistical model of intermediate structure *Phys. Rev. Lett.* **17** 478

[64] Vonach H K and Hille M 1969 Die energieabhängigkeitderniveaudichteimbereich der kernmassen $A = 30$–65 *Nucl. Phys.* A **127** 289

[65] Lang D W 1966 The angular momentum dependence of the nuclear level density *Nucl. Phys.* **77** 545

[66] Cline C K 1972 The Pauli exclusion principle in pre-equilibrium decay *Nucl. Phys.* A **195** 353

[67] Weisskopf V F 1951 Radiative transition probabilities in nuclei *Phys. Rev.* **83** 1073

[68] Axel P 1962 Electric dipole ground-state transition width strength function and 7-MeV photon interactions *Phys. Rev.* **126** 671

IOP Publishing

Pre-equilibrium Emission in Nuclear Reactions
Fundamentals, measurements and analysis
B P Singh, Manoj K Sharma and R Prasad

Chapter 5

Experimental measurements

Details of some important experiments on the measurement of pre-equilibrium and compound nuclear emission components are provided in this chapter. Analysis of the experimental data using model-based computer codes is also discussed.

5.1 Introduction

Deciphering the nuclear reaction mechanism is like finding the cause of an air crash from its debris scattered all around the crash site. In a planned nuclear reaction, all parameters of incident channel are known beforehand as they are set by the one carrying out the experiment, however, measured parameters in the exit channel, like reaction cross-section, the energy and angular distributions of ejectile and/or residual nucleus, excitation functions (EFs) etc, carry valuable information about the mechanism of the reaction. The first theoretical prescription for a binary nuclear reaction was proposed by Bohr [1] in 1936 in the form of the compound nucleus (CN) reaction mechanism. According to this theory a binary nuclear reaction was assumed to take place in two steps; formation of the CN by the fusion of the incident ion with the target nucleus and its subsequent decay. The formation and decay of the CN were assumed to be independent of each other. As has already been mentioned in earlier chapters, the CN theory was experimentally verified for the first time by the pioneering experiment of Ghoshal [2]. Comparison of the results of several independent experiments with predictions of the compound reaction mechanism led to the conclusion that the theory holds good at relatively lower energies. However, with the increase in the energy of the incident ion, deviations of the experimental data from the theoretical predictions were observed. These deviations have been ascribed mainly to the pre-equilibrium emission that is found to compete with CN emission. In order to have a better understanding of the reaction mechanism involved, attempts were made to study the dynamics of the formation of an excited CN and its evolution to the equilibrium states [3, 4]. With the availability of particle accelerators capable of providing high resolution beams of different types of incident particles, experimentally it has become possible to study the reaction dynamics and to separate out the pre-equilibrium emission component from the competing CN emission component. This separation has been facilitated by the characteristic properties of the pre-equilibrium component. As already mentioned, the important signatures of the pre-equilibrium emission are: the presence of larger number of emitted high-energy particles as compared to what is predicted by the statistical CN model [5]; forward-peaked angular distribution of the emitted particles [6]; slowly descending tails of the EFs [7]; and stretched particle distribution in angular momentum space [8, 9] etc. Further, the phenomenon of pre-equilibrium emission has also been studied recently in heavy-ion reactions as well where large linear as well as angular momentums are involved. As such, in heavy-ion reactions a smaller recoil range/linear momentum of the reaction residues after the emission of pre-compound particles as compared to CN particles [10] has also been observed. In this chapter, details of some experiments carried out for the study of pre-equilibrium emission and the analysis of the experimental data are presented.

Experimental details presented here include not only light-ion induced reactions but also reactions induced by heavy ions where the pre-equilibrium emission is found to play an important role in reaction dynamics. In the following, results of neutrons, protons, alpha particles and heavy-ion induced reactions are presented where pre-equilibrium emission has been conclusively established and is found to contribute significantly to the reaction process, particularly at lower energies where CN emission is expected to dominate. In general, two types of experiments, the off-beam and the in-beam experiments have been carried out to decipher the reaction mechanism and to separate out the relative contributions of the pre-equilibrium and equilibrium components. Brief experimental details of some outstanding experiments are presented here before the results, and their comparison with theoretical models is made.

5.2 Neutron induced reactions

5.2.1 Measurement and analysis of cross-sections

The neutron being a neutral particle, it cannot be accelerated directly by any kind of field to generate a collimated primary beam. The beam of neutrons may be obtained either from nuclear reactors or from a neutron generator where neutrons are produced from some nuclear reaction like, $d(t,n)\alpha$. In most of the experiments where neutron induced reactions are studied, the above neutron producing reaction has been used. The above reaction is also popularly referred to as the D–T reaction. Here, D stands for deuteron particle, an isotope of hydrogen (^{2}H), while T stands for tritium, another isotope of hydrogen (^{3}H). In order to obtain neutrons from the above reaction, an accelerated beam of deuterons is made to hit a tritium target. Neutrons produced in the forward direction have different energies depending on the angle of emission with respect to the incident deuteron beam. Energy of neutrons at any desired angle of emission may be computed using the reaction kinematics.

In a typical experimental set-up [11], a set of five target samples mounted on a circular aluminum ring, respectively, at 0°, 30°, 60°, 90° and 120° with respect to the incident deuteron beam were irradiated by a neutron flux obtained by hitting a tritium target with a ≈175 keV deuteron beam. From the kinematics, the energy of neutrons, for the samples mounted at 0°, 30°, 60°, 90° and 120° angular position were found to be 14.77, 14.68, 14.42, 14.07 and 13.73 MeV, respectively. Vishal *et al* [11] used activation technique to measure cross-sections for the reaction ^{140}Ce(*n*, 2*n*)^{139}Ce at 14.07 MeV, 14.42 MeV, 14.68 MeV and 14.77 MeV neutron energies. These cross-section values are presented in figure (5.1). Theoretical calculations of cross-section for corresponding energies have been performed using the codes TALYS-1.8 [12] and EMPIRE-II [13] where both equilibrium and pre-equilibrium emission are considered. Theoretical values for cross-sections are also shown in figure 5.1. As may be seen from this figure, calculation done including pre-equilibrium emission to the CN emission, agrees well with the experimental data.

In a similar way the measurement and analysis of EF for the reaction ^{80}Se(*n*,*p*)^{80}As have been performed by Attar *et al* [14]. The experimental data was analysed using the codes EMPIRE-II and TALYS-1.8. As may be observed in figure 5.2, TALYS-1.8 and EMPIRE-II calculations performed for reaction ^{80}Se(*n*,*p*)^{80}As with the inclusion of pre-equilibrium contribution along with CN contribution agree well with the experimental data.

Further, Pandey *et al* [15] measured cross-sections for reactions ^{nat}Yb(*n*,*x*)^{172}Tm and ^{nat}Yb(*n*, *x*)^{173}Tm at 14.6 MeV using a neutron generator. Their results are shown in figure 5.3 and figure 5.4, respectively. Theoretical calculations for these reactions ^{nat}Yb(*n*,*x*)^{172}Tm and ^{nat}Yb(*n*, *x*) have been done using the model codes EMPIRE and TALYS and are also shown in figures 5.3 and 5.4. In these figures EMPIRE-II calculations are shown by dotted curves, however, TALYS calculations are shown by the solid curve. It has been found that theoretical

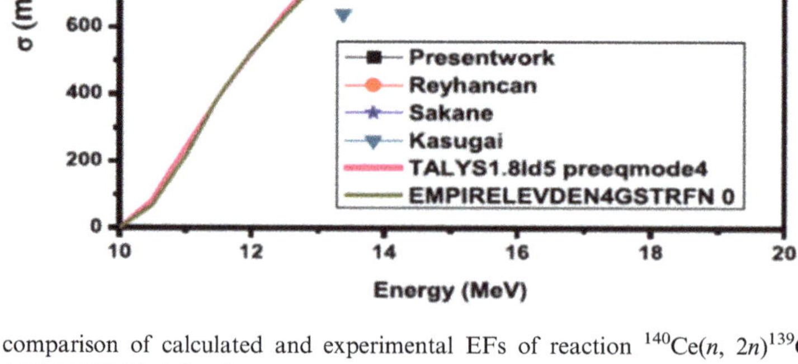

Figure 5.1. A comparison of calculated and experimental EFs of reaction ^{140}Ce(*n*, 2*n*)^{139}Ce using codes TALYS-1.8 and EMPIRE-II. Other literature values of the data are also shown in the figure for comparison (reprinted from [11] with permission from Elsevier).

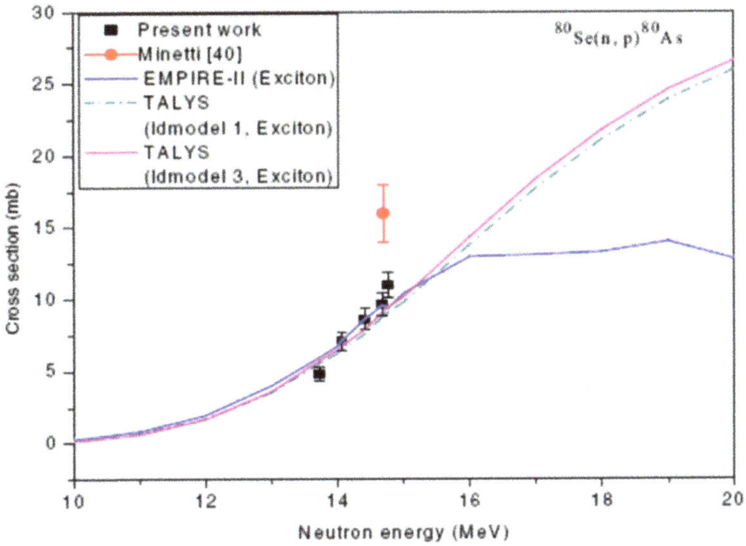

Figure 5.2. A comparison of calculated and experimental EFs for reaction ^{80}Se(n,p)^{80}As using codes TALYS-1.8 and EMPIRE-II. In this figure the reported values of cross-section of different authors are also shown for comparison (reprinted from [14] with permission from American Physical Society).

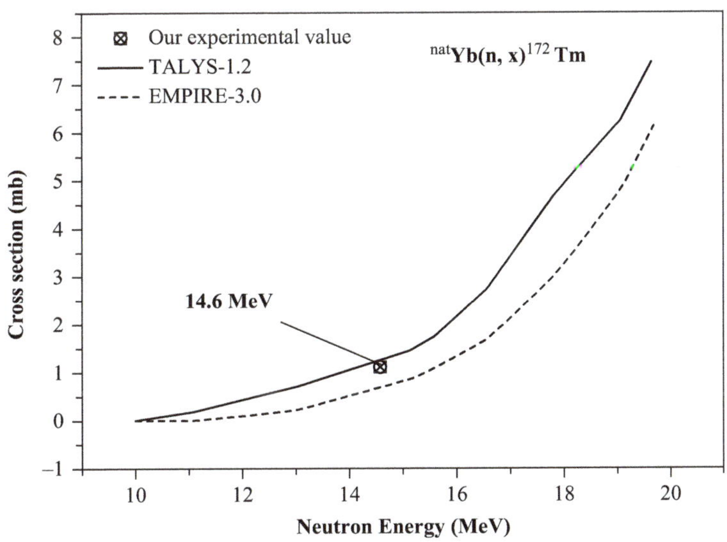

Figure 5.3. A comparison of experimentally measured cross-section and theoretically calculated EFs for the reaction ^{nat}Yb(n, x)^{172}Tm. Calculations performed by TALYS-1.2 (solid line) and EMPIRE-3.0 (dashed line) are also shown (reprinted from [15] with permission from Elsevier).

calculations performed by code TALYS are reasonably in agreement with the measured values of cross-section data, when pre-equilibrium emission is taken into consideration.

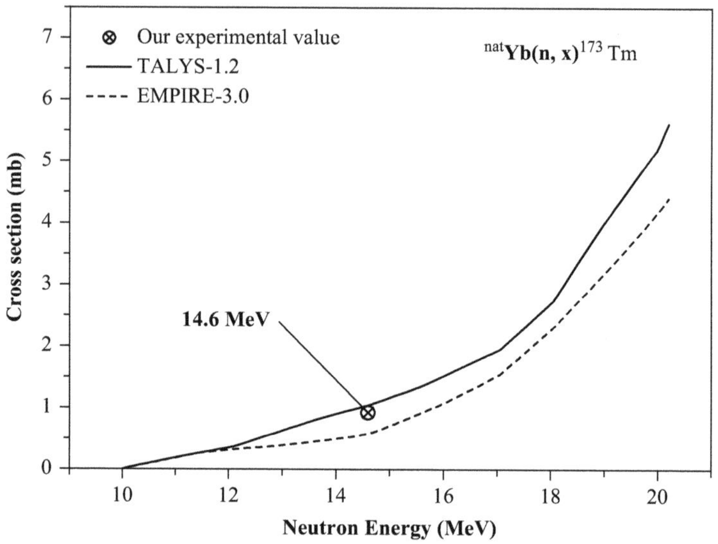

Figure 5.4 A comparison of experimentally measured cross-section and theoretically calculated Efs for the reaction $^{nat}Yb(n, x)^{173}Tm$. Theoretical calculations for the reaction performed by TALYS-1.2 (solid line) and EMPIRE-3.0 (dashed line) are also shown (reprinted from [15] with permission from Elsevier).

In experiments carried out by Birn *et al* [16] cross-sections for the reaction $^{78}Se(n, \alpha)^{75}Ge$ at several neutron energies varying from ≈6 to 14.6 MeV have been measured to study the pre-equilibrium emission. Theoretical calculations for the reaction $^{78}Se(n,\alpha)^{75}Ge$ were performed with the model code STAPRE [17] by considering two different values of rigid body moment of inertia parameter $\eta = 0.5$ and $\eta = 1.0$. It may be pointed out that this code gives the prediction of cross-section considering Hauser-Feshbach theory [18] for CN emission and for pre-equilibrium emission using the exciton model [19]. The experimentally measured and theoretically calculated EFs for the reaction $^{78}Se(n,\alpha)^{75}Ge$ are shown in figure 5.5. In this figure, calculations performed with rigid body moment of inertia parameter $\eta = 1.0$, are shown by dotted curves, however, calculations performed with $\eta = 0.5$ are shown by solid curve. The measurement and analysis of the EFs for reactions $^{181}Ta(n, 2n)^{180}Ta$ and $^{181}Ta(n, 3n)^{179}Ta$ [20] have been performed using activation technique while theoretical calculations are done using code STAPRE as above and are also shown in figure 5.6. It has been observed that theoretical calculations performed by code STAPRE with $\eta = 0.5$ for reactions $^{78}Se(n,\alpha)^{75}Ge$, $^{181}Ta(n, 2n)^{180}Ta$ and $^{181}Ta(n, 3n)^{179}Ta$ are reasonably in agreement with the measured value of cross-sections. Since code STAPRE includes pre-equilibrium emission, a good agreement between theory and experiments clearly brings out the importance of pre-equilibrium emission even at such low energies.

Cross-sections for *(n,p)* reactions with several target nuclei at 14.8 MeV neutron energy were measured using Cockcroft–Walton neutron generator at the Nuclear Physics Lab of Aligarh Muslim University, (AMU), Aligarh, India. Gupta *et al* [21]

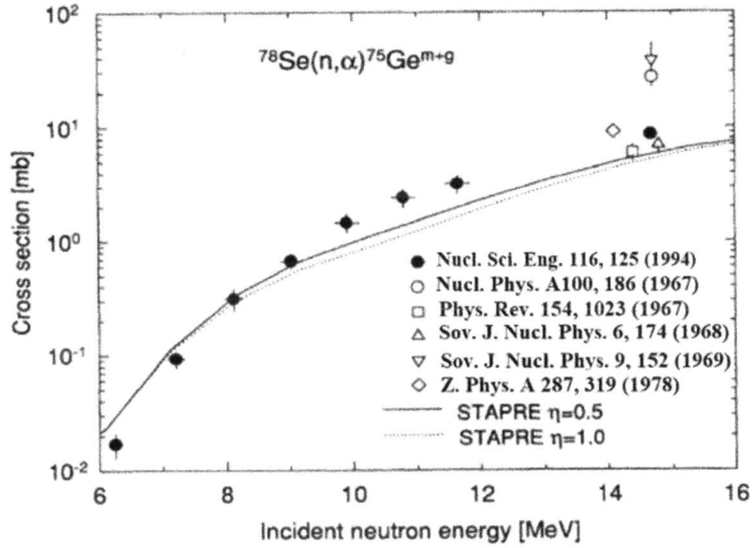

Figure 5.5. Experimentally measured and theoretically calculated EFs for the reaction ^{70}Se(n,α) ^{75}Ge. The calculations performed with code STAPRE with η = 1.0 (solid curve) and η = 0.5 (dotted curve) are shown (reprinted from [16] with permission from American Physical Society).

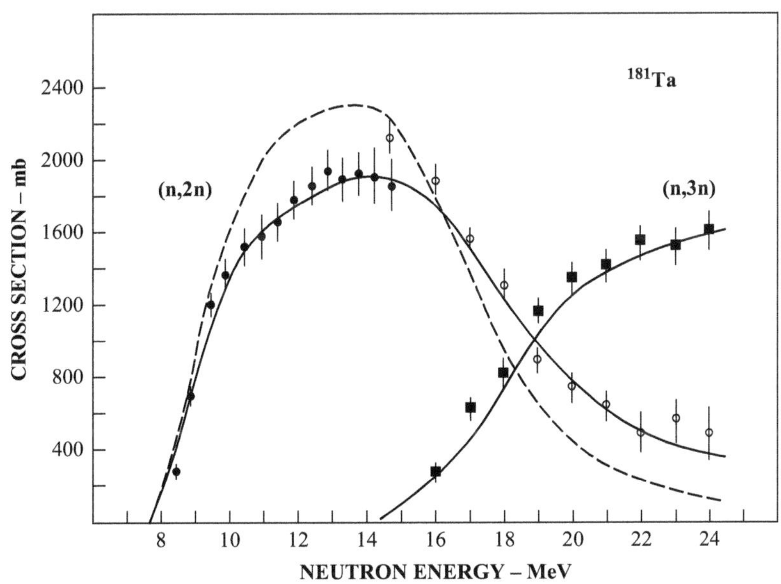

Figure 5.6. Experimentally measured and theoretically calculated EFs for the reactions ^{181}Ta$(n,2n)$ ^{180}Ta and ^{181}Ta $(n,3n)^{179}$Ta. The solid circles represent the experimental data. The dotted curve shows CN calculations, while solid curves include pre-compound emission (reprinted from [20] with permission from Elsevier).

carried out some experiments using the activation technique to determine the (n,p) reaction cross-sections for a large number of target nuclei. The measured cross-section data have been compared with the theoretically calculated values predicted by code ACT [22] with and without considering the pre-equilibrium emission. It may be remarked that code ACT [22] is a modified version of code STAPRE developed by Uhl *et al* [17]. In figure 5.7(a, b) and 5.8(a, b), respectively, the ratio of (*n,p*) reaction cross-section, i.e., $\{\sigma(n,\ p)_{\text{exp}}/\sigma(n,\ p)_{\text{cal}}\}$ obtained experimentally and calculated theoretically using code ACT for 18 nuclei are plotted against atomic mass number A and the asymmetry parameter $(N - Z)/A$, respectively.

In figure 5.7(a) and (b) the ratio $\sigma(n,\ p)_{\text{exp}}/\sigma(n,\ p)_{\text{cal}}$ is plotted against atomic mass number of target nuclei where the $\sigma(n,\ p)_{\text{cal}}$ is the cross-section calculated with and without the inclusion of pre-compound contributions. It is interesting to see in figure 5.7(a) that all the data point (shown by hollow circles) except for ^{24}Mg, ^{52}Cr and ^{64}Zn nuclei are higher than their values calculated without the consideration of pre-equilibrium emission. The higher values of the ratio $\sigma(n,\ p)_{\text{exp}}/\sigma(n,\ p)_{\text{cal}}$ suggest the presence of non-CN process in these $(n,\ p)$ reactions at the energy of interest. On the other hand, when pre-equilibrium emission is also taken into consideration, the ratio of experimental to that of theoretical cross-sections approaches the unity line indicating the presence of pre-equilibrium emission in these reactions at this energy. In figure 5.8, the ratio of experimental and theoretical cross-section, i.e.,

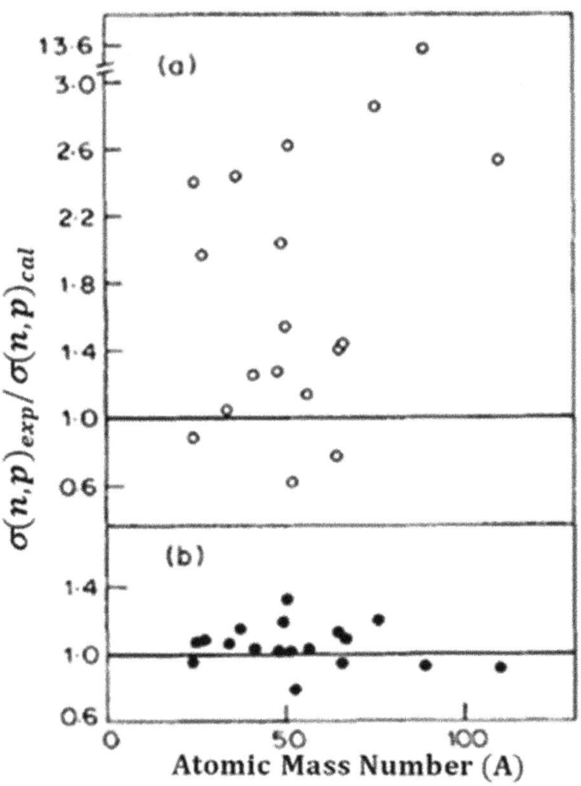

Figure 5.7. Comparison between experimental and calculated (*n,p*) reaction cross-section ratio $\frac{\sigma(n,p)\text{exp}}{\sigma(n,p)\text{cal}}$ versus atomic mass number (*A*) of the target nucleus at 14.8 MeV. (a) Calculation based on equilibrium statistical theory only (b) calculations include the pre-equilibrium emission (reprinted from [21] with permission from Springer).

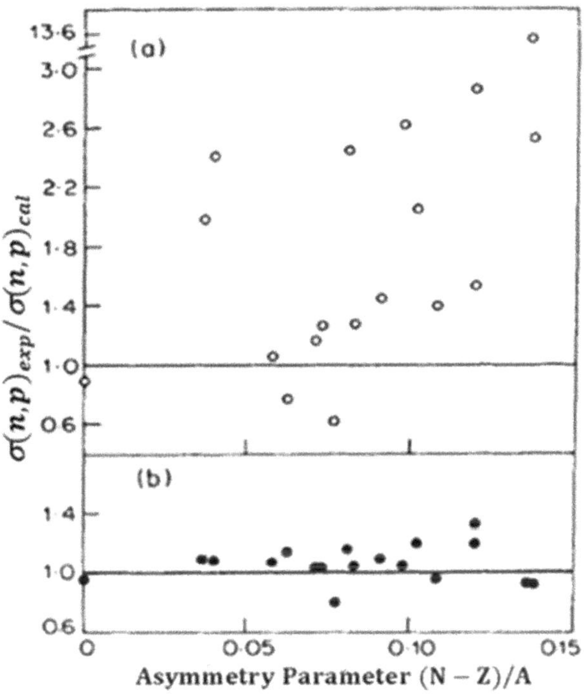

Figure 5.8. Comparison between experimental and calculated (*n,p*) reaction cross-section ratio versus asymmetry parameter $(N{-}Z)/A$ of the target nucleus at 14.8 MeV (reprinted from [21] with permission from Springer).

$\sigma(n,\, p)_{\mathrm{exp}}/\sigma(n,\, p)_{\mathrm{cal}}$for the same target nuclei is plotted against asymmetry parameter $(N{-}Z)/A$ with or without the consideration of pre-equilibrium emission. As can be seen in figure 5.8(b), the experimental data comes very close to the theoretical calculations with the inclusion of the pre-equilibrium emission. This also strengthens the observation and presence of pre-equilibrium emission in neutron induced reactions at low energies.

5.2.2 Measurements and analysis of energy spectra of light charged particles

As mentioned earlier, the measurement and analysis of energy spectra of emitted charged particles may also give information of considerable value regarding pre-equilibrium emission. The energy spectra of light charged particles may be measured in the in-beam experiments. Bevilacqua *et al* [23], measured energy spectra of deuterons, tritons, ^{3}He, and α-particles emitted in the interaction of neutrons with Fe (Iron) and Bi (Bismuth) target nuclei at the Svedberg Laboratory (TSL) in Uppsala, Sweden. In their experiments cyclotron was used to accelerate protons up to 180 MeV, and were extracted to interact with a 23.5 mm-thick ^{7}Li target producing neutrons. The neutrons produced in the reaction were collimated using iron collimators into a beam and then transported inside the Medley reaction chamber. The ‘Medley’ is a spectrometer system consisting of eight three-element telescopes, placed at angles from 20° to 160°, in steps of 20°, and a thin target positioned inside the 90 cm-diameter reaction chamber. Each of the telescopes consists of two fully depleted surface barrier silicon detectors which served as ΔE detector and a *CsI*(*Tl*) scintillator which served as E detector. The particle identification is performed using the ΔE–E technique. The measured energy spectra [22] of deuterons, ^{3}He, triton, and α-particles produced in the interaction of neutron beam with natural bismuth (Bi) target are shown in figure 5.9. The measured energy spectra are compared with the calculations done using TALYS code [12] with default parameter set, which considers both the CN and the pre-equilibrium emission contributions.

As may be seen in figure 5.9, the energy spectra of the light charged particles at forward angles are well reproduced by considering pre-compound emission in the TALYS [12] calculations. However, for backward angles experimental data are reproduced by considering CN emission only. This provides further evidence for the pre-equilibrium emission of light charged particles in the forward direction (a characteristic of pre-equilibrium emission) in neutron induced reactions. It may be pointed out that the above measurements are done using high energy neutrons and are presented to show that such non-compound processes are dominant at relatively high energies.

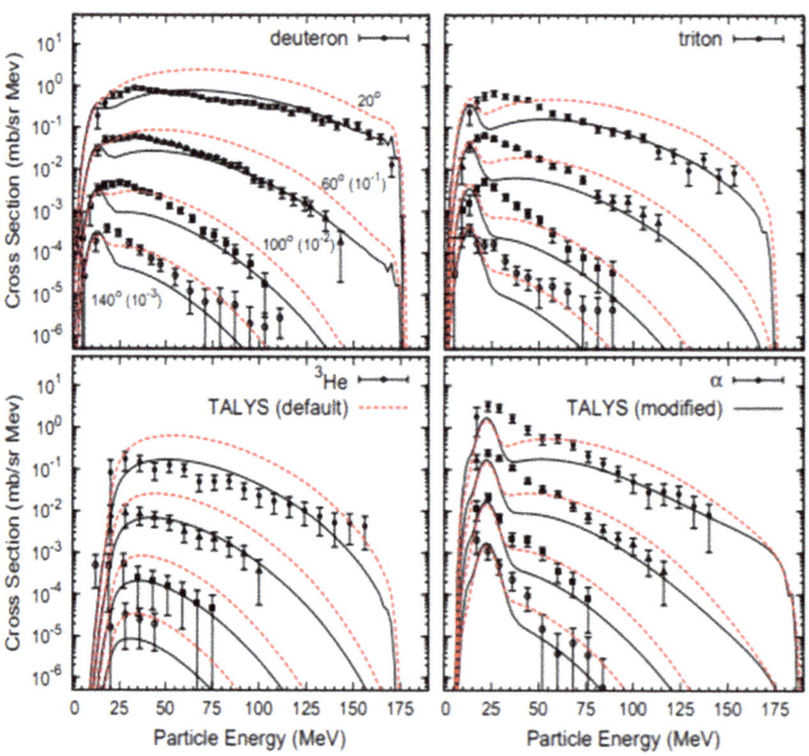

Figure 5.9. Measured light charged particle spectra in the interaction of 175 MeV neutron beam with Bi. Experimental data at 20°, 60°, 100° and 140° in the laboratory system are compared with TALYS calculations with default parameters (dashed line) and with TALYS calculations modified to include Kalbach contribution to the pre-equilibrium emission (solid line). See text for details.

5.3 Proton induced reactions

5.3.1 Measurements and analysis of EFs

Contrary to neutrons, the proton is the lightest charged particle which can be directly accelerated by the cyclotron or linear accelerators, to higher energies for studying the reactions induced by these particles. Details of some of the measurements of EFs carried out at the Variable Energy Cyclotron Centre (VECC), Kolkata, India using the off-beam stacked foil activation technique are presented here. As a typical example, measured [24, 25] EFs for (p,n) reactions on enriched ^{130}Te, and ^{51}V target nuclei are shown in figures 5.10–5.13, respectively. Brief details of the measurement and analysis of EFs for proton induced reactions are given here. Samples of ^{130}Te, and ^{51}V were prepared employing vacuum evaporation technique by depositing target material on aluminum (Al) foils. A stack containing a few targets each followed by a thick Al-catcher foil was irradiated with the proton beam. After the irradiation the activities induced in the irradiated samples were recorded using a High Purity Germanium (HPGe) detector. Intensity of observed gamma activities in the irradiated targets has been used to obtain the cross-section using standard formulations, as presented in chapter 2. Theoretical calculations for these reactions have been performed employing both the semi-classical codes ALICE-91 [26] and ACT [17, 22] as well as a quantum mechanical code EXIFON [27]. The details of these codes and their parameters have already been presented in chapter 4.

Experimentally measured EF for the reaction ^{130}Te$(p,n)^{130}$I is shown [24] in figure 5.10. The experimental data has been analysed within the framework of code ALICE-91 [26]. In this code the CN calculations are performed using the Weisskopf–Ewing model [28] while the PE component is simulated employing the hybrid model [29]. In figure 5.10, the dotted curve represents pure CN calculations, however, calculations done by including pre-compound emission along with the CN emission are shown by the

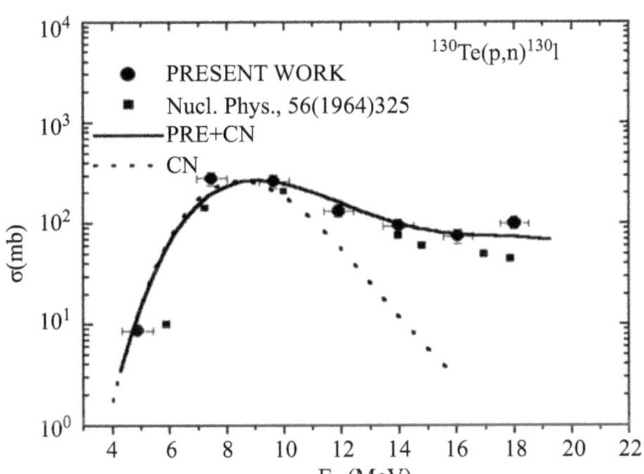

Figure 5.10. Experimentally measured and theoretically calculated EFs for reaction ^{130}Te$(p,n)^{130}$I by using compound (dotted curve) and/or pre-compound model (solid line) ALICE-91 (reprinted from [24] with permission from Elsevier).

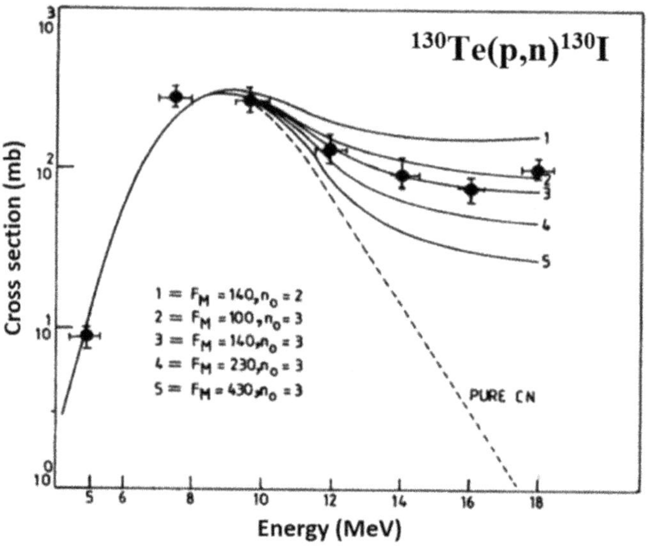

Figure 5.11. Experimentally measured and theoretically calculated EFs for reaction ^{130}Te(p,n)^{130}I. Theoretical calculations were done using the code ACT. Solid curves show compound plus pre-compound contributions for different values of parameters F_M and initial exciton number n_0. The dotted curve shows CN calculations (adapted from [25] with permission from American Physical Society).

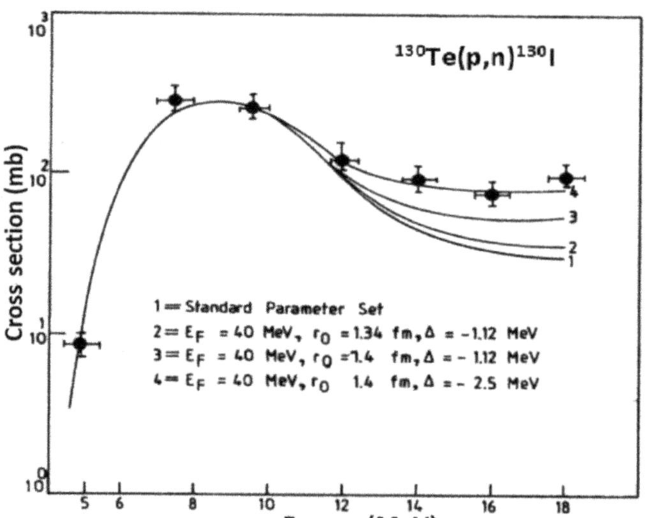

Figure 5.12. Experimentally measured and theoretically calculated EFs using quantum mechanical code EXIFON for reaction ^{130}Te(p,n)^{130}I. The effects of important parameters of this code are shown. As such, both the semi-classical as well as quantum mechanical model codes reproduced the experimental data and indicate the influence of the pre-equilibrium emission particularly towards the tail portion of the EFs or rather at relatively higher energies (adapted from [25] with permission from American Physical Society).

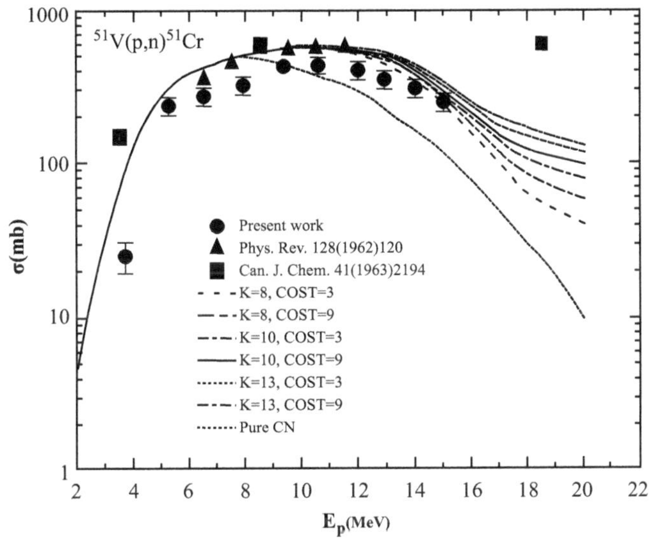

Figure 5.13. Experimentally measured and theoretically calculated EFs for the reaction $^{51}V(p,n)^{51}Cr$. Effect of variation of parameters (K and COST) of the code on theoretical calculations is shown (reprinted from [25] with permission from Elsevier).

dark solid curve. The available literature data is also presented in this figure for comparison. In ALICE-91 calculations, there are basically three parameters. The level density parameter in these calculations is given by $a = \frac{A}{K}\mathrm{MeV}^{-1}$, where A is atomic mass number while K is a parameter which may be used to fit the experimental data. The level density parameter a largely influences the CN calculations. The initial exciton number n_0(which is equal to the sum of initially excited number of particles and holes), and the mean free path multiplier parameter govern the pre-equilibrium component. In figure 5.10, the best fit of the experimental data for $^{130}Te(p, n)^{130}I$ reaction is obtained by taking the value of K=10, the initial exciton number $n_0 = 3(2p + 1h)$ and mean free path multiplier as zero. The initial configuration of the compound system, i.e., the initial exciton number n_0, in these calculations is taken equal to 3 with two particles and one-hole configuration. This configuration may be justified by assuming that the first interaction of the proton with the target nucleus may give rise to the excitation of one particle above the Fermi energy leaving behind a hole in excited state resulting in the two-particle ($2p$) and one-hole ($1h$) state. As can be seen from this figure, the experimental data agrees well with the theoretical calculations done using code ALICE, when the pre-compound emission is included in calculations along with the CN calculations, thus, indicating the importance of pre-compound emission at these energies.

In figures 5.11 and 5.12 the experimentally measured EFs for the reaction $^{130}Te(p, n)^{130}I$ are compared with the calculations done using semi-classical code ACT [17, 22] and quantum mechanical code EXIFON [25], respectively. In code ACT [17, 22] evaluation of the internal transition rates is made by the square of the average matrix element $|\overline{M}|^2 = F_M \bullet A^{-3}E^{-1}$. Here, F_M is treated as an adjustable parameter. Another important parameter of code ACT that governs the pre-equilibrium component is the initial exciton number $n_0 = n_p + n_h$. The calculations have been performed using different values of parameter F_M and the initial exciton number n_0. The effect of variation of parameter F_M and the initial exciton number n_0 is also shown in these figures. As may be observed from figure 5.11, with the increase in the value of parameter F_M the cross-section values in the tail portion of the EF decreases. Further, in figure 5.11 the CN calculations shown by the dotted curve are not able to reproduce the experimental data in the tail portion of the EFs. When calculation was performed using code ACT [17, 22] that includes pre-equilibrium emission and parameters $F_M = 140$ MeV^{-1} and initial exciton number $n_0 = 3(2p + 1h)$ were used, then, it was found that experimental data is well reproduced over the entire energy range. This value of initial exciton number may be justified by assuming that the incident proton on hitting the target nucleus excites a particle above the Fermi level leaving behind a hole in excited state, i.e., in all $3(2p+1h)$ excitons in the first step of reaction. This reflects that pre-equilibrium emission is a competing reaction mechanism in proton induced reactions at these low energies. The experimentally measured EF for the reaction ^{130}Te(p,n)^{130}I is compared with calculations done with quantum mechanical model code EXIFON in figure 5.12. The detailed description of the code EXIFON has already been presented in chapter 4. Theoretical calculations of the EF shown in figure 5.12 have been performed using the code EXIFON [27] with the standard parameter set [27, 30] which includes the strength of surface delta interaction $F = 27.5$ MeV, radius parameter $r_0 = (1.21+4.0A^{-2/3} - 15A^{-4/3})$fm, potential depth $V_0 = (52 - 0.3E_a)$MeV, Fermi energy $E_F = 33$ MeV, and pairing shift $\Delta = 12.8A^{-1/2}$MeV. Optical model potentials given, respectively, by Wilmore and Hodgson [31] for neutron, Perey *et al* [32] for proton and Huizenga and Igo [33] for alpha particle were used in these calculations. Calculations done with these parameters are found to underestimate the experimental data, particularly in the tail portion of the EF. In order to match the experimental data with theoretical calculations done with quantum mechanical code the values of some of the parameters of this code have been varied from that of the standard set. As such, the value of the pairing correction term (Δ) has been changed from −1.12 (standard value) to −2.5 which is in agreement with the value used by Kalbach-Cline *et al* [34] and Coryell [35]. The Fermi energy E_F is related to the single-particle state density g and through it to the level density parameter. The different values of the radius parameter r_0 within the permissible range have also been considered. These values of E_F(=40 MeV) and r_0(=1.4fm) are used in the present calculations and are found to give satisfactory reproduction of the experimentally measured E_F in the entire energy range.

In the same way, the EF for the reaction ^{51}V(p,n)^{51}Cr has been measured [25] by activation technique and the analysis has been performed using the model code ALICE-91 with and without the inclusion of pre-equilibrium emission. The comparison is shown in figure 5.13. Available literature data is also shown in this figure. Calculations have been done with different values of level density parameter K and also by varying the value of mean free path multiplier COST. As can be seen from this figure, a value of $K = 10$ along with the value of mean free path multiplier COST equal to three reproduces the experimental data satisfactorily, only when pre-equilibrium emission is also taken into consideration with CN calculations.

Parashari *et al* [36] measured EFs for proton induced reactions on ^{93}Nb target, using the stacked foil activation technique followed by off-line γ-ray spectrometry at the 14UD BARC-TIFR Pelletron accelerator, Mumbai, India. In this experiment, a stack of five niobium foils was irradiated with proton beam of energy 22 MeV. The proton beam is successively degraded along the stack of five niobium foils. The energy falling on the mid of the each ^{93}Nb target sample was calculated to be 21.85, 19.53, 16.98, 14.13 and 11.50 MeV, respectively. The cross-sections of the identified reaction residues ^{93m}Mo, ^{92m}Nb and ^{89g}Zr populated through (p,n), (p,pn) and $(p,\alpha n)$ reaction channels have been measured and compared with the existing literature data available in EXFOR [37] library and are found to be in good agreement. To provide a clear insight into the pre-compound reaction dynamics, the measured EFs for these reactions have also been compared with the theoretical predictions of TALYS-1.9 code using suitable input parameters for the pure CN and compound plus pre-equilibrium reaction channels. Analysis of the experimental data indicates that the pre-equilibrium emission dominates over the compound process at higher proton energies. As a typical example the experimentally measured and theoretically calculated EFs for reaction ^{93}Nb$(p,n)^{93m}$Mo calculated using code TALYS-1.9 are shown in figure 5.14.

In this figure, the solid curve shows CN calculations including pre-compound emission, while the dotted curve shows pure CN calculations. The theoretical

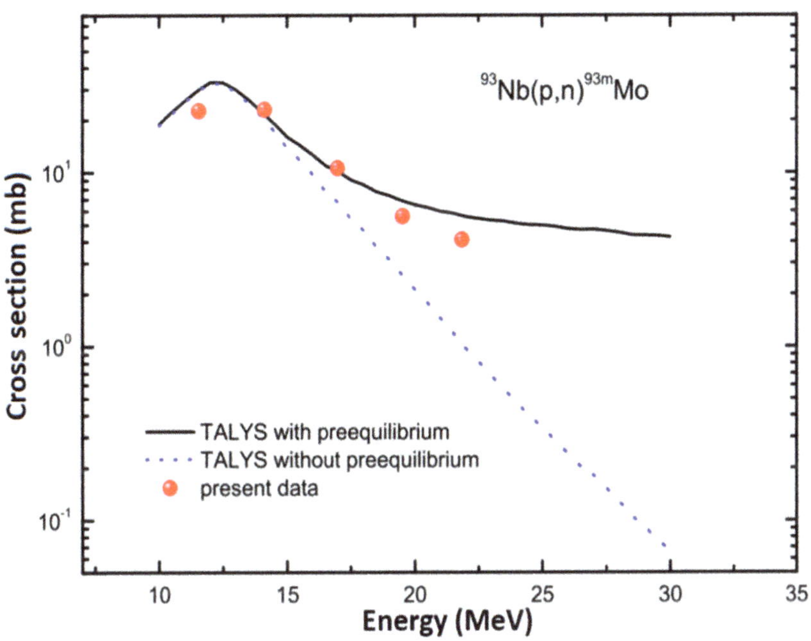

Figure 5.14. Experimentally measured and theoretically (TALYS-1.9) calculated EFs for reaction ^{93}Nb$(p, n)^{93m}$Mo. The solid curve shows the calculations done including both the compound and the pre-equilibrium emissions, while the dotted curve shows the pure CN calculations (reprinted from [36] with permission from Elsevier).

calculations have been found to be consistent with the experimental data particularly in the tail portion of the EF, a characteristic of the pre-compound emission.

5.3.2 Measurement and analysis of energy spectra of neutrons

In order to study pre-compound emission process in proton induced reactions, some in-beam experiments by using the time-of-flight (TOF) technique have also been carried out, where energy spectra of emitted neutrons have been measured. Holbling *et al* [38] measured energy spectra of neutrons in the ^{104,105,106,108,110}Pd (p,xn) reactions at $E = 26.1$ MeV. The standard Hamburg TOF set-up [38] consists of eight liquid scintillator detectors with appropriate shielding and a target chamber surrounded by two bending magnets which produce a total deflection of the beam of 34°, was used in these experiments. In this TOF set-up, there are three optional target positions: the first located in front of the first magnet, the second between the two magnets, and the third behind the second magnet.

The spectra from (p, xn) reactions on a whole series of Pd isotopes have been measured and compared with pre-equilibrium-model calculations. The measured double differential cross-sections for emitted neutrons in ^{106}Pd$(p$,xn) reactions at forward angles was found to have a typical shape like the one shown in figure 5.15. As can be seen from this figure, there are two regions in the spectra, one smooth (except for isobar analogue resonances) at lower outgoing energies, the other structured at high energies. Moreover, the structures observed at different angles are correlated suggesting their connection with the structure of the residual nucleus. The observed structure can be reasonably well described by pre-equilibrium-model calculations with two-quasi-particle state densities for proton–particle neutron–hole configurations based on Nilsson-model.

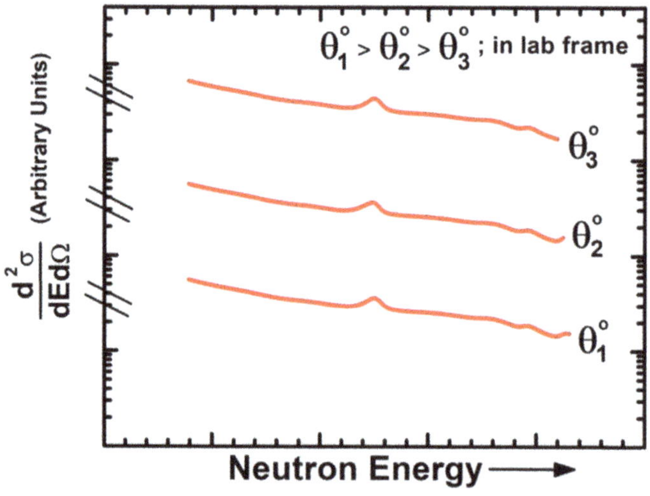

Figure 5.15. A typical double differential cross-section measured at forward angles.

5.4 Alpha induced reactions

5.4.1 Measurement and analysis of EFs

In the case of alpha induced reactions, it is pertinent to mention that due to relatively larger mass as compared to that of the proton a relatively large angular momentum is imparted to the composite system. Further, the measured cross-section data for the reactions induced by alpha particles are of prime importance because the features of these EFs at relatively low, medium and high energies may give information of considerable value about the reaction mechanism involved. As was observed in proton induced reactions, the low energy portion of the EF is dominated by the CN reaction mechanism, however, with the increase in projectile energy the pre-equilibrium component gains strength and starts to dominate. Experimental details and measured EFs for some alpha induced reactions are given in the following. Comparison of measured EFs with their theoretical counterparts obtained from model-based computer codes is also included. In these experiments of alpha induced reactions, for the systems, $\alpha + {}^{141}_{59}\mathrm{Pr}$, ${}^{93}_{41}\mathrm{Nb}$, ${}^{\mathrm{nat}}_{28}\mathrm{Ni}$, and ${}^{128,130}_{52}\mathrm{Te}$, the activation technique has been used for the measurement of cross-sections. A well collimated beam of alpha particles has been obtained from the Variable Energy Cyclotron Centre (VECC), Kolkata, India, which is a Department of Atomic Energy Facility of the Government of India. Spectroscopically pure, ${}^{141}\mathrm{Pr}$, ${}^{93}\mathrm{Nb}$, ${}^{\mathrm{nat}}\mathrm{Ni}$ target material of thicknesses ≈2.20–3.32 mg cm^{-2}, have been deposited on Al foils of thicknesses ≈6.75 mg cm^{-2} using vacuum evaporation technique. However, in the case of tellurium ($Z = 52$), the enriched isotopes of ${}^{128}\mathrm{Te}$(87%) and ${}^{130}\mathrm{Te}$(61%) have been used for making sample targets for irradiation. The Al foils served both as energy degrader as well as catcher/backing foils, where the recoiling residues from the composite system may be trapped. The thickness of each target sample/Al-catcher foil was determined precisely using alpha transmission method. In the stacked foil activation technique, an energetic beam of alpha particles traverses through the stack of samples. The incident beam loses energy in each target sample as well as the backing Al foils. Thus, it is possible to bombard successive samples of the stack at decreasing energies in a single irradiation. The experimental details for these systems $\alpha + {}^{141}_{59}\mathrm{Pr}$, ${}^{93}_{41}\mathrm{Nb}$, ${}^{\mathrm{nat}}_{28}\mathrm{Ni}$, and ${}^{128,130}_{52}\mathrm{Te}$, referred to above were almost similar. As such, here the experimental details of only one system i.e., $\alpha + {}^{141}_{59}\mathrm{Pr}$ are presented. A stack consisting of eight ${}^{141}_{59}\mathrm{Pr}$ target with Al-catcher assemblies was irradiated for ≈12 h with alpha-beam of 40 MeV [39] energy, obtained from the Variable Energy Cyclotron Centre (VECC), Kolkata, India. The beam current typically of ≈100 nA in each case was monitored using the ORTEC's current integrator count rate. The calculations for average beam energy on a given target of the stack were done using the stopping power program SRIM [37]. The post-irradiation analysis of the data has been carried out using ORTEC's high resolution large-volume (100 c.c.) HPGe detector coupled to a PC based multichannel analyzer. It may be pointed out here that the residues were identified by their characteristic gamma-rays and measured half-lives. The measured intensities of the characteristic gamma lines of the residues of interest have been used to determine the cross-sections for the corresponding reaction channels employing a FORTRAN program based on standard formulations described in section 2.8 of chapter 2.

The analysis of experimentally measured EF for the reaction ^{141}Pr(α,n)^{144}Pm has been performed using different theoretical model codes viz., PACE4 [40, 41] and ACT [17, 22], respectively. These are statistical model codes and employ the semi-classical approach. The details of these codes have already been presented in chapter 4. In figure 5.16 the experimentally measured EF for the reaction ^{141}Pr(α,n)^{144}Pm is compared with its theoretically calculated counterpart. Theoretical excitation functions have been obtained using code PACE4, which uses a free parameter K to get the value of the level density parameter 'a'. Calculated EFs for different values of free parameter K are also shown in these figures. Code PACE4 [40, 41] calculates cross-sections using Bass formula [42] based on Hausher-Feshbach [18] formalism and uses Monte Carlo simulation. In this code, level density parameter $a = A/K$ is an important parameter which governs the shape of EF in the low energy region where CN emission is important. Here, A is the mass number of the residual nucleus and K is a free parameter which may be varied to match the experimental data. In figure 5.16, along with the present measurements the cross-sections obtained from literature are also shown [43]. The calculations done using code PACE4, for different values of parameter K =8, 9 and 10, are shown in the figure.

As can be seen from the figure, a value of K = 8 (default value) is found to satisfactorily reproduce the experimentally measured reaction cross-sections, particularly up to the peak region. The experimentally measured values by Ansari *et al* [43] for the reaction have also been indicated in figure 5.16. An enhancement in the experimental cross-sections compared to PACE4 calculations has been observed

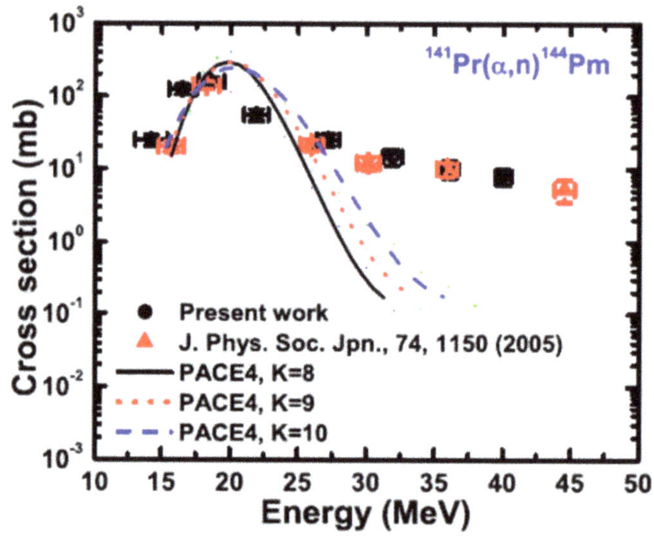

Figure 5.16. Experimentally measured and theoretically (PACE4) calculated EFs for reactions ^{141}Pr(α, n)^{144}Pm. The effect of the variation of parameter K (= 8, 9 and 10) on PACE4 calculations is shown (reprinted from [39] with permission from American Physical Society).

towards higher projectile energies, i.e., in the tail portion of the EF. It may be pointed out that this observed enhancement in the experimental EF for the reaction ^{141}Pr(α,n) ^{144}Pm above projectile energy of 25 MeV could not be reproduced even by changing the values of parameter K in the PACE4 calculations. This is expected, since, code PACE4 takes into account evaporation reaction channels proceeding only via CN process, therefore, the observed enhancement in the measured EF as compared to PACE4 calculations indicate the presence of some other mechanism apart from the CN evaporation one that becomes significant at higher energies. Obviously, it is the pre-equilibrium channel. It may be remarked that the contribution of the direct reaction to the same reaction residues is expected to be negligible at these energies. Hence, observed enhancement of the experimental cross-sections in the tail portion of EF for reaction ^{141}Pr(α,n)^{144}Pm as compared to the theoretical calculations may be attributed to the pre-equilibrium emission process, which is expected to be a dominant mode of reaction mechanism in one neutron emission channel.

The theoretical calculations for the EF for the reaction ^{141}Pr(α,n)^{144}Pm have been performed using the code ACT [17] which is based on the lines of code STAPRE [22]. In this code, each evaporation step is treated within the framework of statistical model with consideration of angular momentum and parity conservation. For the emission of the first particle, pre-equilibrium emission is taken into account. In this code the CN calculations are performed using the Hauser-Feshbach model, while the exciton model is employed for the simulation of pre-equilibrium emission.

Two important parameters in exciton model that govern the strength of pre-equilibrium component are the average matrix element F_M and the initial exciton number n_0. A value $F_M = 430$ MeV^{-1}, that has some justification for alpha induced reactions is often used. Thus, the only parameter that may be varied to match the experimental data is the initial exciton number n_0. Effect of the variation of n_0 on theoretical calculations for reactions ^{141}Pr(α,n)^{144}Pm is shown in figure 5.17. It may be observed from this figure that a value of $n_0 = 4$ satisfactorily reproduces the experimental data over the complete energy range.

In the case of (α+^{93}Nb) system, the EFs for ^{93}Nb(α,n)^{96m}Tc, ^{93}Nb(α, n)^{96}Tc, ^{93}Nb (α, 2n)^{95m}Tc, ^{93}Nb(α, 2n)^{95g}Tc and ^{93}Nb(α,3n)^{94}Tc reactions have been measured [44] in the energy range from threshold to ≈10MeV/nucleon, using the activation technique. The measured EFs have also been compared with theoretical predictions based on the semi-classical code ACT [17, 22], which takes into account CN as well as pre-equilibrium emission. One of the most important advantages of using code ACT [17, 22] is that it can predict the cross-section values of the isomeric states of the reaction residues as well. The measured EFs for the reactions ^{93}Nb(α,n)^{96m}Tc and ^{93}Nb(α, n)^{96}Tc are shown in figures 5.18 and 5.19, respectively. In these figures the reported values of the cross-sections, given in literature [44], have also been shown. The isomer ^{96m}Tc produced in (α,n) reaction decays (98%) to ^{96}Tc via internal transition. The half-life of ^{96}Tc is 4.35 d which is much larger than that of the metastable state ^{96m}Tc ($t_{1/2}$ = 51.5 min). As such, the observed activity of the residue ^{96}Tc shown in figure 5.19, has contributions from both the isomeric as well as the ground states. As can be seen from figure 5.18, theoretical calculations done

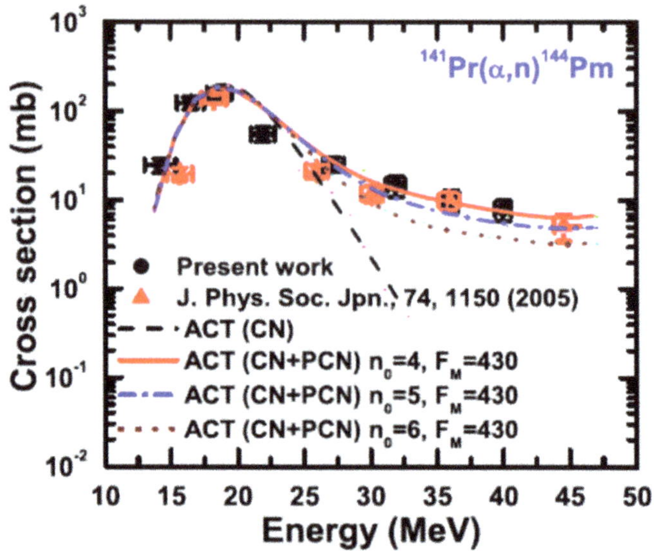

Figure 5.17. Experimentally measured and theoretically calculated EFs for reactions ^{141}Pr(α,n)^{144}Pm using code ACT. Effect of variation of initial exciton number is also shown in this figure (reprinted from [39] with permission from American Physical Society).

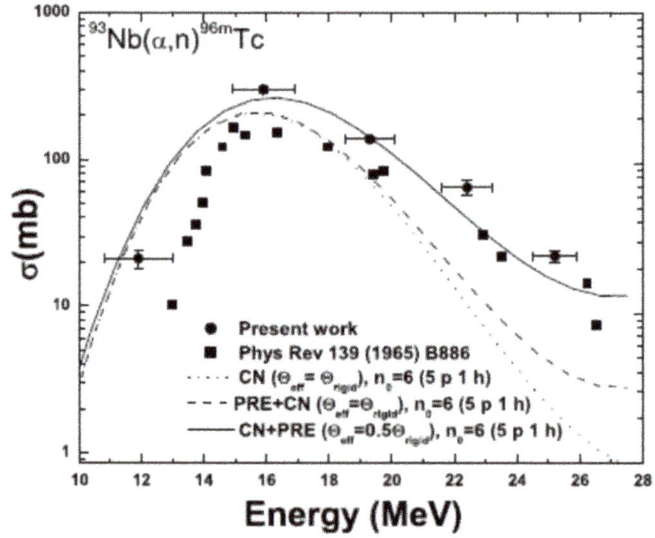

Figure 5.18. Experimentally measured and theoretically calculated EF for the reaction ^{93}Nb(α,n)^{96m}Tc. Theoretical calculations done using code ACT and the effect of variation of parameter effective moment of inertia parameter on calculated EFs is also shown (reprinted from [44] with permission from Springer).

using code ACT considering only the pure CN emission with moment of inertia parameter $\Theta_{\text{effective}} = \Theta_{\text{rigid}}$ along with initial exciton number $n_0 = 6(5p + 1h)$ are not able to reproduce the tail portion of the EF for this reaction. Further, when pre-compound emission is included with the CN calculations, keeping $\Theta_{\text{effective}} = \Theta_{\text{rigid}}$

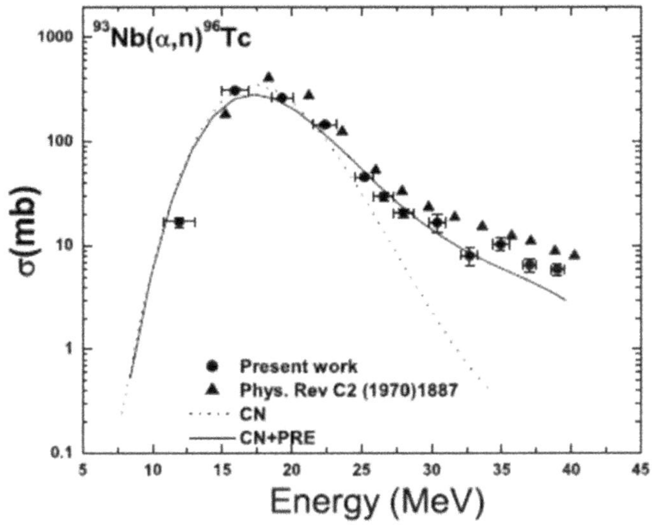

Figure 5.19. Experimentally measured and theoretically calculated EFs for reaction $^{93}Nb(\alpha,n)^{96}Tc$. Theoretical calculations are done using code ACT (reprinted from [44] with permission from Springer).

with $n_0 = 6(5p + 1h)$, theoretical calculations do not reproduce the measured EF in the high-energy tail portion of the EF. However, when pre-compound emission is included along with the CN calculations for the value of $\Theta_{\text{effective}} = 0.5\Theta_{\text{rigid}}$ with $n_0 = 6(5p + 1h)$ the code ACT satisfactorily reproduces the measured EF, even in the high energy tail portion of the EF. This indicates that the moment of inertia parameter $\Theta_{\text{effective}}$ equal to 50% of the Θ_{rigid} better explains the data. In figure 5.19, the experimentally measured EF for the reaction $^{93}Nb(\alpha, n)^{96}Tc$ is compared with theoretical predictions of the code ACT, and found to explain the data satisfactorily when pre-compound emission is taken into consideration along with CN emission.

The EFs for the $^{58}Ni(\alpha, p)^{61}Cu$, $^{58}Ni(\alpha, pn)^{60}Cu$, $^{60}Ni(\alpha, p2n)^{61}Cu$, $^{60}Ni(\alpha, n)^{63}Zn$, $^{60}Ni(\alpha, 2n)^{62}Zn$, $^{61}Ni(\alpha, 3n)^{62}Zn$, and $^{61}Ni(\alpha, 2n)^{63}Zn$ reactions have been measured in the energy range from threshold to ≈40 MeV [45]. The experimentally measured cross-sections for the production of residues in (α + ^{nat}Ni) system have been compared with the predictions of the theoretical model code ALICE-91 [26] with and without the inclusion of pre-equilibrium emission. This code calculates the CN reaction cross-section on the basis of Weisskopf–Ewing model [28], while pre-compound component is simulated employing the hybrid/geometry-dependent hybrid model [29]. In the ALICE-91 calculations, there are mainly three important parameters. The level density in these calculations is given by $a = \frac{A}{K}\text{MeV}^{-1}$. Here, A is the atomic mass number of the residual nucleus while K is a parameter which may be varied to fit the experimental data. The level density parameter K essentially determines the strength of the CN emission component which largely influences the peak portion of the EF. The initial exciton number n_0, which is the sum of initially excited number of particles and holes, is an important parameter that mainly governs the pre-compound emission. The mean free path multiplier COST may also be varied to match the experimental data.

In the hybrid model calculations, the intermediate states of the system are characterized by the excitation energy E and number n_p of excited particles and number $n_h$of excited holes. The states of these particles and holes are defined relative to the ground state of the nucleus and are called excitons. In the pre-equilibrium reaction, the number of excitons will be different for different projectiles and does not depend on the structure of the target nuclei. As has been pointed out in the description of the code ALICE-91, the initial configuration of the composite system is defined by initial exciton number $n_0 = (n_p + n_h)$, which is an important parameter to determine the pre-equilibrium contribution in the reaction. Calculations with different values of initial exciton number n_0, have been performed. As a representative case, theoretically calculated EFs for different values of n_0 ranging from 4 to 6 with configurations $n_0 = 4(2p + 2n + 0h)$, $n_0 = 5(3p + 2n + 0h)$, and $n_0 = 6(3p + 2n + 1h)$ for the reaction ^{58}Ni(α,p)^{61}Cu are shown in figure 5.20. In the case of α-particle induced reactions, values of initial exciton number $n_0 = 4$ or 5 have been justified, as a lower value of n_0 means a larger number of two-body interactions prior to the equilibration of the composite system and hence larger pre-equilibrium contribution in the reaction. By fixing the value of $n_0 = 4$, the calculations have been performed with a set of other parameters $a = A/8$ and mean free path multiplier equal to one, for reaction ^{58}Ni(α, p)^{61}Cu and the calculated values are found to agree reasonably well with the experimental data. As may be seen from figure 5.20, a satisfactorily reproduction of the experimental data is found, in general, for reactions ^{58}Ni(α,p)^{61}Cu when

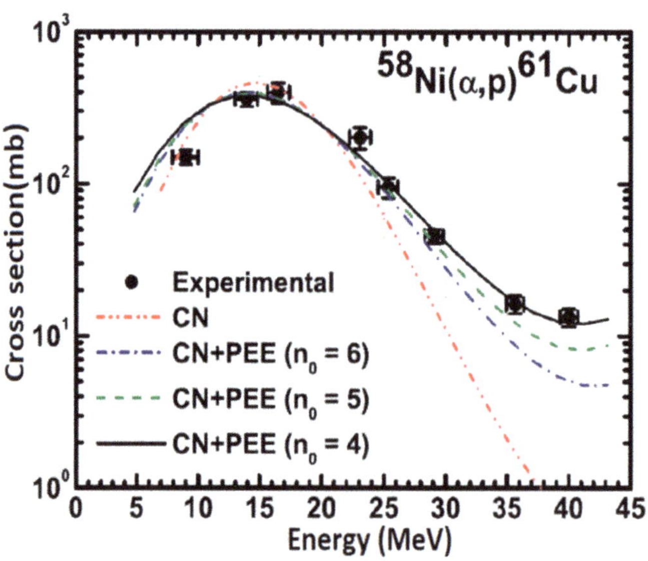

Figure 5.20. Experimentally measured and theoretically calculated EFs for reaction ^{58}Ni(α,p)^{61}Cu. Theoretical calculations are done using code ALICE-91. The effect of different values of initial exciton number (n_0) ranging from 4 to 6 with configurations $n_0 = 4(2p + 2n + 0h)$, $n_0 = 5(3p + 2n + 0h)$, and $n_0 = 6(3p + 2n + 1h)$ on calculated EFs is also shown (adapted from [45] with permission from American Physical Society).

pre-compound emission is taken into consideration in calculations, indicating the importance of non-equilibrium processes. With this consistent set of parameters, i.e., the level density parameter $a = A/8$, the initial exciton number $n_0 = 4$ and mean free path multiplier equal to unity, for the code ALICE-91, the calculations have also been performed for the EFs for the reactions ^{60}Ni(α,n)^{63}Zn, ^{60}Ni(α, 2n)^{62}Zn and ^{61}Ni (α,2n)^{63}Zn, respectively [45]. As a representative case, experimentally measured and theoretically calculated EFs have been shown in figure 5.21 for the reaction ^{61}Ni (α,2n)^{63}Zn, where the data could be explained by taking pre-equilibrium emission into consideration. It has been observed that the experimental EFs are satisfactorily reproduced by the CN calculations plus the pre-equilibrium calculations based on the exciton model both for neutron as well as proton emission channels.

From the analysis done using the code ALICE-91, for the alpha induced reactions on nickel isotopes, it has been observed that there is a significant contribution from the pre-compound emission processes at the studied range of energies. Thus, the pre-compound emission plays an important role in reaction dynamics of α-induced reactions at these energies and measured EFs for most of the reactions are well described by the theoretical predictions of the model codes provided pre-equilibrium emission is included in calculations.

With the motivation to study the pre-equilibrium emission in α-particle induced reactions, EFs for three reactions ^{128}Te(α,np)^{130}I, ^{130}Te(α,np)^{132}I, and ^{130}Te(α, np)^{132m}I have been also measured [46]. The semi-classical code ACT [17, 22] has been used to calculate cross-sections for reactions ^{128}Te(α,np)^{130}I, ^{130}Te(α,np)^{132}I,

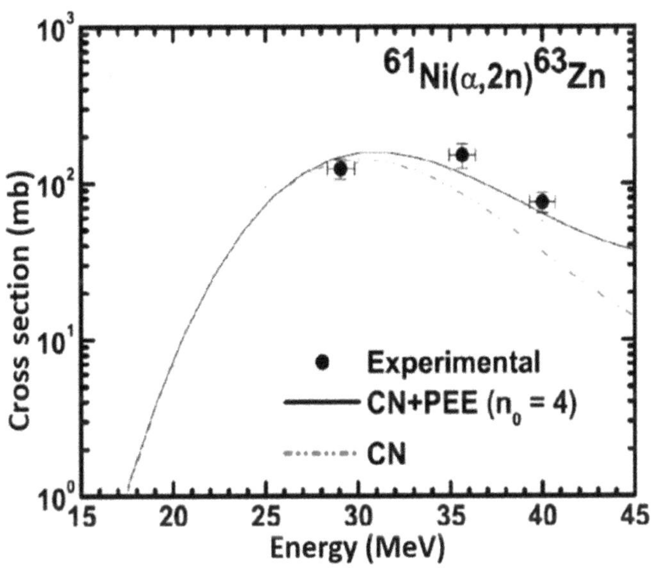

Figure 5.21. Experimentally measured and theoretically calculated EFs for reaction ^{61}Ni(α, 2n)^{63}Zn using the code ALICE-91 with $a = A/8$, $n_0 = 4$ and MFP = 1. Dashed line represents pure CN calculations while the solid line represents CN calculations with the inclusion of pre-equilibrium emission (adapted from [45] with permission from American Physical Society).

and ^{130}Te(α,np)^{132m}I, respectively [46]. In code ACT [17, 22] the CN calculations are performed using the Hauser-Feshbach model [18], while the pre-equilibrium component is simulated employing the exciton model of Griffin [19] consistently using the same set of parameters. In the analysis of the data, it is found that the same residual nucleus may be populated via two different sequences of particle emissions, i.e., there may be first chance emission of either neutron or proton. In the computer code ACT [17, 22] it is possible to consider sequential emission of particles as well. Thus, it is possible to calculate the EFs using different sequences of particle emissions in reactions like ^{128}Te(α,np)^{130}I, ^{130}Te(α,np)^{132}I, and ^{130}Te(α,np)^{132m}I. In these reactions, the residual nucleus may be produced by either the first chance emission of the neutron followed by emission of proton or vice versa. In the present analysis the calculations were performed for both the possible sequences of particle emission, i.e., (α, np) and (α, pn) for the same target nucleus. Since both of the reaction channels are energetically open, the experimental data is compared with the sum of the two reaction paths (first chance emission of neutron followed by proton emission and vice versa) for each reaction, and is shown in figure 5.22.

Using the provisions in the code ACT, the calculations of the EFs for the reactions ^{128}Te(α, np)^{130}I, ^{130}Te(α, np)^{132}I, and ^{130}Te(α, np)^{132m}I have been performed using both the first chance neutron and the first chance proton emission sequences [46]. In figure 5.23, the dashed-dotted line represents the calculated EF with first chance neutron emission for reaction ^{130}Te(α, np)^{132}I, while the dotted line is with the first chance emission of the proton for the same reaction. The solid line is the calculated EF with the first chance emission of the neutron for reaction

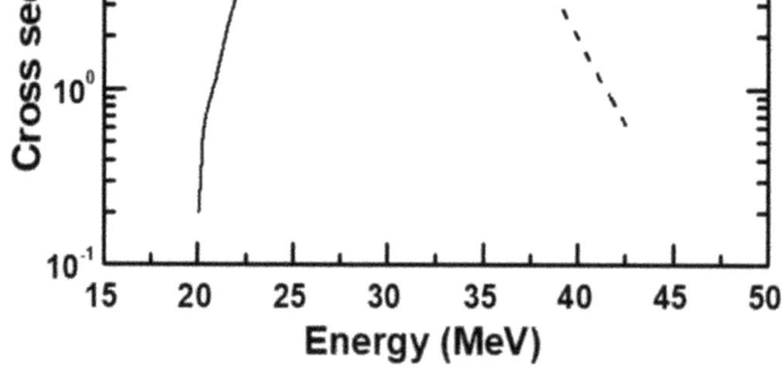

Figure 5.22. Experimentally measured and theoretically calculated EFs for reaction ^{128}Te(α,np)^{130}I using the code ACT. Hauser-Feshbach calculation is shown by the dashed line, while the solid line indicates the sum of compound and pre-equilibrium (exciton model) calculations (adapted from [46] with permission from American Physical Society).

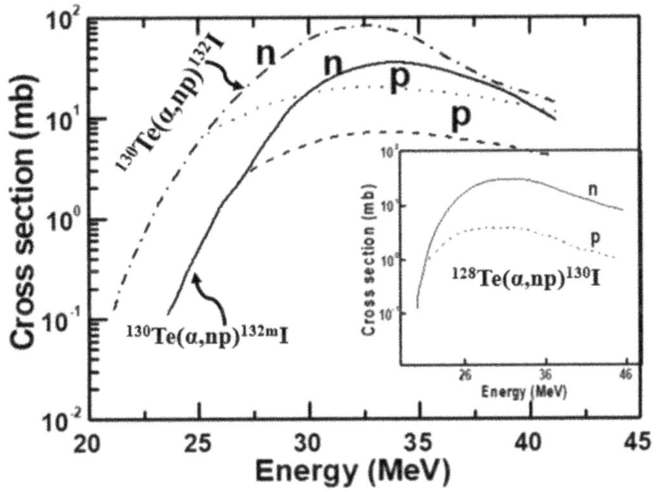

Figure 5.23. Theoretically calculated EFs including pre-equilibrium and CN calculations with two different particle emission sequences (adapted from [46] with permission from American Physical Society).

^{130}Te(α, np)^{132m}I, while the dashed line is with the first chance emission of proton for the same reaction. The figure in the inset (of figure 5.23) is for the reaction ^{128}Te(α, np)^{130}I with the solid line indicating calculations for the first chance emission of neutron and the dashed line for the first chance emission of proton. It has been observed that the EFs calculated assuming the first chance emission of the neutron, in general, have higher cross-sections. A possible reason for the observed enhancement of first chance neutron emission over first chance proton emission may be the Coulomb barrier which inhibits the emission of charged particles. It may be remarked that in both the cases the pre-equilibrium emission has been observed.

5.4.2 Measurements of neutron energy spectra in (α, xn) reactions

The reaction mechanism responsible for producing fast neutrons as pre-compound particles in alpha induced reactions is, somehow, complex. Analysis of experimentally measured EFs for several (α, xn) reactions presented in previous sections, indicates significant contribution from pre-compound emission at the energy range of interest. Fields *et al* [47] carried out systematic study of (α,2$n\gamma$) reactions using alpha particles of energies ranging from 28 to 35 MeV and target nuclei with atomic mass number from 80 to 210. They measured exclusive neutron spectra and angular distributions. Alpha particle beams of energies 28, 32 and 35 MeV were obtained from the 132 cm AVF cyclotron of University of Colorado. Either enriched or isotopically pure samples were used for the experiments. The n–γ–Δt coincidences were recorded using an array of four NE224 liquid scintillation detectors which were placed at 30° with respect to the beam direction, coupled to the Ge(Li) detector at 90° to the beam.

The usual (n–γ) pulse shape discrimination was employed for neutron detectors while the energies of the neutrons were determined employing the TOF technique over a flight path of 60 cm. Using (n–γ) coincidence data the neutron energy spectra were obtained [47]. The neutron spectra were obtained only for those cases which gave a clean gamma ray spectrum. As a representative case the inclusive neutron spectra measured at 30° for reactions $^{112,118,124}Sn(\alpha, xn\gamma)$ at 35 MeV are shown in figure 5.24. The spectra were fitted with a two-temperature model representing the yields with a functional form given by:

$$Y(E_n, \theta) = N\left[E_n e^{\left(-\frac{E_n}{kT_e}\right)} + (kT_e/kT_p)\{a(\theta)\}^{-1}H(E_n, T)(E_n - T)e^{\{-(E_n-T)/kT_p\}}\right]$$

Here, the term N is used for the normalization, the terms kT_e and kT_p refer to the equilibrium and pre-equilibrium nuclear temperatures, respectively. The term $a(\theta)$ represents the square of the ratio of the height of first maximum to that of second maximum in the neutron spectrum, the term $H(E_n, T)$ is a function describing two regions of the distributions such that $H(E_n, T) = 0$, if $E_n < T$, while $H(E_n, T) = 1$, if $E_n > T$. Thus, the spectra are parameterized as indicated above by the two Maxwellian distributions fitted by two-phase de-excitation model [48]. Analysis of the experimental data in terms of the compound and the pre-equilibrium components is shown in figure 5.24. It may be observed in figure 5.24, the strength of pre-equilibrium component increases with the increase of the target mass number.

The observed enhancement in the pre-compound part of the neutron spectrum was satisfactorily explained by theoretical calculations done using Blann's Geometry-dependent

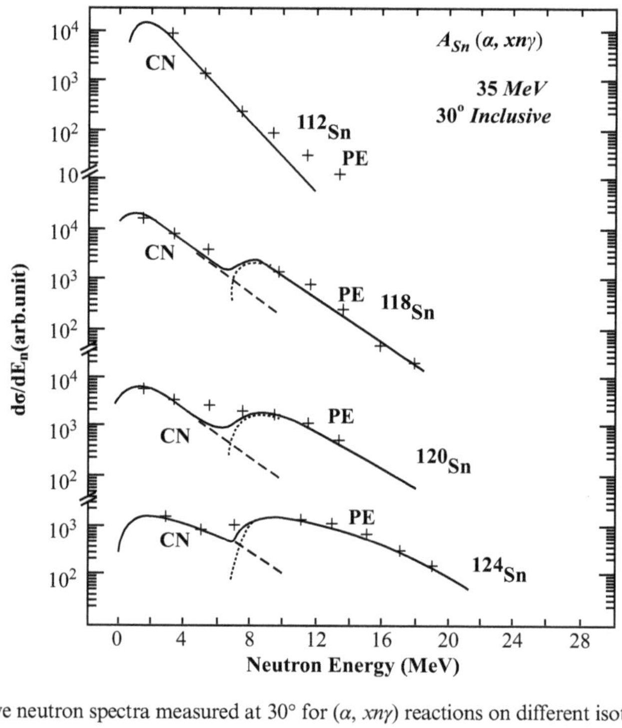

Figure 5.24. Inclusive neutron spectra measured at 30° for $(\alpha, xn\gamma)$ reactions on different isotopes of Sn (tin). The curves are fits with two-component decay model; one component representing CN part while the high energy part is attributed to pre-equilibrium (PE) emission as indicated (adapted from [47] with permission from Elsevier).

hybrid model which takes into account pre-equilibrium emission. It may thus be concluded that pre-equilibrium emission is a dominant reaction mechanism in statistical nuclear reactions even at low and moderate energies.

5.5 Heavy-ion induced reactions

The study of heavy-ion induced reactions is complex in nature as compared to the light-ion induced reactions because of the fact that both the projectile ion and the target nucleus are many nucleon systems. The large Coulomb barrier between the interacting heavy ions, the relatively large mass of projectile nuclei and hence the angular momentum carried by it make the heavy-ion induced reactions quite special. However, because of the small de Broglie wavelength of the relative motion as compared to the size of the target nucleus the heavy-ion reactions may be treated semi-classically. In the semi-classical approach, the radial motion may be treated classically and the angular motion in the central force field quantum mechanically. The semi-classical nature of the heavy-ion interaction makes it possible to give their overall description in terms of minimal distance between the interacting ions, i.e., r_{minimum}, which is related to the impact parameter 'b' by the relation;

$$r_{\text{minimum}} = \frac{b}{\sqrt{\left\{1 - \frac{V(r_{\text{minimum}})}{E_{\text{C.M.}}}\right\}}}$$

Here, $V(r_{\text{minimum}})$ is the potential between the two interacting heavy ions, while $E_{\text{C.M.}}$ is the center of mass energy of interacting ions. It may be pointed out that, for the sake of simplicity, the nuclear field of the target nucleus may be assumed to have a sharp boundary. In this sharp cut-off approach depending on the different values of impact parameter b, the elastic scattering, deep inelastic scattering, incomplete fusion, transfer reactions, fusion reactions etc, are considered to take place at energies from near the Coulomb barrier to well above it. It may be pointed out that in the scattering process the identity of the target and the projectile ions remains unaltered, while in the transfer or in fusion reactions their identities are completely lost.

In the collision of two heavy ions, when the center of mass energy is greater than the Coulomb barrier, the interacting partners may fuse together giving rise to the complete fusion. Once the fusion takes place the energy and angular momentum of the projectile are shared among all the nucleons of the fused composite system leading to the equilibration of the system. However, depending on whether an amount of energy more than the binding energy of a particular nuclear particle is concentrated on it, the particle may escape out of the system well before the equilibration of the compound system has taken place. Such emission of nuclear particles is referred to as the pre-compound or pre-equilibrium emission. The mechanism of such reactions is called pre-equilibrium or pre-compound reaction mechanism. In recent years with the availability of accelerators producing beams of variety of particles other than protons, and alpha particles, pre-compound emission in heavy-ion induced reactions has also been studied. In order to study such reactions, in most of the experiments the energy distribution, angular distribution etc, of light emitted particles or gamma rays produced in a nuclear reaction are studied. However, significantly important information regarding the nuclear reaction mechanism may be obtained from the study of properties of heavy residues produced in the reaction. These heavy residues may be identified from their characteristic mass and charge measured using a suitable analyzing system like recoil mass separator coupled with suitable detector. These heavy residues may also be identified by their characteristic gamma rays and by measuring their half-lives, if radioactive. From the intensity of characteristic gamma rays the population cross-section of the residues produced in a reaction may be determined using the formulations described earlier in chapter 2. Important information regarding reaction dynamics may be obtained through the measurement and analysis of EFs of reactions over a wide range of energy. It may be remarked that features of measured EF at low, medium and high energies may give significant information regarding the reaction mechanism involved in the nuclear reaction. The measured EFs for neutron emission channels in reactions induced by ^{7}Li, ^{11}B, ^{12}C, ^{14}N, ^{16}O and ^{19}F ions etc, in different target nuclei at energies ≈4–7 MeV/nucleons, have been analysed, to obtain information regarding pre-equilibrium emission. The analysis of data has been done within the framework of three different statistical model codes PACE4 [40, 41], ALICE-91 [26] and EMPIRE [13]. Brief descriptions of these codes are already presented in chapter 4. The analysis of relevant data using the above-mentioned codes within the framework of compound and pre-equilibrium emission is presented in the following sub-sections.

5.5.1 Measurement and analysis of EFs with ^{7}Li and ^{11}B beams: ^{7}Li+^{93}Nb; ^{7}Li+^{89}Y and ^{11}B+^{181}Ta systems

Experiments to study the reaction dynamics of heavy-ion collisions for the systems ^{7}Li+^{93}Nb and ^{7}Li+^{89}Y and ^{11}B+^{181}Ta have been carried out [49–51] in the energy range nearly up to 45 MeV energy. It may be pointed out that both the stable isotopes of lithium, i.e., ^{6}Li and ^{7}Li have relatively low binding energy per nucleon (~5.3 MeV) as compared to the adjacent α-particle (~7.1 MeV) and beryllium (~6.5 MeV). As such, the ^{7}Li is also sometimes referred to as the loosely bound nucleus. In the above experiments, the stacked foil activation technique followed by off-line γ-ray spectroscopy has been used. The experiment has been performed at the BARC-TIFR Pelletron accelerator facility, Mumbai, India. Spectroscopically pure (99.99%) samples of natural niobium and ytterbium (^{93}Nb) and tantalum (^{181}Ta) of thickness ≈2.3–3.9 mg cm^{-2} were prepared using a rolling machine. A stack of the target foils each followed by an Al-catcher foil was used for irradiation. The total charge collected during the irradiation was measured by an electron-suppressed Faraday cup placed at the rear of the target assembly. The calculations of energy loss (ΔE) of the incident beam in each target as well as in catcher foil were done using the code SRIM [37] based on the stopping power of incident ion in target material. After the irradiation the stack was taken out from the scattering chamber and the activities induced in each target foil together with respective aluminum catcher foil was measured separately using an off-line γ-spectrometer based detector set-up. The residue ^{97}Ru ($t_{1/2}$ = 2.83d) populated in reaction ^{93}Nb(^{7}Li,3n) ^{97}Ru was identified on the basis of its characteristic gamma-rays and measured half-life. The observed intensities of the characteristic gamma-rays of the reaction residues have been used to determine their cross-section using the formulations presented in chapter 2. The measured EF for the reaction ^{93}Nb (^{7}Li,3n) ^{97}Ru is shown in figure 5.25 [49]. Experimentally measured EF for this reaction has been analysed within the framework of three statistical model codes viz., PACE4 [40, 41], ALICE-91 [26] and EMPIRE [13]. Brief accounts of these codes and analysis done is presented here. The code PACE4 [40, 41] calculates the cross-section for the population of reaction residue using Bass formulations [42] based on Hausher-Feshbach [18] theory. The level density parameter $a = A/K$ is one of the important parameters of the code PACE. Here, A is the atomic mass number of the residual nucleus and K is a free parameter which may be varied to match the measured EF. In the present calculations a value of $K = 10$ has been used, which is the default value in this code and is used frequently for heavy-ion induced reactions. A comparison of the measured EF with calculations performed using code PACE4 [40, 41] with level density parameter, $K = 10$ for the reaction ^{93}Nb(^{7}Li,3n)^{97}Ru is shown in figure 5.25 by the solid curve. The experimental data is shown by solid circles with error bars.

It may be seen from this figure that the calculations done with code PACE4 underestimate the experimentally measured cross-section data for this reaction particularly in the tails portion of the EF. The enhancement in the measured EF (see figure 5.25) with respect to the calculations done with code PACE4 may be due

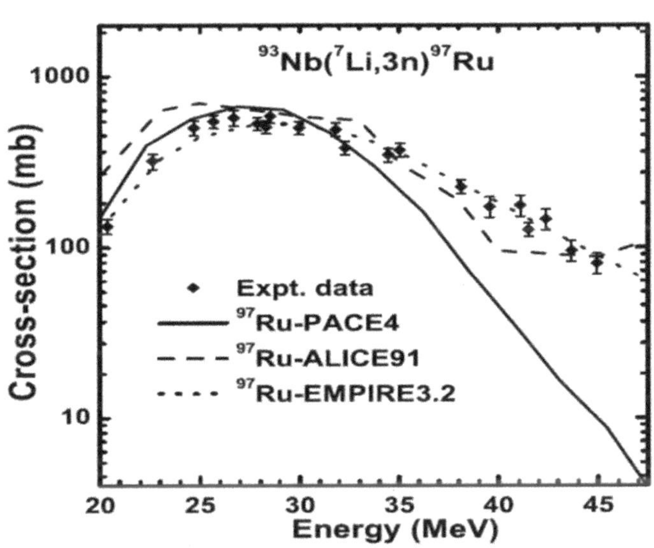

Figure 5.25. Comparison of experimentally measured EF (solid circles with error bars) for reaction ^{93}Nb (^{7}Li, 3n)^{97}Ru with those obtained theoretically using codes PACE4 (solid curve), ALICE91 (dashed curve), and EMPIRE3.2 (dotted curve) (reprinted from [49] with permission from American Physical Society).

to the contribution of pre-equilibrium emission, which is not considered in this code, as it is based on the pure CN model. In order to verify the above, calculations done by codes ALICE-91 [26] and EMPIRE [13] are also shown in figure 5.25, as these codes take into account the contribution due to compound as well as pre-equilibrium emissions. In code ALICE-91 [26], the CN calculations are based on the Weisskopf–Ewing model [28], while pre-equilibrium emission is simulated employing the hybrid/geometry-dependent hybrid model [29]. In ALICE-91 calculations, the level density related parameter *K*, initial exciton number n_0, and mean free path multiplier COST are three parameters that may be assigned different values. Although the parameter *K* influences the CN contribution, the initial exciton number n_0, and the mean free path multiplier COST largely govern the pre-equilibrium contribution. It may be seen from figure 5.25, that calculations done with code ALICE-91 including both the compound and pre-compound emissions give a better description of experimental data. It indicates that pre-compound emission is a dominant reaction mechanism at relatively higher energies. Further, the calculations have also been performed with code EMPIRE3.2, which also takes into account both the contributions of compound and pre-equilibrium emission. As can be seen from figure 5.25, both codes ALICE-91 as well as EMPIRE3.2 satisfactorily reproduce the experimental data in the high energy tails portion, where the contribution of pre-equilibrium emission is expected. It may, therefore, be concluded that there is a significant contribution of pre-equilibrium emission and this process is a dominant reaction mechanism at relatively higher energies.

The measured EF for the reaction ^{89}Y (^{7}Li,3*n*)^{93m}Mo [50] is shown in figure 5.26 and has been analysed using the codes ALICE-91 [26] and EMPIRE3.2 [13]. Both these codes take into account the CN and pre-equilibrium emissions. A comparison

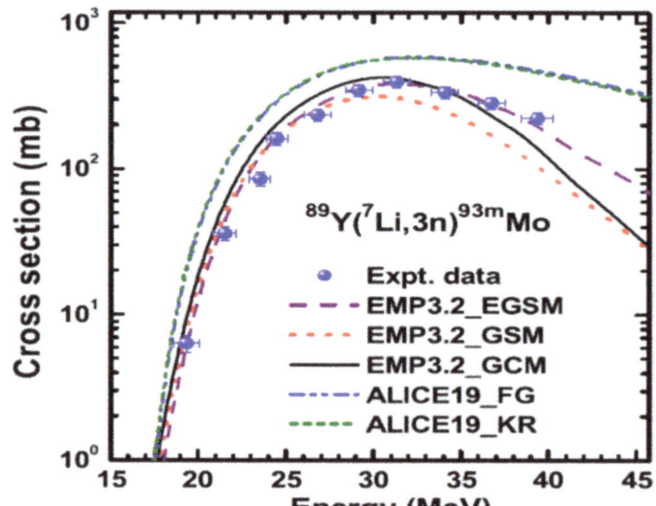

Figure 5.26. Comparison of experimentally measured EF for reaction ^{89}Nb (^{7}Li,3*n*)^{93m}Mo with those of theoretically calculated using codes ALICE-91 and EMPIRE3.2, with different level density parameter prescriptions. For details see the text (adapted from [50] with permission from American Physical Society).

of experimentally measured and theoretically calculated (ALICE-91) EFs for the residues ^{93m}Mo produced in the reaction ^{89}Y(^{7}Li,3n) is also shown in figure 5.26. The ALICE-91 calculations have been performed by two different prescriptions of level densities; Fermi gas (FG) with backshifted pairing and Kataria–Ramamurthy (KR) formalism [26]. In these calculations the level density related parameter $K = 10$, initial exciton number $n_0 = 7$ and mean free path multiplier COST equal to zero have been taken. It has been observed that the calculations done with code ALICE-91 overestimate the measured EF. This may be due to the fact that the code ALICE-91 does not calculate the cross-section of the metastable state separately, however, it gives total cross-section of the reaction residue including both the meta plus ground states.

Further, the calculations for the same reaction ^{89}Nb(^{7}Li,3n)^{93m}Mo have also been performed with the theoretical model code EMPIRE3.2 [13], which also takes into account the contributions of pre-equilibrium along with CN emission. Further, the code EMPIRE3.2 [13] is also capable of calculating the cross-section of metastable and ground states of the reaction residues separately in a given reaction channel. In this code the CN calculations are done on the basis of Hauser-Feshbach [16] formalism, while the contribution of the pre-compound emission is simulated either by the quantum mechanical models considering multi-step direct (MSD) or multi-step compound (MSC) processes [52] or by phenomenological pre-equilibrium (PEQ) models [53] (exciton model or hybrid Monte Carlo simulation). To investigate the effect of nuclear level density on cross-sections, three phenomenological level density models: the Gilbert–Cameron Model (GCM) [54]; the generalized superfluid model (GSM) [55], and the enhanced generalized superfluid model (EGSM) [56] may be used. The effect of these different options of level density parameter sets of code EMPIRE3.2 on the production cross-section of the residue ^{93m}Mo in the reactions ^{89}Y(^{7}Li, 3n) are shown [50] in figure 5.26. In general, the shape of the experimental EF is satisfactorily reproduced over the entire energy range by EMPIRE calculations with all the three different level-density models: EGSM, GSM, and GCM. Analysis of the measured data within the framework of this code shows that an admixture of compound and pre-equilibrium processes is needed to reproduce the experimental data of this reaction channel over the entire energy range, especially towards the high energy side where pre-equilibrium contribution is likely to be significant.

Studies have also been done to investigate pre-equilibrium emission in reactions induced by ^{11}B (boron), which is also referred to as a loosely bound projectile with average binding energy per nucleon around ~7.5 MeV. In view of the above, measurement of EF for the production of residues in a three-neutron emission channel from the CN formed in the ^{11}B+^{181}Ta system has been carried out [51]. An off-line γ-ray spectroscopy based stacked-foil activation technique has been used to determine the production cross-sections of radio nuclides populated in the reaction ^{181}Ta(^{11}B,3n)^{189}Pt at ≈5 MeV/nucleon energies. The beam of ^{11}B (charge state 5^+) delivered by 14UD BARC-TIFR Pelletron facility, Mumbai, India, was used to bombard the stack of self-supporting spectroscopically pure (99.99%) samples of natural tantalum (thicknesses ≈ 1.2–2.0 mg cm^{-2}) foils, each backed by a thin

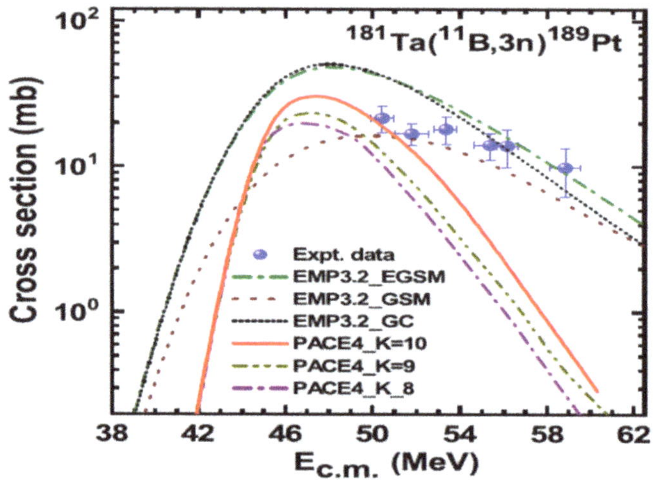

Figure 5.27. Comparison between experimentally measured and theoretically calculated EFs for the production of the residues ^{189}Pt in the reaction ^{181}Ta (^{11}B, 3n) with codes PACE4 and EMPIRE3.2, using different level density parameter prescriptions. See details in text (adapted from [51] with permission from American Physical Society).

aluminum catcher foil [51]. The experimentally measured EF for ^{181}Ta(^{11}B,3n)^{189}Pt reaction has been compared with the predictions of the theoretical model codes viz., PACE4 [40, 41] and EMPIRE3.2 [13], and is shown in figure 5.27. Details of these codes have already been presented. As can be seen from this figure, the calculations done with code PACE4 [40, 41] are not able to reproduce the experimental data towards the higher energy side of the EF, with different values of the level density related parameters K = 8,9, and 10. It is mainly because of the fact that the code PACE4, takes into account only the calculations done using the pure CN model. The enhancement towards the tail portion of EF may be due to the contribution expected from pre-equilibrium emission which is important at relatively higher energies.

In order to test the above, calculations of the EFs for reaction ^{181}Ta(^{11}B, 3n)^{189}Pt have also been done, including pre-equilibrium emission, with the code EMPIRE with three different level density models: the GCM [54], the GSM [55], and the EGSM [56]. As may be seen from figure 5.27, the trends as well as magnitude of measured cross-sections are satisfactorily reproduced by these calculations that include pre-equilibrium emission along with the CN contribution. In general, for simulating the EF of 3n-channel at relatively higher energy, the contribution of pre-compound emission has to be included for satisfactory reproduction of the experimental data, indicating the importance of on-equilibrium reaction mechanism.

5.5.2 Measurement and analysis of EFs with ^{12}C beam: ^{12}C+^{169}Tm, ^{12}C+^{128}Te and ^{12}C+ ^{197}Au systems

In recent years, with the availability of particle accelerators capable of producing heavy-ion beams with energies around 4–7 MeV/nucleon, several experiments have been carried out to study the reaction dynamics. Efforts have been made to study the heavy-ion fusion reactions where the pre-equilibrium emission plays an important role. An adequate description of the mechanism of pre-equilibrium emission may be achieved by generating more and more precise experimental data on this process. With this view, experiments to investigate the reaction dynamics of light-particle emission in heavy-ion fusion reactions for systems $^{12}C+^{169}Tm$, $^{12}C+^{128}Te$ and $^{12}C+^{197}Au$ have been carried out in the energy range ≈5–7 MeV/nucleon [57]. The stable isotope of carbon ^{12}C has relatively high binding energy per nucleon (≈7.68 MeV), as such these are generally treated as strongly bound projectiles. It is emphasized that such projectiles are less likely to break up at the time of interaction. As such, the beams of ^{12}C may be considered to be suitable to study the pre-equilibrium emission at low projectile energies, where, the probability of formation and decay of a pure CN process is relatively high. Thus, it may be quite interesting to see the interplay between the compound and pre-equilibrium emission processes in reactions induced by low energy stable ^{12}C ion beam in fusion reactions. Experimental observations have indicated that as beam energy increases, ^{12}C ions break up into alpha-clusters and fusion of one of alpha particle and/or cluster of alpha particles with target nucleus leads to the phenomenon of the incomplete/breakup fusion processes in heavy-ion reactions. Consequently, in such reactions at relatively higher projectile energies, the probability of alpha particle emission from breakup or from fusion evaporation as compared to the particle emission in the pre-equilibrium processes may decrease.

Recently, some experiments have been carried out using the 15 UD Pelletron accelerator of the Inter-University Accelerator Centre (IUAC), New Delhi, India, where, the beam of ^{12}C is obtained to perform experiments to study the pre-equilibrium emission process [57]. Experiments have been performed in the General Purpose Scattering Chamber (GPSC) beam line of IUAC, New Delhi, where a big scattering chamber of diameter 1.5 m is particularly used for low energy reaction dynamics studies. Typical inside and outside views of the GPSC facility are shown in figure 5.28(a) and (b). In figure 5.28(c) the in-vacuum transfer facility is shown which is employed to take out and insert the samples inside the scattering chamber without

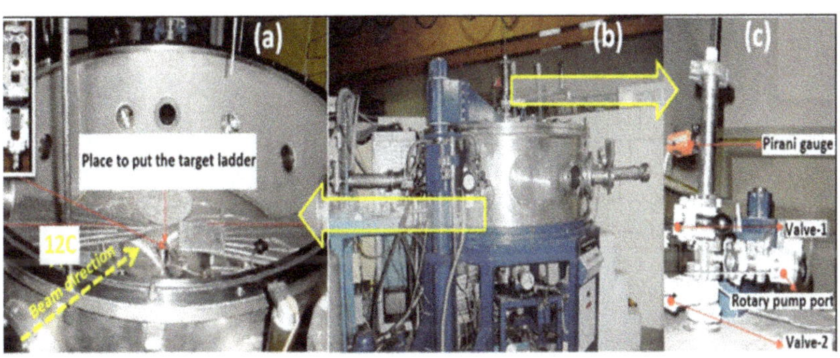

Figure 5.28. Inside and the outside views of the GPSC facility. Figure (c) shows in-vacuum transfer facility for targets without disturbing the vacuum of the scattering chamber.

disturbing its high vacuum. Thus, time spent to recover vacuum for about 6–8 h between two consecutive irradiations is saved. The important experimental details of such target preparation, measurements of thickness of the target and calibration of detector etc for the GPSC set-up of IUAC are discussed in the following.

Isotopically pure ^{169}Tm (abundance = 100%) targets of thickness ≈900 μg cm^{-2} and Al-catcher foils of thickness ≈2 mg cm^{-2} were prepared by rolling machine. Here, thick Al foils served a two-fold purpose; one as an energy degrader for producing energy difference between two successive target samples and the other as a catcher/backing foil for trapping the residues recoiling out from the target as the result of momentum transfer. The thickness of each target sample/Al-catcher was determined precisely prior to its use in the stack. The thickness of the samples and each catcher foil was measured using an α-transmission method that is based on the measurement of the energy lost by 5.487 MeV α-particles (obtained from a standard ^{241}Am source), while passing through the target sample or backing foil. In the stacked foil activation technique, the energetic beam of ^{12}C particles traverses through the samples with degrading beam energies. Three (sample plus catcher) stacks were individually irradiated at different incident energies of ^{12}C beam so as to cover the incidence energy range of ≈54–90 MeV. Irradiations were carried out in GPSC with beam currents ≈20–30 nA. Precise energy of the incident ^{12}C beam at the middle of each sample was determined using the computer code SRIM [37]. After the irradiation, the irradiated stack was taken out from the scattering chamber with the help of a fast extraction system of in-vacuum transfer, which reduces the lapse time between the end of irradiation and the start of counting of the activities produced in the samples. The radio-activities produced in different samples of the stack and the catcher foil due to the irradiations were followed off-line using pre-calibrated HPGe detectors. A typical γ-ray spectrum of a standard source ^{152}Eu of known strength, recorded prior to actual experiment for ^{12}C+^{169}Tm system, to determine the efficiency of the gamma spectrometer set-up, at source–detector distance 3 cm is shown in figure 5.29.

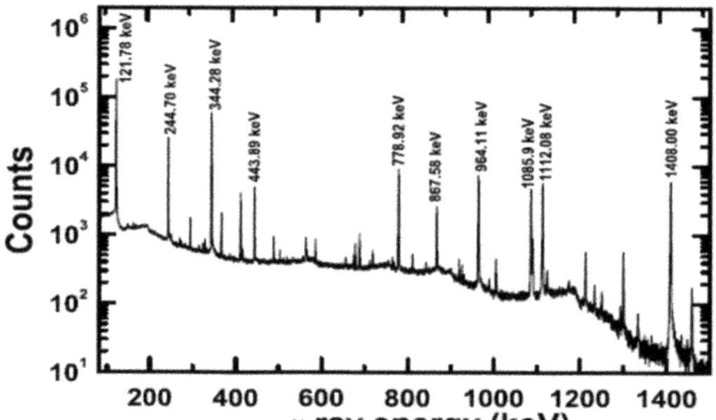

Figure 5.29. A typical γ-ray spectrum of a standard source ^{152}Eu of known strength covering energy ranging from ≈121 to 1408 keV. This spectrum recorded prior to actual experiment for ^{12}C+^{169}Tm system at a definite source–detector distance.

The dead time of the spectrometer was kept around 10% by adjusting the sample–detector distance, and necessary dead time corrections in the geometry-dependent efficiency of detector were applied. The uncertainty in geometry-dependent efficiency of detectors due to the dead time factor was estimated to be $\leqslant 2\%$. Figure 5.30 sows the variation of geometry-dependent efficiency with γ-ray energy in the energy range 121.7 to 1408 keV for a standard ^{152}Eu gamma source. After the irradiation the samples were taken out from the scattering chamber and counting of each sample plus catcher foil was done using a high-resolution large-volume (100 c.c.) HPGe detector coupled to a PC based multichannel analyzer. A typical sketch of the data acquisition system used to record the gamma-ray activities in the irradiated target-catcher assembly is shown in figure 5.31. The reaction residues were identified by their characteristic

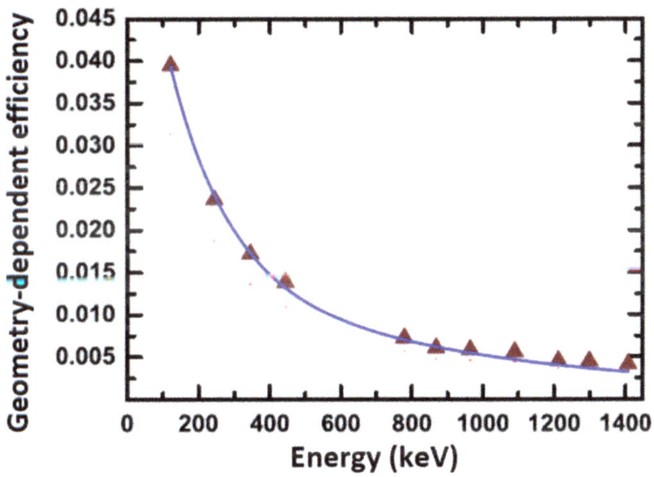

Figure 5.30. A typical geometry-dependent efficiency of the spectrometer as a function of gamma ray energy obtained for a standard gamma source ^{152}Eu.

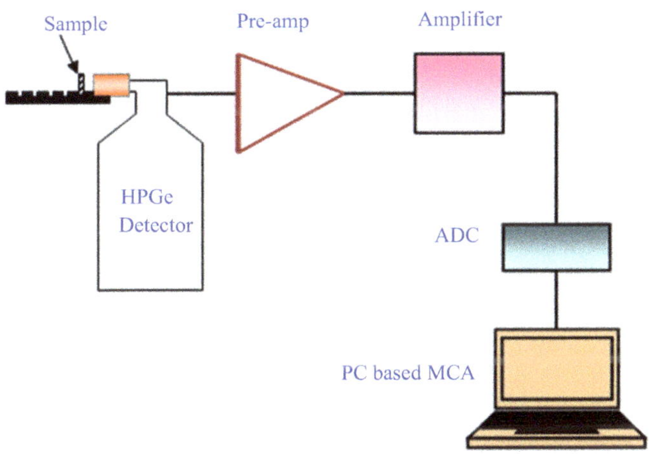

Figure 5.31. Sketch of the experimental set-up and data acquisition system used to record gamma-ray activities in the irradiated target-catcher assembly.

γ-lines, and verified by decay-curve analysis. The measured intensities of the characteristic γ-lines of the identified residues have been used to determine the cross-sections for the corresponding reaction channels employing a FORTRAN program based on standard formulations given in section 2.8 of chapter 2.

The measured EF of the reaction ^{169}Tm(^{12}C,3n)^{178}Re has been compared with theoretical predictions based on pure statistical model code PACE4 [40, 41] and the geometry-dependent hybrid (GDH) [29] model-based code ALICE-91 [26]. Details of code PACE4 [40, 41] along with its important parameters have already been discussed in reference [57]. The de-excitation of the CN in code PACE4 [40, 41] is followed by Monte Carlo procedure. Theoretical calculations for cross-section for the population of residues over the measured energy range have been performed using code PACE4 [40, 41]. In this code the angular momentum projections are calculated at each stage of de-excitation, which enables the determination of angular distribution of the emitted particles. Theoretical calculation of cross-section for reaction ^{169}Tm(^{12}C,3n) has also been performed employing the code ALICE-91 [26], which considers the pre-equilibrium emission as well. The code ALICE-91, uses the GDH model [26] for simulating pre-compound emission. In this code the level density parameter 'a', the mean free path multiplier 'COST' and the initial exciton number 'n_0' are the important parameters and may influence the calculations. The level density parameter 'a' affects mainly the equilibrium component while the initial exciton number 'n_0' and the mean free path multiplier govern, predominantly, the pre-equilibrium component. The level density in this code is calculated from the expression 'a' = A/K, where, A is the mass number of the residual nucleus while, K is an adjustable parameter. The intermediate states of the CN are characterized by the excitation energy E^*, and number n_p, of excited particles and n_h of excited holes. Particles and holes are defined relative to the ground state of the nucleus and are called excitons. The initial configuration of excited compound system is represented by exciton number $n_0 = (n_p + n_h)$,which is an important parameter of pre-equilibrium formalism. For the reactions induced by ^{12}C beam, a value of $n_0 = 13$ with configuration ($6p$+$6n$+$1h$) has been found to satisfactorily reproduce the measured EF data, where, p, n and h represent the number of excited protons, neutrons and holes, respectively. The configurations of $n_0 = 13$ (12 particles + 1 hole) for ^{12}C may be justified, assuming that the first interaction gives rise to the excitation of one particle above the Fermi energy leaving behind a hole in the excited state. Keeping the value of $n_0 = 13$, the effect of variation of parameter K on the EFs has also been studied. This is shown in figure 5.32 for the reaction ^{169}Tm(^{12}C,3n)^{178}Re by comparing the experimentally measured EFs with calculations done using ALICE-91 code with different values of parameter K = 10, 12, 14, and 16. In this figure, the measured cross-sections for the residues ^{178}Re by Chakrabarty *et al* [58] are also shown for comparison. The dashed, dotted, dash doted and solid curves are calculations done corresponding to the parameter K =10, 12, 14, and 16, respectively. It can be seen from this figure that a value of K = 16 satisfactorily represents the peak portion of the experimental data. It may be observed from figure 5.32 that the peak of the measured EF is slightly shifted towards the higher beam energies than that of the calculations done with ALICE-91 [26]. In ALICE-91 calculations,

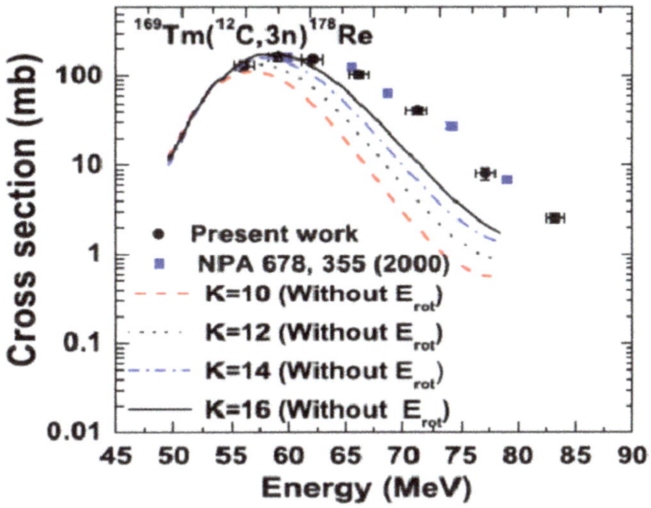

Figure 5.32. Experimentally measured and theoretically calculated EFs for reaction ^{169}Tm(^{12}C,3n)^{178}Re. Calculations done with code ALICE-91 with the value of initial exciton number $n_0 = 13$, the effect of different values of parameter $K = 10$, 12, 14, and 16 are shown. Cross sections measured by Chakrabarty *et al* [58] are also shown in the figure for comparison (reprinted from [57] with permission from American Physical Society).

the angular momentum effects have not been taken into account. As a matter of fact, in the case of heavy-ion induced reactions, the incident particle imparts relatively larger angular momentum to the composite system as compared to the light projectile. The higher angular momentum inhibits particle emission more than it does gamma emission in the last stages of the nuclear de-excitation process, and hence, the peak of EF corresponding to the particle emission mode gets shifted to higher energies [55]. The possible energy shift due to angular momentum effects may be estimated from the rotational energy given by $E_{\text{rot}} \approx (m/M)E_{\text{lab}}$. Here, m/M is the ratio of the projectile and the target nucleus masses and E_{lab} is the incident energy in laboratory frame.

It may be observed from these figures that with the inclusion of rotational energy in calculations, theoretical curve with $K = 16$, reproduces the experimental data for reaction ^{169}Tm(^{12}C,3n)^{178}Re (figure 5.33) Satisfactorily. Thus, an admixture of compound and pre-equilibrium emission explains the experimental data satisfactorily.

Further, in order to study the pre-equilibrium emission in 3n-channel in the system ^{12}C+^{128}Te an experiment was performed using activation technique [55] employing enriched isotope of tellurium-128 (abundance 87%). The thickness of the samples was measured by the standard alpha transmission method as already described. The measured thickness of the ^{128}Te deposition was found to be ≈0.92 mg cm^{-2}. Here, the thick Al backing of 6.75 mg cm^{-2}, served as energy degrader as well as backing to trap the recoiling nuclei from the target foil. The experiment was done at the 15 UD Pelletron accelerator facility of the IUAC, New Delhi, India. The experimentally measured EF for reaction ^{128}Te (^{12}C,3n) ^{137}Ce [57] is shown in figure 5.34 and has been analyzed using the Monte Carlo statistical model code

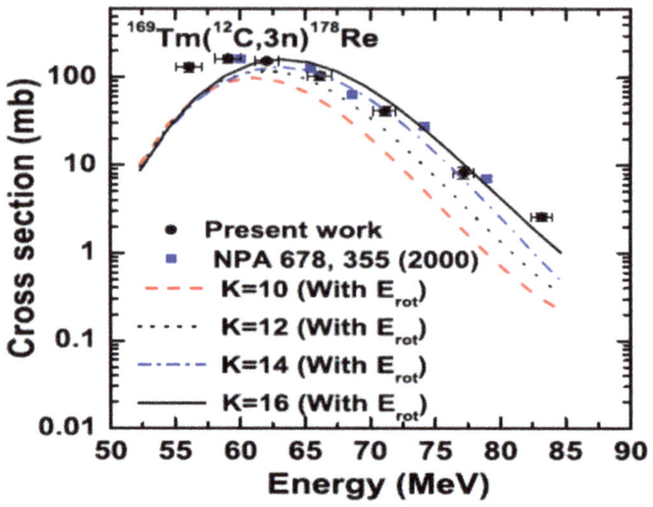

Figure 5.33. Experimentally measured and theoretically calculated EFs for reaction ^{169}Tm(^{12}C,3n)^{178}Re. Keeping the value of initial exciton number $n_0 = 13$, the effect of rotational energy (E_{rot}) with the different values of parameter K =10, 12, 14, and 16 in ALICE-91 calculations is shown. Cross-sections measured by Chakrabarty *et al* [58] are also shown in the figure for comparison (reprinted from [57] with permission from American Physical Society).

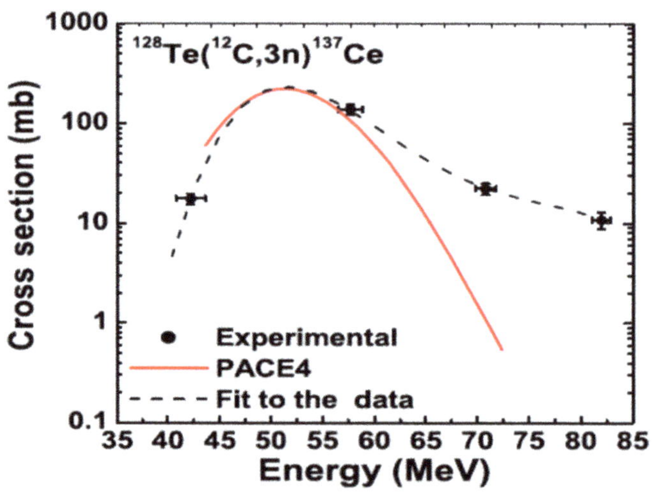

Figure 5.34. Experimentally measured and theoretically calculated EFs for reaction ^{128}Te(^{12}C,3n)^{137}Ce, using Monte Carlo–based code PACE4 (solid line). Dashed line is the fit to experimental data (adapted from [57] with permission from American Physical Society).

PACE4 [40, 41]. A comparison of experimentally measured and theoretically calculated (PACE4) EFs with level density parameter, $K = 8$ (by solid curve) for the reaction ^{128}Te (^{12}C,3n) ^{137}Ce is also shown in figure 5.34.

It may be observed from figure 5.34, that the PACE4 calculations done for this reaction reproduce the experimental data nearly up to 60 MeV, however, the calculations are found to underestimate the experimental cross-sections at relatively higher energies. The enhancement in the measured EF with respect to the calculations done with code PACE4 has been attributed to the contribution from pre-equilibrium emission, as observed for some other reactions as well.

One of the early experiments where pre-equilibrium emission was observed in heavy-ion collisions, was carried out for ^{12}C + ^{197}Au system by Vergini *et al* [59] at the MP Tandem of the Laboratorio Nazionale del Sud, Catania, Italy. Activation technique was used for measuring the cross-sections for the production of reaction residues at nine energies between 57 and 97 MeV, in steps of 5 MeV. The close data points provided a better test of the predictions of pre-equilibrium models. Thicknesses of the Au (Gold) foil targets used were ≈800 μg cm^{-2} and those of Al-catchers were ≈810 μg cm^{-2}. Here, Al-catchers served as backing to collect the recoiling nuclei from target and also degraded energy falling on the next target sample. A ^{12}C beam of energy 97 MeV with an average beam current ≈600 nA was used. The experimentally measured and theoretically calculated EFs with and without including the pre-equilibrium emission are shown in figures 5.35 and 5.36. In these figures, experimental data of Vergini *et al* [59] is shown by solid circles while experimental data given in reference [59] for the same reactions by Thomas *et al* [60] (crosses), Bimbot, Lefort, and Simon [61] (open squares), and Baba *et a*l [62] (open circles) are also shown. The results of Vergini *et al* [59] are found to be in reasonable agreement with those of Bimbot, Lefort, and Simon [61]. Also, they are found to be reasonably well in agreement with the results of Baba *et al* [62] except for the production cross-section of At isotopes around 90 MeV or so. The values given by Baba *et al* [62] disagree with those reported by Bimbot, Lefort, and Simon [61],

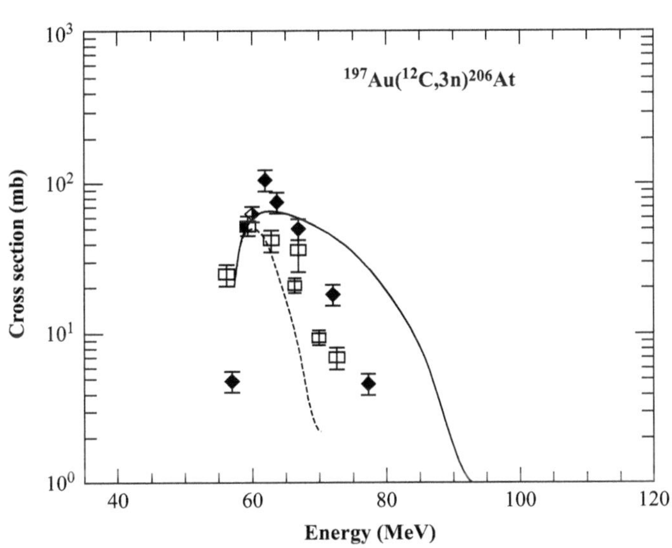

Figure 5.35. EFs for $^{197}Au(^{12}C,3n)$ ^{206}At reaction. Experimental data of Vergini *et al* along with other literature values are also included. In this figure, the full and dashed lines give the result of calculations made by, respectively, considering and neglecting the pre-equilibrium emission (adapted from [59] with permission from American Physical Society).

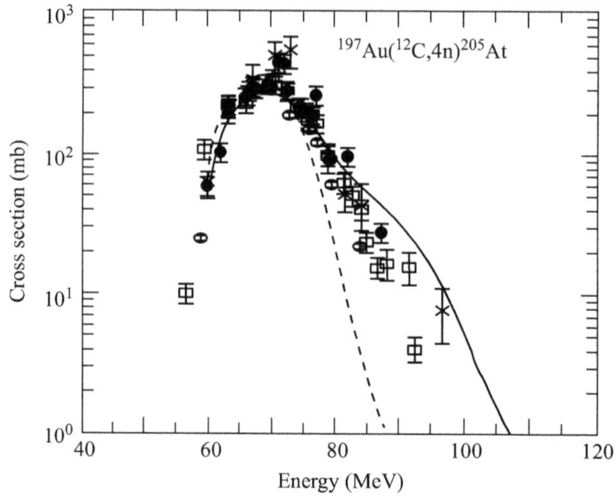

Figure 5.36. EFs for reaction $^{197}Au(^{12}C,4n)\ ^{205}At$. The experimental data of Vergini *et al* along with other literature values are included in the figure. In this figure, the full and dashed lines give the result of calculations made, respectively, by considering and neglecting the pre-equilibrium emission (adapted from [59] with permission from American Physical Society).

however, these are found to agree with results obtained by Vergini *et al* [59]. The results of Baba *et al* [62] are quite small and cause a sudden and inexplicable decrease in the cross-section given by these authors (see figure 5 of reference [59]). The cross-sections measured by Thomas *et al* [60] are also relatively much less in terms of absolute value than those measured in the experiment of Vergini *et al* [59] and those of Bimbot, Lefort, and Simon [61] and Baba *et al* [62]. On the other hand their observed energy dependence is the same as that observed and these data points may be normalized with a certain factor that turns out to be nearly equal to to the data of Vergini *et al* [59]. The analysis of the measured EFs for the identified residues has been made by estimating production cross-section with the help of the Boltzmann master equation (BME) model [63]. It may be pointed out that the BME model has a provision to calculate cross-section of the production residues with or without including pre-equilibrium emission. In the BME theory [63] one uses the fact that when the projectile and the target nucleus fuse together, the momentum distribution of nucleons is significantly different from that corresponding to the case of statistical equilibrium condition as a result of relatively higher degree of correlation giving rise to the translatory motion of the projectile and the target with respect to the center of mass. The momentum distribution undergoes a sort of mixing through a series of nucleon–nucleon interactions, and in between these interactions the emission of nucleons with energy higher than that of particles evaporated by an equilibrated system may take place. In this way the CN proceeds toward the statistical equilibrium condition, and the fusion of the incident particle and the target nucleus is described by solving coupled master equations to get, as functions of time, the occupation probabilities of the states of the composite nucleus from nucleon–nucleon interactions and emission into the continuum of both the nucleons or the combinations of particles like clusters. It can been seen from figures 5.35 and 5.36 that in the case of production of ^{206,205}At isotopes via 3*n*, and 4*n* emission reaction channels, respectively, an appreciable contribution of pre-equilibrium emission of nucleons (shown by solid curves) is observed with average energy substantially higher than that of nucleons evaporated by a fully equilibrated nucleus (shown by dotted curves). The results of analysis made by considering the emission of pre-equilibrium nucleons in the BME model are found to satisfactorily reproduce the measured experimental data of the EF for reaction residues produced in the interaction of ^{12}C beam with ^{197}Au target nuclei. As such, it may be concluded that there is a significant contribution of pre-equilibrium emission in these reaction channels as well.

5.5.3 Measurement and analysis of EFs with ^{14}N, ^{16}O, and ^{19}F beams: ^{14}N+^{181}Ta, ^{16}O+^{159}Tb, ^{16}O+^{169}Tm, ^{16}O+^{181}Ta and ^{19}F+^{169}Tm systems

The results of experiments with ^{12}C ion beams discussed in section 5.5.2, have clearly indicated the presence of the pre-equilibrium emission at relatively higher energies, where, the enhancement in the experimental data as compared to the calculated cross-sections based on CN models was attributed to the pre-equilibrium emission process. The pre-equilibrium emission has also been investigated with a non-cluster ^{14}N beam on relatively heavy target ^{181}Ta at energies around 4–7 MeV/nucleon using the off-line γ–ray spectroscopy [64]. The EFs for the population of radioactive residues ^{192}Hg produced via 3*n* reactions channel in a ^{14}N +^{181}Ta system has been measured and analyzed within the framework of statistical model code PACE4 [40, 41]. Details of the beam and the target used as well as the experimental set-up employed in these measurements of EF are discussed here in brief. The experiment was carried out at the ion-beam facility of the IUAC, New Delhi, India. It may be pointed out here that negative ions of elements having negative electron affinity like ^{14}N cannot be produced in elemental form. Electron affinity of nitrogen is −0.07 eV. As such, these beams may be produced in the molecular form containing the element of the required beam. In view of this, for the case of ^{14}N, the C–N molecules are first injected in the accelerator column and are pre-accelerated. As this C–N molecule beam passes through the stripper, the accelerated beam in the molecular form gets dissociated and get stripped off forming positive ions of interest. These positive ions are then accelerated downwards in the accelerator column. The final selection of beam of ^{14}N and its energy is made with the help of bending magnet and the analyser system. For the experiment, pure ^{181}Ta targets (purity 99.98%) and aluminum catcher/energy-degrader foils (thickness ≈1.2–1.7 mg cm^{-2}) were prepared by rolling technique Three stacks each consisting of four target-catcher foil assemblies were irradiated separately at energies ≈88, 85 and 83 MeV so that energy interval between two successive samples was less than 2 MeV and covers the data points at twelve energies between ≈65 and 88 MeV [64]. Irradiations were carried out in the GPSC having in-vacuum transfer facility for ≈8–10 h duration each. The beam current was monitored during the irradiation using a Faraday cup installed downstream the beam line connected to an ORTEC's current integrator; and was maintained nearly constant ≈3–5 pnA. The activities induced in each target-catcher foil assembly were recorded using a pre-calibrated HPGe detector set-up. The measured EF for the reaction ^{181}Ta(^{14}N,3n)^{192}Hg is shown in figure 5.37 and has been compared with the predictions of the statistical model code PACE4 [40, 41]. In this figure, the effect of variation of different values of the level density parameter constant $K = 8–10$, is also shown. As may be seen, the high-energy tail portion of the EF could not be reproduced with the statistical model predictions of code PACE4. Since this code is based on CN model calculations, the observed enhancement in the measured EF as compared to the calculations may be attributed to the presence of pre-equilibrium emission process. As can be seen from figure 5.37, the pre-equilibrium contribution, represented with vertical arrows increasing with beam energy, indicates its importance at relatively higher energies. The other stable heavy-ion beams of ^{16}O and ^{19}F having relatively larger values of binding energy per nucleon have also been used to perform the experiments at the 15 UD Pelletron accelerator of IUAC, New Delhi, India. The measurement of the EFs for the 3n-reaction channels have been carried out for ^{16}O+^{159}Tb, ^{16}O+^{169}Tm, ^{16}O+^{181}Ta systems [57] as well. Analysis of data indicated significant pre-equilibrium contribution in the 3n emission channels, as confirmed by ALICE-91 calculations.

In order to study the pre-equilibrium emission, the EFs of neutron emitting reaction channels in the interaction of ^{19}F +^{169}Tm system have also been measured using off-line γ-ray spectroscopy. The experimental details are similar to those already described above. Three stacks of target-catcher assemblies were irradiated. Each stack consisted of three targets each followed by Al-catcher foil of

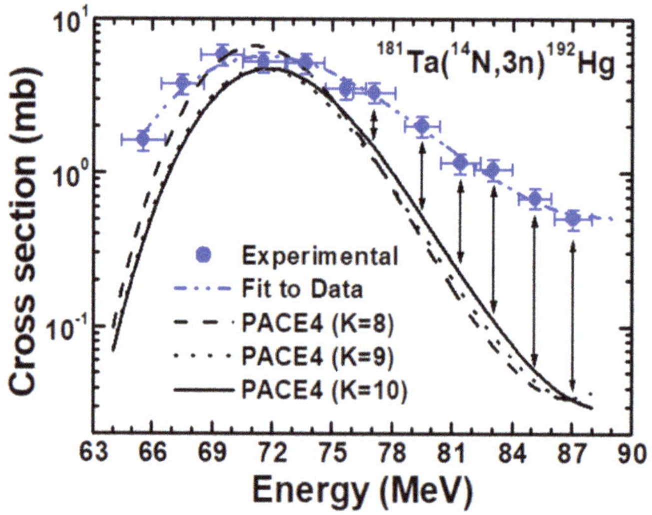

Figure 5.37. Experimentally measured and theoretically calculated EFs for reaction ^{181}Ta(^{14}N,3n)^{192}Hg using the code PACE4. The effect of different values of the level density parameter constant $K = 8$, 9 and 10, on theoretical calculations is shown in the figure (adapted from [64] with permission from American Physical Socety).

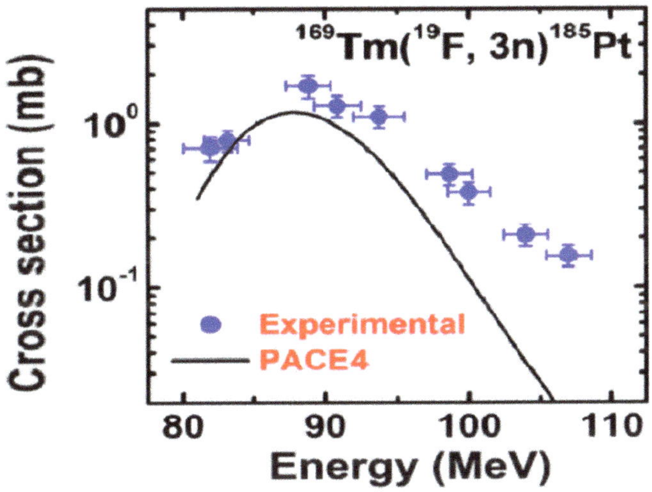

Figure 5.38. Experimentally measured EF for reaction ^{169}Tm(^{19}F,3n)^{185}Pt and theoretical calculations performed with code PACE4 (adapted from [65], copyright IOP Publishing, all rights reserved).

suitable thickness placed normal to the beam direction. The irradiations of stacks were done separately at energies ≈108, 105, 102 MeV [65], respectively. After the irradiation the activities induced in the samples were recorded using HPGe detectors and intensities of characteristic gamma lines have been used to determine the cross-sections for the residues of interest. The measured EF for the reaction ^{169}Tm (^{19}F,3n)^{185}Pt and the corresponding theoretical calculation has been done within the framework of statistical model code PACE4 [40, 41] and is shown in figure 5.38. The solid circles with error bars in this figure represent the experimental data, while the solid curve represents calculations done with the code PACE4. As may be seen from this figure, the calculations done with code PACE4 are not able to represent the experimental data in the higher energy tail portion of the EF. It is expected, as the code takes into account only the pure CN model calculations. The observed enhancement of cross-sections towards the tail portion of EF has been attributed to the contribution from pre-equilibrium emission, which is likely to be a dominant mode of reaction at relatively higher energies.

5.5.4 Pre-equilibrium emission from recoil range distribution: ^{16}O+^{159}Tb system

At low energies, in heavy-ion interactions the complete, incomplete fusion and/or transfer reactions have been studied in detail. However, it is likely that at these energies, reactions may also proceed via pre-equilibrium emission as well. From the measurement and analysis of EFs, significant contribution from pre-equilibrium emission has been observed in reactions induced by both light as well as heavy-ion induced reactions. In case of the EFs, the pre-equilibrium component has been deduced by comparing the measured cross-sections with the predictions of the CN theory. Any enhancement over and above the calculated values has been attributed to the pre-equilibrium emission. Although every effort is made to suitably choose the parameters of the compound and pre-equilibrium models, such contributions may sometime become questionable in the sense that such contributions are said to be model dependent. Since both the compound and pre-equilibrium reactions compete with each other, experimentally it is very difficult to separate out their contribution in a given reaction channel. As such, to get direct evidence, precise measurements are required to get these two distinctly different reactions first identified and then to separate their contributions.

It may be remarked that in the case of pre-equilibrium emission the high-energy nuclear particles are emitted well before the establishment of thermodynamical equilibrium of the composite nucleus. These high-energy particles, generally referred to as the pre-equilibrium particles, are emitted mostly in the forward direction and take away a significant amount of linear momentum from the system. As a result, the reaction residues of pre-equilibrium emission are left with a relatively small amount of linear momentum and if passed through an absorbing medium, travel a smaller distance in the medium as compared to the residues populated via CN process. Thus, the measurement of the linear momentum transferred to the reaction products or the range distribution may be taken as one of the most direct and irrefutable methods to disentangle the CN and pre-equilibrium emission residues. As already mentioned, the linear momentum transferred to the reaction residue in a reaction will be largest for the residues populated via CN process and will have a reduced value for residues populated through pre-equilibrium emission. Recently, measurements of the recoil ranges have been used to derive information about the momentum transfer in the compound and pre-equilibrium emission process. In order to characterize the pre-equilibrium emission process on the basis of the linear momentum involved in the reaction, experiments have been carried out using ^{16}O ion beams delivered from the 15UD-Pelletron accelerator at the Inter-University Accelerator Centre (IUAC), New Delhi, India. In these experiments recoil range distributions of residues in some reactions that are likely to be populated by both the compound and pre-equilibrium emission channels have been measured for ^{16}O+^{159}Tb, ^{16}O+^{169}Tm and ^{16}O+^{181}Ta systems [66]. Brief experimental details of measurements of recoil range distribution for ^{16}O+^{159}Tb system used to obtain information regarding pre-equilibrium emission are given below.

The beam of oxygen ions ($^{16}O^{7+}$) obtained from the 15 UD Pelletron accelerator facility of the IUAC, New Delhi, India has been used for the experiments employing recoil-catcher activation technique. The target of ^{159}Tb was prepared by the rolling method and thin Al-catcher foils were prepared by vacuum evaporation technique [66]. The thickness of target sample and Al-catcher foils were determined by the α-transmission method. The thickness of the target was $\approx$190 μg cm^{-2}, however, the thicknesses of Al-catchers were ranging from $\approx$25–50 μg cm^{-2}. A thin ^{159}Tb target followed by a number of Al foils was pasted on rectangular aluminum holders (2.5 × 2.1 cm^2) each having aconcentric hole of 10 mm diameter, and a stack was prepared. Separate irradiations at different energies were carried out in the GPSC having an in-vacuum transfer facility. A series of thin Al-catcher foils were placed normal to the beam direction behind the target sample, so that the heavy recoiling residues populated via compound and pre-compound reaction processes, depending on their linear momentum, may be trapped at the respective ranges in Al-catcher foil. The duration of irradiation in the experiment was typically about 24 h with a beam current of ≈3–4 pnA. Attempts were made to keep the beam current as nearly constant as possible all through the duration of irradiation. After the irradiation, activities produced in each Al-catcher foil due to the trapping of recoiling radioactive heavy residues of interest were recorded separately using a pre-calibrated high-resolution HPGe gamma ray spectrometer of 100 c.c. active volume coupled to the CAMAC-based CANDLE software. The resolution of the γ-spectrometer was ≈2 keV for 1.33 MeV γ-ray of ^{60}Co source. The geometry-dependent efficiency of the detector for various γ-ray energies at different source–detector separations was determined using the standard ^{152}Eu source of known activity.

The identification of populated reaction residues was made on the basis of their characteristic γ-ray energies, and was further confirmed by measuring the half-life. The cross-sections (σ) in different catcher foils for the reaction residues of interest was determined from the intensity of their characteristic γ-rays. In order to obtain the recoil range distribution of the reaction products, the measured values of cross-section of the reaction residues in each catcher foil was divided by its thickness, the resulting yield of the reaction product was then plotted against cumulative catcher foil thickness to get experimental distribution of ranges of the reaction products in successive catcher foils. As typical examples, experimentally measured recoil range distributions for the residues produced in the reactions ^{159}Tb(^{16}O,2n)^{173}Ta and ^{159}Tb(^{16}O,3n)^{172}Ta at 90 MeV are shown [66] in figures 5.39 and 5.40, respectively. The experimental data of the measured recoil range distributions were fitted using the ORIGIN software and are shown by solid lines to guide the eye. The bumps (peaks) in the experimental range distributions arise due to the overlap of the heavy residues produced via two different

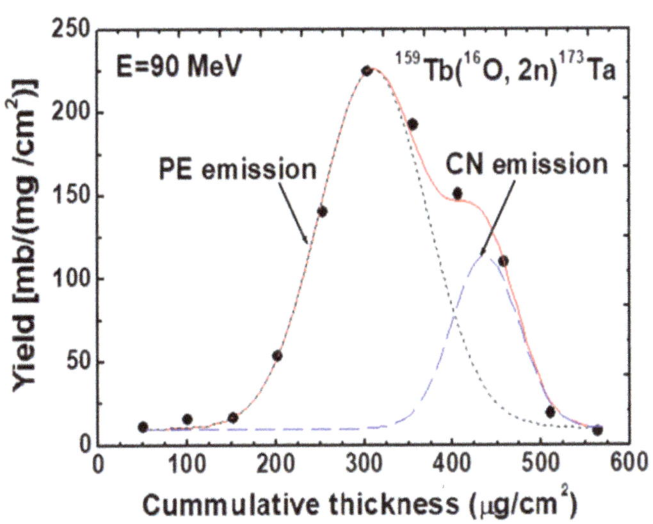

Figure 5.39. Measured recoil range distribution for the ^{173}Ta residues in the reaction ^{159}Tb(^{16}O,2*n*) at ≈90 MeV. In this figure Gaussian peak at higher cumulative thickness represents the CN contribution while the other Gaussian peak at lower thickness represents the contribution due to pre-equilibrium emission as detailed in the text (adapted from [66] with permission from American Physical Society).

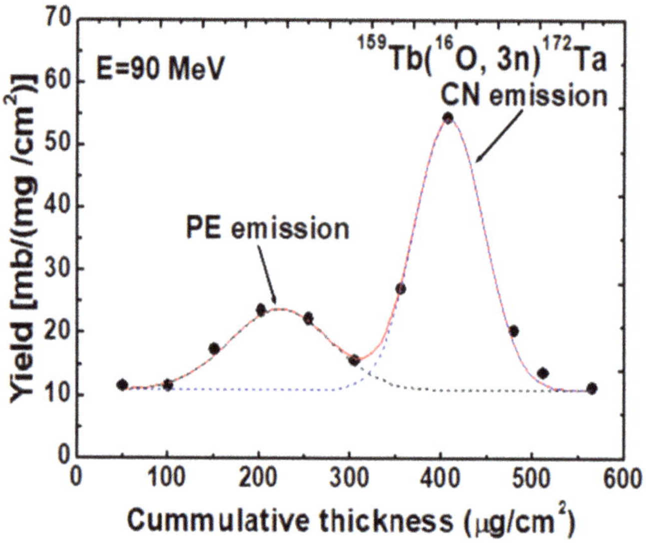

Figure 5.40. Experimentally measured recoil range distribution for the reaction $^{159}Tb(^{16}O,3n)^{172}Ta$ at energy ≈90 MeV. In this figure Gaussian peak at higher cumulative thickness represents the CN contribution while the other Gaussian peak at lower thickness represents the contribution due to pre-equilibrium emission (adapted from [66] with permission from American Physical Society).

reaction mechanisms, i.e., the compound and the pre-equilibrium emission of particles. As may be seen from these figures, the experimental range distribution data for the reactions $^{159}Tb(^{16}O,2n)^{173}Ta$ and $^{159}Tb(^{16}O,3n)^{172}Ta$ show two prominent peaks. One peak is at a relatively lower value (≈300 μg cm^{-2}) of cumulative catcher thickness while the other is at the higher range value ≈450 μg cm^{-2}. The peak at higher value (≈450 μg cm^{-2}) of cumulative thickness corresponds to the fraction of residues produced through the CN process and is consistent with the full momentum transfer events that are produced in the complete fusion reactions. On the other hand, the observed peak at relatively smaller range (≈300 μg cm^{-2}) may be attributed to the fact that the residues ^{173}Ta and ^{172}Ta may also be produced via the pre-equilibrium emission process. In such a case of a pre-equilibrium reaction, the emission of high-energy neutrons and/or protons may take place prior to the establishment of thermodynamical equilibrium taking away significant part of the linear momentum, leaving the residue with smaller linear momentum and hence smaller range in a stopping medium. The observed well-separated peaks in the recoil range distribution plots give a clear separation of compound and pre-equilibrium emission processes. Such a disentanglement of the two processes is direct experimental separation and there is no model dependency involved in it.

Similar measurements for the reaction channel $^{169}Tm(^{16}O,2n)^{183}Ir$ at 88 MeV and $^{181}Ta(^{16}O,2n)^{195}Tl$ at three energies, i.e., 81, 90 and 96 MeV [66], carried out using the recoil-catcher activation technique also indicate significant pre-equilibrium contributions that can be directly separated out on the basis of observed area of the corresponding peaks in the recoil range distribution plots. The measurements at three energies clearly indicate that the contributions of these processes are energy dependent and as the beam energy increases the pre-equilibrium component starts increasing while CN component decreases, as expected.

5.6 In-beam studies of the pre-equilibrium emission

5.6.1 Forward-to-backward yield ratios of reaction residues

Investigations on the pre-equilibrium-emission have also been made through in-beam particle-gamma coincidence technique by measuring forward-to-backward yield ratios [67] of the reaction products produced in the different de-excitation processes. In these in-beam experiments, the coincidences between prompt gamma-rays of the residue and the charged particles emitted in various angular (mostly forward and backward) zones have been recorded to obtain the yield of the reaction products. Several gating conditions viz., p-γ, α-γ i.e., particle (Z = 1, 2) gated gamma spectra have been projected onto the energy spectra to achieve the information about the particle emission via different de-excitation modes of reaction mechanism. Since, in the pre-equilibrium emission, emitted energetic particles are likely to be concentrated in the forward cone, the yield of residues corresponding to forward emitted particles is expected to be relatively large as compared to those emitted in the backward cone. In the case of CN emission, the isotropic emission of nuclear particles is expected and hence, the yield corresponding to such events may be nearly the same in the forward and backward directions. The enhanced forward cone yield, i.e., the difference of forward-to-backward yields may be attributed to the pre-equilibrium-emission.

With this motivation the forward-to-backward yield ratios have been obtained in the experiments performed at the Inter-University Accelerator Centre (IUAC), New Delhi, India. Isotopically pure (abundance ≈100%), self-supporting ^{169}Tm target of thickness ≈0.93 mg cm^{-2} was bombarded by $^{16}O^{+7}$ ≈ 5.6 MeV/nucleon energy with a beam current of ≈3–4 pnA. The target was placed at the center of the Charged Particle Detector Array (CPDA) set-up, in conjunction with the Gamma Detector Array (GDA) set-up [67], as shown in figure 5.41.

The GDA set-up consists of twelve Compton suppressed, high resolution HPGe detectors installed in three angular zone rings (i.e., 45°, 99°, 153°, four detectors in each ring) with respect to the direction of beam axis. While, the array of CPDA is an assembly of 14 Phoswich detectors arranged in two truncated hexagonal pyramid shapes. The bases of these pyramids lie in a horizontal plane with each other having trapezoidal shape. However, the top and bottom spaces are filled by two hexagonal detectors which together with trapezoids cover nearly 90% of the total solid angle. The array of fourteen charged particle detectors has been divided into three angular zones, i.e., (i) forward (F) 10°–60°, (ii) sideways (*S*) 60°–120° and (iii) backward (B) 120°–170° zones. Depending on the fast and slow components of the CPDs particles (a sum of protons and α's) and α's in each angular zone may be identified. To remove the scattered beam particles reaching the detectors, the charged particle detectors were covered by Al-absorbers of appropriate thickness. The coincidences were performed between particles (Z = 1, 2) and prompt γ-rays employing three gating conditions corresponding to a given angular zone for each value of Z of the

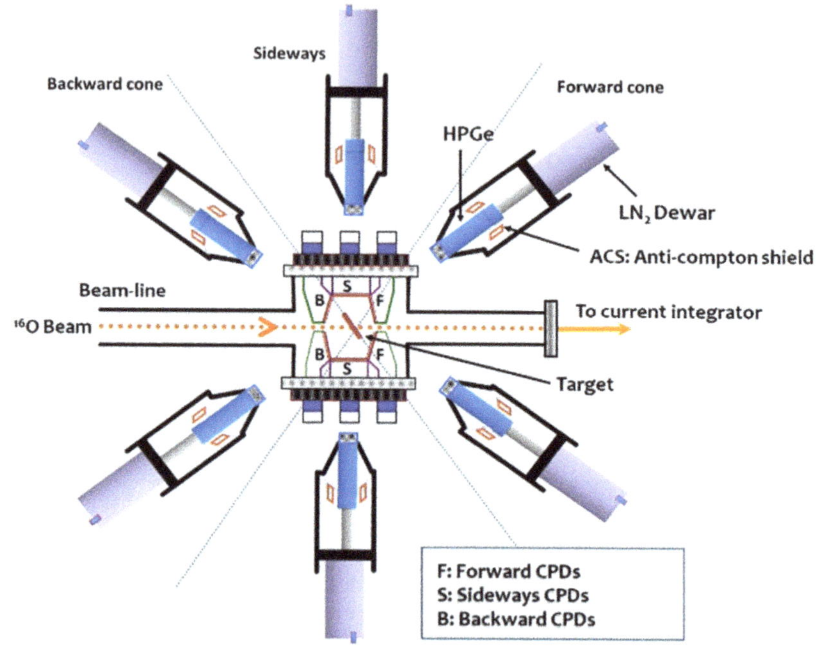

Figure 5.41. Schematic representation of the Charged Particle Detector Array (CPDA), and gamma detector array (GDA) set-ups used to record forward and backward yield of particles.

particle. Multi-parameter data in event-by-event mode have been collected. The intensity and area under the efficiency corrected photo-peak of characteristic prompt γ-transitions were used to determine the relative production yield for the observed spin (J_{Obs}) of different reaction products. The yields of different reaction products identified from forward (F), backward (B) and sideways (S) proton-gated spectra have been plotted as a function of observed spin (J_{Obs}). The yield of the reaction product ^{181}Os populated via *p*3*n*-channel and determined with proper efficiency corrections in forward (F), backward (B) and sideways (S) angular zones were plotted [67] as a function of (J_{Obs}). The errors from different factors in the experiment data were found to be less than 10%. It was observed [67] that the yield values for ^{181}Os identified from backward (B)-proton-gated spectra are almost similar to that identified from sideways (S)-proton-gated spectra within the experimental uncertainties. However, a significant enhancement in the yields of production residues is observed in forward (F)-proton-gated spectra. This enhancement of yield in the forward (F) cone has been attributed to the contribution coming from pre-equilibrium emission of particles. It is because of the fact that the particles emitted via the pre-equilibrium emission process may be focused predominantly in the forward cone, and are expected to show relatively high yield as compared to that formed via CN process. Further, the same residues populated via two de-excitation processes may have more than unity value in the forward-to-backward yield ratio. On the other hand, emission of particles by only one de-excitation process, i.e., CN emission may have unity value in the forward-to-backward yield ratio values.

As a representative case, the forward (F)-to-backward (B) yield ratios $R_{Y(F/B)}$ of different reaction products viz., ^{183}Re. ^{183}Re and ^{181}Re populated via $2p$, $2pn$, and $2p2n$ channels, respectively and the reaction products ^{182}W, ^{181}W and ^{179}W populated via $3p$, $3pn$ *and* $3p3n$ reaction channels, respectively, have been shown as a function of J_{Observed} in figure 5.42 and figure 5.43, respectively. As can be seen

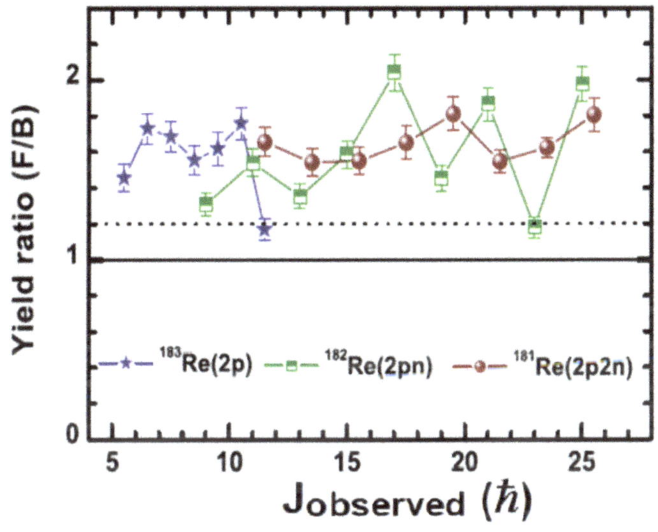

Figure 5.42. Experimentally measured forward-to-backward yield ratio $R_{Y(F/B)}$ of ^{183,182,181}Re residues as a function of spins of the residues indicating enhancement in the forward yield as compared to backward (adapted from [67] with permission from World Scientific).

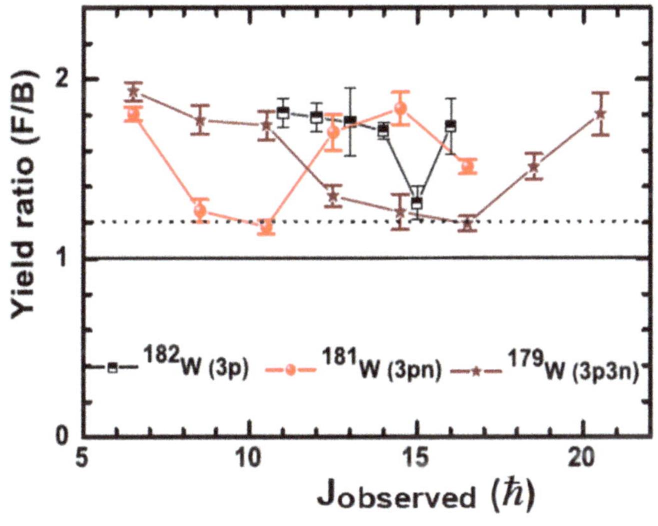

Figure 5.43. Experimentally measured yield ratios $R_{Y(F/B)}$ of ^{182,181,179}W residues as a function of observed spins indicating enhancement in the forward yield as compared to backward (adapted from [67] with permission from World Scientific).

from these figures, the $R_{Y(F/B)}$ is found to be higher than unity and shows a significant enhancement above horizontal dotted lines, which indicates the contribution from pre-equilibrium emission in the production of these reaction products. It may also be noticed that the enhancement in the yield ratios is large for higher values of J_{Observed}, in general. This seems to indicate the origin of different reaction products from both, the central or non-central (peripheral) interactions. In non-central interactions, the centrifugal field may increase the driving input angular momenta (l-values) imparted into the system [67]. As such, the composite(s) system formed at large l-values may undergo the emission of pre-equilibrium nucleons leaving behind the composite nucleus in high spin states.

5.6.2 Spin distribution and angular momentum

The knowledge of input angular momentum involved in the interaction process may provide important information to characterize the reaction mechanism. The angular momentum involved in the reaction may be obtained by measuring the spin distributions of residues produced in a given reaction channel. In the pre-equilibrium emission, the emitted particles take away a significant part of angular momentum; as such the angular momentum associated with the residual nuclei is expected to be relatively lower than that associated with the CN process. As a result, the reaction residues left after pre-equilibrium emission are expected to be populated with relatively less high spin states as compared to the spin states of the residues populated after CN emission. Therefore, the measurement of spin distribution may be one of the appropriate methods to decipher the pre-equilibrium from CN process. In order to determine the involvement of input angular momentum in the pre-equilibrium emission and the CN processes, the experiments based on particle-gamma coincidence technique have been performed at the IUAC, New Delhi, India [66]. Here, the population of spin states during de-excitation of reaction residues produced in the reactions ^{16}O+ ^{159}Tb and ^{16}O+ ^{169}Tm have been studied [66]. The details of the various components used in the experiment have already been discussed in the previous section. The system-wise details of the results obtained and their interpretation for the above-mentioned systems are given here.

In order to obtain the driving input angular momentum associated with pre-equilibrium reaction channel, a particle-γ coincidence experiment was performed for the ^{16}O+^{159}Tb system at $\approx$93 MeV and for ^{16}O+^{169}Tm system at 88 MeV. Spectroscopically pure, self-supporting thin natural ^{159}Tb and ^{169}Tm targets (purity 99.99%) of thickness $\approx$1.72 mg cm^{-2} and 0.93 mg cm^{-2}, respectively, were prepared by the rolling technique. In two separate experiments these targets were bombarded, respectively, with beams of ^{16}O of $\approx$95 MeV and $\approx$88 MeV. The experimental set-up consists of a GDA along with Charged Particle Detector Array (CPDA). As a representation the top flange of the CPDA set-up with photomultiplier tube (PMT) fitted are shown in figure 5.44, while the arrangement of complete CPDA set-up fitted in-beam line is presented in the picture shown in figure 5.45.

The GDA has 12 Compton suppressed, high resolution HPGe γ-spectrometers arranged at 45° (forward), 99° (side upper), and 153° (backward) angles with respect

Figure 5.44. A picture of the top flange of the CPDA set-up with PMT fitted.

to the beam direction and there are four detectors at each of these angles. The CPDA is a set of 14-phoswich detectors housed in a 14 cm diameter scattering chamber, covering nearly 90% of total solid angle. As a matter of fact, the CPDs at forward (F) angles are expected to detect both compound and pre-equilibrium components, while backward (B) angle detectors may detect only particles produced via CN process because of isotropic nature of emission in this process. Forward and backward particle gated gamma ray spectra were used to identify the reaction residues.

The values of relative production yields of the residues (observed area under the peak of the experimentally measured prompt gamma lines) have been obtained from the singles spectra and plotted as a function of observed spin J_{Observed} corresponding to prompt gamma-transitions. In order to get information about the involved input angular momenta and to investigate the entry state spin population of the reactions ^{169}Tm(^{16}O,2n)^{183}Ir and ^{159}Tb(^{16}O,2n)^{173}Ta the relative yields have been normalized [66] with minimum observed spin $J_{\text{Observed}}^{\text{Minimum}}$, at highest yield $Y_{\text{Observed}}^{\text{Maximum}}$. The experimentally measured spin distributions obtained from prompt γ-rays recorded in forward and backward directions for the reaction ^{169}Tm(^{16}O,2n)^{183}I are shown in figures 5.46 and 5.47, respectively.

Figure 5.45. A picture of the arrangement of CPDA inside the small chamber in the beam line.

It may be observed from these figures that the measured spin distributions and their decay patterns for reaction $^{169}Tm(^{16}O,2n)^{183}Ir$ obtained in the forward and backward directions are distinctly different. The distinctly different decay patterns in forward and backward directions indicate that the two processes are quite different from each other. A close inspection of these figures indicate that in the case of CN process, it is expected that higher spin states in the residues are populated giving rise to linearly increasing population of lower spins due to their feeding from the higher spin states. On the other hand, in the case of the pre-equilibrium process, residues are left with relatively lower excitation energy after emitting the energetic neutrons and hence with lower spin entry state, therefore, the feeding of lower spin sates is not linear like the CN process. The value of the mean input angular momentum involved in each case is obtained by the spin where the normalized yield falls to half of its maximum value obtained at lowest observed spin. The mean input angular momenta, i.e., the spin at half yield deduced from experimental spin distributions of figures 5.46 and 5.47 are found to be $\approx 8.5\hbar$ and $\approx 9.5\hbar$ in forward and backward directions, respectively, for the reaction $^{169}Tm(^{16}O,2n)^{183}Ir$. Similar observation [66] has also been made in the case of $^{159}Tb(^{16}O,2n)^{173}Ta$ reaction channel where the mean input angular momentum as obtained from the spin distributions was found to be $10.5\hbar$ and $11.5\hbar$ in forward and backward directions, respectively. The observed lower values of the mean input

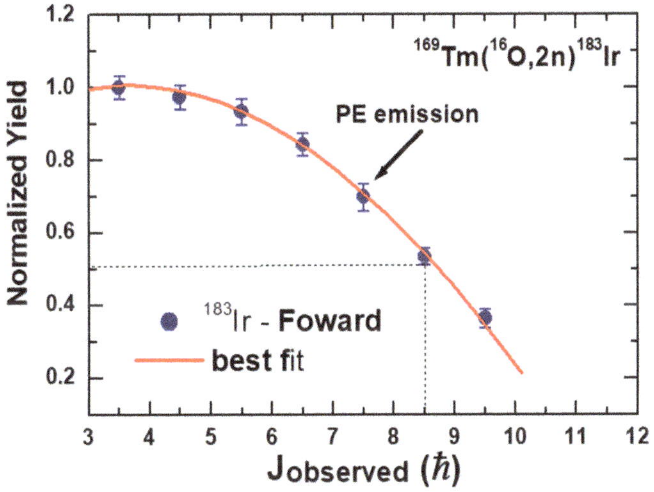

Figure 5.46. Experimentally measured spin distributions for reaction $^{169}Tm(^{16}O,2n)^{183}Ir$ in forward direction. The curve represents best fit to the measured spin distribution (adapted from [66] with permission from American Physical Society).

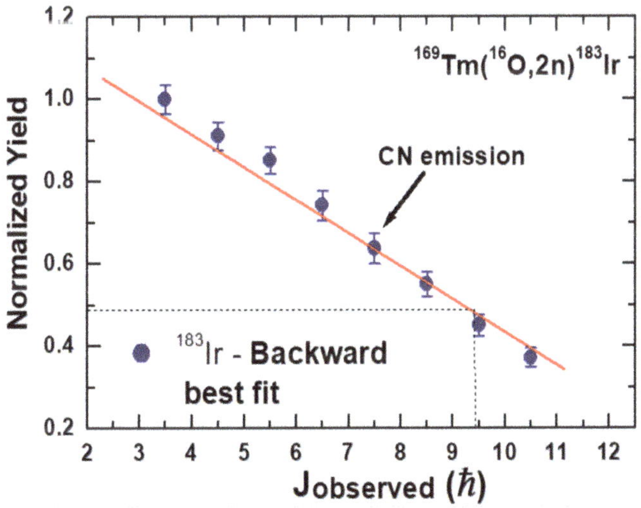

Figure 5.47. Experimentally measured spin distributions for reaction $^{169}Tm(^{16}O,2n)^{183}Ir$ in backward direction. The curve represents the best fit to the measured spin distribution (adapted from [66] with permission from American Physical Society).

angular momentum in forward direction are due to the fact that emission of pre-equilibrium particles takes away a significant part of the angular momentum. On the other hand, a relatively higher observed value of the mean input angular momentum in backward direction is because of the emission of two equilibrated neutrons which takes away relatively lesser angular momentum. As such, it may be observed that the measured spin distributions give direct evidence of the pre-equilibrium emission process. Thus, the results of the measurements of the spin distributions further supplement the conclusions drawn from the recoil range and EF measurements. Further, these measurements provide additional information on the mean input angular momentum associated with the pre-equilibrium and the CN processes as well.

As may be observed, from a large amount of data analyzed within the framework of compound and pre-equilibrium emission and presented above, that these reactions dominate in their respective energy regimes. At low energies the CN emission is important; however, as the energy of the incident particle increases the pre-equilibrium emission becomes dominant. Several types of experiments have been described to study the pre-equilibrium emission over a wide range of energy and projectile target systems viz., (i) the measurement of cross-sections in neutron induced reactions, (ii) the measurement of EFs for proton induced reactions over a large number of nuclei and wide energy region, (iii) the measurement of EFs in alpha induced reactions on a large number of target nuclei and over wide energy region, then (iv) the measurement of EFs in heavy-ion induced reactions involving beam of ^{7}Li, ^{11}B, ^{12}C, ^{14}N, ^{16}O, ^{19}F etc. In all these reaction systems a significant contribution of pre-equilibrium emission has been observed particularly at relatively higher energies. The other kind of measurements giving direct and irrefutable method to decipher the compound from pre-equilibrium emission in the case of heavy-ion collisions were of the recoil range distribution method. The details of these measurements were also presented. Here, the compound and pre-equilibrium emission contributions have been separated out on the basis of linear momentum involved in the reactions. The observation of larger number of emitted nuclear particles in the forward cone were also used to indicate the observation of pre-equilibrium emission through the particle-gamma coincidence experiment involving the charge particle detector coupled to the gamma detector array systems. The mean input angular momentum involved in the CN and pre-equilibrium reactions has also been obtained using the spin distributions measured in particle-gamma coincidence experiments. In the case of pre-equilibrium emission, the mean input angular momentum was found to be less than that in case of CN emission, as expected. In conclusion, it may be mentioned that a variety of parameters may be used to describe a given reaction dynamics of pre-equilibrium emission. In the next chapter, an attempt has been made to parameterize the pre-equilibrium emission that may be useful for the study of such reactions.

Bibliography

[1] Bohr N 1936 Neutron capture and nuclear constitution *Nature* **137** 344

[2] Ghoshal S N 1950 An experimental verification of the theory of compound nucleus *Phys. Rev.* **80** 939

[3] Koning A J and Akkermans J M 1993 Computational comparison of quantum-mechanical models for multistep direct reactions *Phys. Rev.* C **47** 724

[4] Dimitrova S S, Cowley A A, Zemlyanaya E V and Lukyanov K V 2014 Pre-equilibrium mechanisms in the ^{93}Nb(p,α) inclusive reaction at incident energies from 65 to 160 MeV *Phys. Rev.* C **90** 054604

[5] Wu J R, Chang C C and Holmgren H D 1979 Charged-particle spectra: 90 MeV protons on ^{27}Al, ^{58}Ni, ^{90}Zr, and ^{209}Bi *Phys. Rev.* C **19** 698

[6] Awes T C, Poggi G, Gelbke C K, Back B B, Glagola B G, Breuer H and Viola V E 1981 Pre-compound emission of light particles in the reaction $^{16}O+^{238}U$ at 20 MeV/nucleon *Phys. Rev.* C **24** 89

[7] Sharma M K, Kumar M, Shuaib M, Sharma V R, Yadav A, Singh P P, Singh B P and Prasad R 2018 Systematic study of precompound neutron emission in α-particle-induced reactions *Phys. Rev.* C **98** 054607

[8] Kishimoto T, Sasao M, Suzuki H, Sakai H and Ejiri H 1980 Spin alignment of yrast level populated by pre-equilibrium (α, xnγ) reactions *Nucl. Phys.* A **337** 493

[9] Sakai H, Shimizu A, Ejiri H, Shibata T, Okada K, Kishimoto T and Maeda K 1980 Pre-equilibrium process for ^{175Lu}(*p*, xn γ), ^{176}Yb(^{3}He, xn yp γ), and ^{174,176}Yb(*α*, xn γ) reactions *Phys. Rev.* C **22** 002

[10] Luo J, Jiang L and Li S 2017 Activation cross section and isomeric cross section ratios for the (*n*,2n) reaction on 153Eu *Phys. Rev.* C **96** 044617

[11] Bharud V D, Attar F M D, Dahiwale S S, Dhole D and Bhoraskar N 2019 Cross sections for formation of ^{139m}Ce radioisotope through the ^{140}Ce(n, 2n) reaction over 13.73–14.77MeV neutrons *Appl. Radiat. Isot.* **146** 10

[12] Koning A J, Hilaire S and Goriely S 2015 *TALYS-1.8. Nuclear Reaction Program, Nuclear Data Section* (Vienna: IAEA)

[13] Herman M *et al* 2005 *EMPIRE-II, Nuclear Reaction Model Code 3.2, Nuclear Data Section* (Vienna: IAEA)

[14] Attar F M D, Dhole S D and Bhoraskar V N 2014 Cross sections of the (*n*,*p*) reaction on the ^{78}Se and ^{80}Se isotopes measured for 13.73 MeV to 14.77 MeV and estimated for 10 MeV to 20 MeV neutron energies *Phys. Rev.* C **90** 064609

[15] Pandey B, Agrawal H M and Pepelnik R 2014 . Neutron activation cross-sections for ytterbium isotopes at (14.6±0.3) MeV *Appl. Radiat. Isot* **85** 128

[16] Birn *et al* 1995 Isomeric cross-section ratios for the formation of 75Gem,g in (*n*,*p*), (*n*, *n*), and (*n*,2n) reactionsfrom 6 to 15 MeV, *Phys. Rev.* C **52** 2546

[17] Uhl M and Strohmair B 1981 Report IRK 76/01, NEA Data Bank. France

[18] Hauser W and Feshbach H 1952 The inelastic scattering of neutrons *Phys. Rev.* **87** 366

[19] Griffin J J 1966 Statistical model of intermediate structure *Phys. Rev. Lett.* **17** 478

[20] Herman M, Marcinkowski A and Stankiewicz K 1984 Statistical multi-step compound emission in (*n*,2n) reactions, *Nucl. Phys.* A **430** 69

[21] Gupta J P, Bhardwaj H D and Prasad R 1985 Pre-equilibrium emission effect in (*n*, *p*) reaction cross-sections at 14.8 MeV *Pramana* **24** 637

[22] Bhardwaj H D 1985 *PhD Thesis* A. M.U., Aligarh. India unpublished

[23] Bevilacqua R, Pomp S and Hayashi M *et al* 2011 Study of pre-equilibrium emission of light complex particles from Fe and Bi induced by intermediate energy neutrons, *J. Phys. Conf. Ser.* **312** 082013

[24] Singh B P *et al* 2006 A study of pre-equilibrium emission in some proton and alpha-induced reactions *Nucl. Inst. Meth. Phys. Res.* A **562** 717–20

[25] Musthafa M M, Kumar Sharma M, Singh B P and Prasad R 1995 Measurement and analysis of cross sections for (*p*,*n*) reactions in ^{51}V and ^{113}In *Appl. Radiat. Isot.* **62** 419–28
Musthafa M M *et al* 2005 Semi classical and quantum mechanical analysis of the excitation function for the ^{130}Te(*p*, *n*)^{130}I reaction *Phys. Rev. C* **52** 3174

[26] Blann M 1991 *Report PSR-146, NEA Data Bank, Gif-sur-Yvette* (France)

[27] Kalka H 1991 *EXIFON. A Statistical Multistep Reaction Code, NEA Data Bank* (Saclay. France)

[28] Weisskopf V F and Ewing D H 1940 On the yield of nuclear reactions with heavy elements *Phys. Rev.* **57** 472
[29] Blann M 1971 Hybrid model for pre-equilibrium decay in nuclear reactions *Phys. Rev. Lett.* **27** 337
[30] Singh B P and Prasad R 1995 Measurement and analysis of excitation functions for the reactions 165Ho(α,xn) (x = 1–3) in the energy range N = 10–40 MeV *Phys. Scr.* **51** 440
[31] Wilmore D and Hodgson P E 1964 The calculation of neutron cross-sections from optical potentials *Nucl. Phys.* **55** 673
[32] Perey C M and Perey F G 1976 Compilation of phenomenological optical-model parameters 1954–1975 *At. Data Nucl. Data* **17** 1
[33] Huizenga J R and Igo G 1962 Theoretical reaction cross sections for alpha particles with an optical model *Nucl. Phys.* **29** 462
[34] Kalbach-Cline C, Huizenga J R and Vonach H K 1974 Isospin conservation and pre equilibrium decay in (p, ṕ) reactions on neutron rich tin isotopes *Nucl. Phys.* A **222** 405
[35] Coryell C D 1953 Beta-decay energetics *Annu. Rev. Nucl. Sci.* **2** 331
[36] Parashari S, Mukherjee S, Nayak B K, Makwana R, Suryanarayana S V, Naik H and Sharma S C 2018 Excitation function of the p + ^{93}Ag reactions in the energy range 10–22 MeV *Nucl. Phys.* A **978** 160
[37] http://srim.org
[38] Holbling S, Caplar R, Stamer S, Langkau R and Scobel W 1991 Preequilibrium emission in *(p, xn)* reactions on palladium isotopes *Z. Phys. A (Hadrons Nucl.)* **338** 11
[39] Sharma M K, Musthafa M M, Shuaib M, Kumar M, Sharma V Raj, Yadav A, Singh P P, Singh B P and Prasad R 2019 Semiclassical and quantum mechanical analysis of α-particle-induced reactions on praseodymium: a study relevant to precompound emission *Phys. Rev.* C **99** 014608
[40] Gavron A 1980 Statistical model calculations in heavy ion reactions *Phys. Rev.* C **21** 230
[41] code http://lise.nscl.msu.edu/pace4
[42] Bass R 1977 Nucleus-nucleus potential deduced from experimental fusion cross sections *Phys. Rev. Lett.* **39** 265
[43] Afzal Ansari M, Sathik N P M, Singh D and Rasid M H 2005 Measurement and analysis of alpha particle induced reactions on praseodymium *J. Phys. Soc. Japan* **74** 1150
[44] Sharma M K, Bhardwaj H D, Unnati, Singh P P, Singh B P and Prasad R 2007 A study of pre-equilibrium emission of neutrons in ^{93}Nb(α, xn) reactions *Eur. Phys. J.* A **31** 43
[45] Yadav A, Singh P P, Sharma M K, Singh D P, Unnati, Singh B P and Prasad R 2008 Large pre-equilibrium contribution in α + ^{nat}Ni interactions at ≈8–40 MeV *Phys. Rev.* C **78** 044606
[46] Singh B P, Sankaracharyulu M G V, Afzal Ansari M, Bhardwaj H D and Prasad R 1991 Equilibrium and preequilibrium emission in some α-induced reactions on enriched isotopes ^{128,130}Te at moderate excitation energies, *Phys. Rev.* C **47** 2055
[47] Fields C A, De Boer F W N, Ristinen R A, Smith P A and Sugarbaker E 1982 Systematic investigation of the (α, 2n γ) reaction on medium-heavy nuclei *Nucl. Phys.* A **377** 217
[48] Ejiri H, Itahashi T, Nagai Y, Shibata T, Nakayama S, Kishimoto T, Maeda K, Sakai H and Hoshi M 1978 Gamma rays and neutrons following preequilibrium ^{165}Ho(p, $x$cn$y$pγ) reactions at E_p = 60 MeV *Nucl. Phys.* A **305** 167
[49] Kumar D, Maiti M and Lahiri S 2016 Experimental probe for the production of ^{97}Ru from the ^{7}Li+^{93}Nb reaction: a study of precompound emissions *Phys. Rev.* C **94** 044603
[50] Prajapat R and Maiti M 2020 Measurement of residual cross sections in the ^{7}Li + ^{89}Y reaction up to 5.7 MeV/nucleon and analysis of fusion mechanisms *Phys. Rev.* C **101** 024608

[51] Prajapat R, Maiti M and Kumar D 2021 Fusion-fission events in the ^{11}B+^{181}Ta reaction up to 5.7 MeV/nucleon energy *Phys. Rev.* C **103** 014608

[52] Lenske H and Wolter H H TRISTAN and ORION codes private communication

[53] Chadwick M B DDHMS code private communication

[54] Gilbert A and Cameron A G W 1965 A composite nuclear-level density formula with shell corrections *Can. J. Phys.* **43** 1446

[55] Ignatyuk A V, Weil J L, Raman S and Kahane S 1993 Density of discrete levels in ^{116}Sn *Phys. Rev.* C **47** 1504

[56] D'Arrigo A, Giardina G, Herman M, Ignatyuk A V and Taccone A 1994 Semi-empirical determination of the shell correction temperature and spin dependence by means of nuclear fission *J. Phys.* G **20** 365

[57] Sharma M K, Singh P P, Singh D P, Yadav A, Sharma V R, Bala I, Kumar R, Unnati, Singh B P and Prasad R 2015 Systematic study of pre-equilibrium emission at low energies in ^{12}C- and ^{16}O-induced reactions *Phys. Rev.* C **91** 014603

[58] Chakrabarty S, Tomar B S, Goswami A, Gubbi G K, Manohar S B, Sharma A, Bindukumar B and Mukherjee S 2000 Complete and incomplete fusion reactions in the ^{12}C + ^{169}Tm *Nucl. Phys.* A **678** 355–66

[59] Vergani P, Gadioli E, Vaciago E, Fabrici E, GadiohErba E and Galmarini M Complete and incomplete fusion and emission of preequilibrium nucleons in the interaction of ^{12}C with^{197}Au below 10 MeV/nucleon *Phys. Rev.* C **48** 1815 1993

[60] Thomas T D, Gordon G E, Latimer R M and Seaborg G T 1962 Spallation-fission competition in astatine compound nuclei formed by heavy-ion bombardment *Phys. Rev.* **126** 1805

[61] Bimbot R, Lefort M and Simon A 1968 Nuclear reactions induced by heavy ions: study mechanism of de-excitation of fissile compound nuclei and treatment of high angular momentum *J. Phys.* **29** 563

[62] Baba S *et al* 1988 Evaporation residue formation competing with the fission process in the ^{197}Au+^{16}O,^{12}C reactions and fission barriers at a specified, J window *Z. Phys.* A **331** 53

[63] Cervesato I, Fabrici E, Gadioli E, GadioliErba E and Galmarini M 1992 Light particle emission in Boltzmann master equation theory of pre-equilibrium reactions *Phys. Rev.* C **45** 2369

[64] Shariq Asnain M *et al* 2021 Effect of non-alpha cluster projectile on incomplete fusion dynamics: An experimental study of ^{14}N+^{181}Ta system *Phys. Rev.* C **104** 034616

[65] Shuaib M, Sharma V R and Yadav A *et al* 2017 Influence of incomplete fusion on complete fusion at energies above the Coulomb barrier *J. Phys. G Nucl. Part. Phys.* **44** 105108

[66] Sharma M K, Singh P P, Raj Sharma V, Shuaib M., Singh D P, Yadav A, Unnati, Kumar R, Singh B P and Prasad R 2016 Pre-compound emission in low-energy heavy-ion interactions from recoil range and spin distributions of heavy residues: a new experimental method *Phys. Rev.* C **94** 044617

[67] Yadav A, Singh P P, Sharma V R, Singh D P, Unnati, Sharma M K, Singh B P and Prasad R 2011 Signature of pre-equilibrium emission in forward-to backward yield ratio measurement *Int. J. Mod. Phys.* E **20** 2133

[68] Dilg W, Schantl W, Vonach H and Uhl M 1973 Level density parameters for the back-shifted fermi gas model in the mass range $40 < A < 250$ *Nucl. Phys.* A **217** 269

Chapter 6

Data analysis, parameterisation of pre-equilibrium fraction

Analysis of experimental data on pre-equilibrium emission is presented in this chapter. In most of the reactions induced both by heavy and light ions and at almost all energies, it has been observed that the reaction cross-section may be split into two distinct components; the compound and the pre-equilibrium component. An attempt is made here to develop systematics for pre-equilibrium component with various parameters of the interacting system. The analysis is expected to further enhance the understanding of the complex process of pre-equilibrium emission.

6.1 Introduction

In a typical nuclear reaction, when a projectile nucleus of suitable energy hits a stationary target nucleus, amalgamation of the two nuclei may take place forming a composite nucleus. The energy and angular momentum of the incident particle is shared among all the nucleons of the composite system, and the system moves towards the establishment of thermodynamic equilibrium. Once the state of thermodynamic equilibrium is attained, the accumulation of sufficient energy on a single nucleon or on a cluster of nucleons may occur in a random sequence of events and hence may require much longer emission times, favouring the evaporation of particles. However, it is likely that during the equilibration process depending on whether an amount of energy more than the separation energy of a particular nucleon/cluster of nucleons is concentrated on it, it may get emitted well before the equilibration of the compund nucleus (CN) takes place. Both the intuition and the results of experiments have indicated that emission of such nuclear particles may take place even before the equilibrium of the CN gets materialised. Such emission of nuclear particles is referred to as the pre-equilibrium emission and the particles emitted are called pre-equilibrium particles [1]. Some of the important signatures of such pre-equilibrium emission reactions are the observation of high energy particles

doi:10.1088/978-0-7503-5077-8ch6

as compared to that predicted by CN models, forward peaked angular distribution of emitted particles, observation of high energy tail portion in the excitation function of the measured excitation functions, stretched particle distribution in angular momentum space etc. As such, these characteristics may be used to understand the dynamics of pre-equilibrium reactions. The measurement of reaction time is the most direct and irrefutable method of getting information about the reaction mechanism, but the time scales involved in the nuclear reactions are very short, typically $\approx 10^{-16} - 10^{-21}$ s, and cannot be measured directly with the presently available electronics and the data acquisition systems. Theoretical models are therefore proposed and used to explain the reaction mechanism. Although the CN [2] and the direct reaction theories are more or less well established, recently various semi-classical [3] as well as quantum mechanical theories [4] have also been put forward to explain the pre-equilibrium processes which are intermediate between the two extremes.

The quantum mechanical calculations, at present, have mostly been performed for the nucleon induced reactions because for a complex particle (e.g., α-particle) the quantum mechanical description of initial projectile–target interaction becomes very much complex. Also, the multiparticle emission, i.e., the cases with more than one particle in the continuum have not been treated in the quantistic approach as yet. No doubt, the strong complications presented by such cases of multiparticle emission would make these calculations practically impossible to be carried out. Otherwise, one should use approximations, therefore, approaching the philosophy of the semi-classical models. Considering the above facts, it may be concluded that for such systematic calculations the quantistic approach, if not impossible is impracticable as has also been mentioned by Gruppelaar *et al* [5], As such, most of the analysis of the measured excitation functions is, generally, performed within the framework of semi-classical descriptions. A brief account of these models has already been presented in chapter 4.

As mentioned above, a typical reaction proceeds with the formation of a CN after the amalgamation of projectile and the target nucleus. Once the CN is formed it has some excitation energy E^*_{CN} and the angular momentum l_{CN} equal to the angular momentum in the entrance channel. The excitation energy of the CN depends on the incident energy as well as on the type of projectile–target combination. The excitation energy of the CN E^*_{CN}, is given as;

$$E^*_{\mathrm{CN}} = E^{\mathrm{CM}}_{\mathrm{incident}} + Q \tag{6.1}$$

The first term on the right-hand side $E^{\mathrm{CM}}_{\mathrm{incident}}$ is the center of mass energy of collision while the second term is the Q-value of the reaction. In the case where both the incident particle as well as target nuclei are relatively heavy, it is likely that the composite system formed may be somewhere near the super heavy region. Thus, heavy-ion collisions may give the possibility of forming super heavy nuclei. It may be mentioned that the fission barriers of super heavy nuclei are rather low and, thus, fission is the dominant decay process of excited CN. To survive in competition with fission, an excited super heavy nucleus may evaporate light particles (mostly neutrons) and emit several γ-quanta which are likely to remove a significant amount of excitation energy and angular momentum. This is possible from pre-equilibrium emission of particles that takes away a significant amount of energy leaving behind the residual nucleus with relatively less excitation energy, and may not undergo fission. Thus, heavy nuclei so formed are likely to be populated with a reasonable cross-section due to such processes. As such, pre-equilibrium emission may be a route to populate super heavy elements (SHE).

6.2 Pre-equilibrium fraction

Several types of experiments may be used to study the dynamics of pre-equilibrium emission in nuclear reactions. Many experimental techniques have been developed to obtain and detect neutrons and charged particles of different energies and to measure the cross-sections of different particle-induced reactions. Broadly, two different types of experiments viz., off-beam and in-beam experiments are used to measure cross-sections and study such reactions. The magnitude of pre-equilibrium contribution in a reaction depends not only on the incident beam energy but also on the projectile–target combination. Since equilibrium and pre-equilibrium emissions compete with each other it is, in general, difficult to discern the sole contribution of these processes in an experiment. One of the ways to obtain the individual contribution due to pre-equilibrium emission is to compare the measured cross-section with the predictions of the CN model. The enhancement in the measured cross-section with respect to that predicted by CN model may be taken as the sole contribution due to pre-equilibrium emission. Such analysis was initially done mostly for neutron induced reactions, where a significant pre-equilibrium emission was observed. Because most of the earlier data for neutron induced reactions was available at ≈14 MeV, the experimentally measured and theoretically calculated cross-sections using CN models were compared. The enhancement in the measured cross-section was attributed [6] as an observation due to pre-equilibrium emission, as the direct reactions are unlikely at these energies. On the other hand, in the case of charged particle-induced reactions, the cross-sections were measured over a broad energy range, and the pre-equilibrium contribution was found to be significant and distributed over relatively higher energies. Hence, in order to get energy dependence of the pre-equilibrium emission such difference is obtained at each energy from the measured and the theoretically predicted calculated excitation function due to CN models. This is mostly done for proton, alpha, and heavy-ion induced reactions. The measured excitation function is first reproduced with theoretical calculations using the pre-equilibrium models [7, 8] including the compound contributions. A separate calculation is then done using the CN model. The difference of the two is taken to be as the sole contribution due to pre-equilibrium emission. As a typical example, such difference in measured excitation function is indicated by vertical arrows, and shown in figure 6.1, at different energies.

As such, one defines what is referred to as the pre-equilibrium fraction f_R, which gives the relative strength or contribution of pre-equilibrium emission with respect to

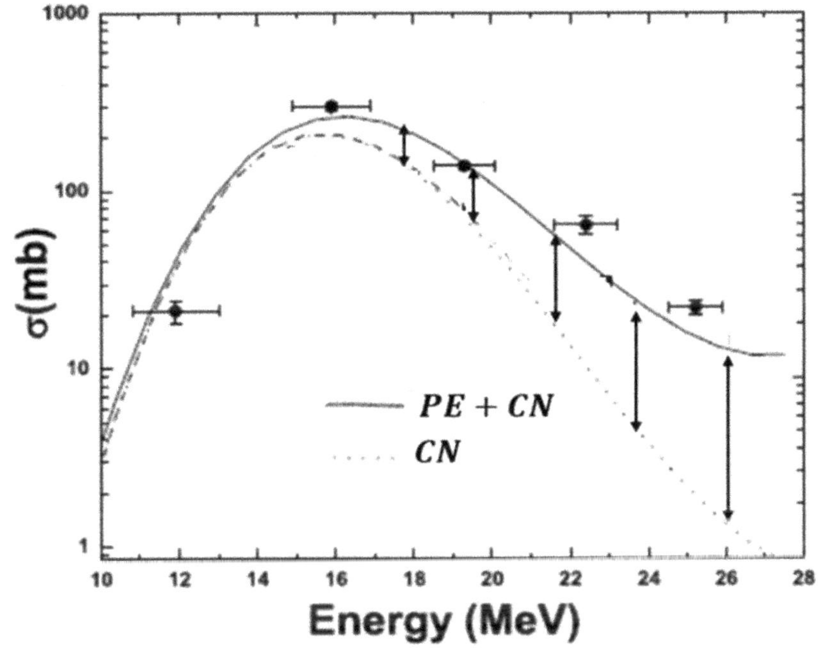

Figure 6.1. Typical representation of pre-equilibrium (PE) contribution deduction from the measured excitation function (solid circles) as the difference of measured cross-sections simulated with PE + CN calculations minus the CN contribution obtained theoretically. Such a PE contribution is represented with vertical arrows in the figure.

the sum of compound and pre-compound emission. The pre-equilibrium fraction f_R, at a given energy may be obtained as;

$$f_R = \frac{\sigma(\exp)_{(CN+PE)} - \sigma_{(CN)}}{\sigma_{(CN+PE)}} \tag{6.2}$$

The value of $\sigma(\exp)_{(CN+PE)}$ is taken from the best fit of the experimental data over the entire energy range using the theoretical calculations which includes both the CN as well as pre-equilibrium emission into consideration consistently with the same set of parameters of the code. However, the $\sigma_{(CN)}$ values are the calculated values of cross-section employing the pure CN model.

6.3 Energy dependence in light-particle-induced reactions

6.3.1 In-proton-induced reactions

As already mentioned, the comparison of theoretically calculated and experimentally measured excitation functions, shows that pre-equilibrium emission must be taken into consideration in the theoretical calculations to reproduce the experimental data. The pre-equilibrium fraction f_R, calculated, using code ACT [9], which is based on the lines of code STAPRE [10] from the best theoretical fits to the measured excitation functions, are plotted for the p + ^{51}V, ^{65}Cu, ^{115}In, ^{130}Te, and

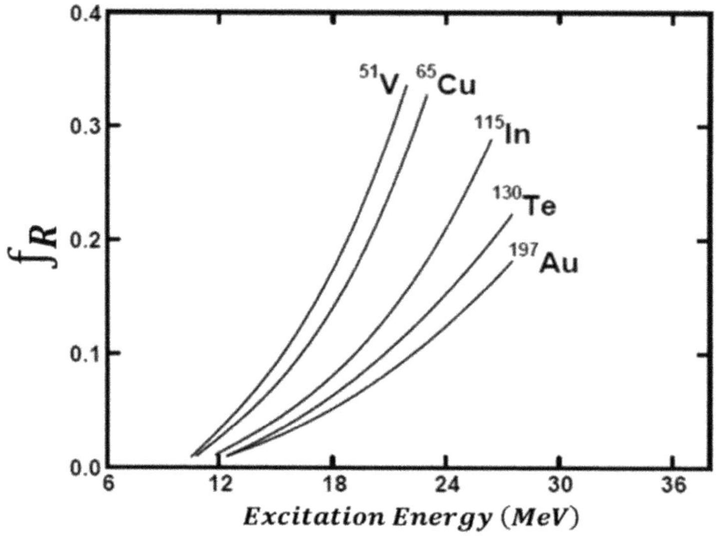

Figure 6.2. Variation of deduced pre-equilibrium fraction f_R, as a function of the excitation energy for proton induced reactions on different targets.

^{197}Au systems, against the excitation energy of the CN in figure 6.2. It is interesting to note from this figure 6.2, that the pre-equilibrium contribution starts building up from the excitation energy of ≈10–12 MeV for almost all the cases presented here for proton induced reactions [11].

Further, as expected, the pre-equilibrium fraction is found to increase with excitation energy, and hence with beam energy. Further, the probability of pre-equilibrium emission at a given excitation energy is larger in lighter nuclei as compared to the heavier ones. This may be due to the reason that at a given excitation energy more energy per nucleon is available in lighter systems, while in heavier systems due to relatively large number of available nucleons, per particle energy distribution is less, resulting in relatively small pre-equilibrium contribution.

6.3.2 In α-induced reactions

In order to study the energy dependence of pre-equilibrium emission, the pre-equilibrium fractions may be deduced at each energy as detailed above. Although, as shown in figure 6.1, the pre-equilibrium fraction in the case of individual reaction channels is possible to obtain easily, in the calculations done with code ACT [9], the total pre-equilibrium contribution as obtained for all the allowed channels may be deduced, similar to that mentioned above for alpha-induced reactions. The pre-equilibrium fractions have been deduced using the code ACT [9] for alpha-induced reactions on several targets. The f_R is taken to be as the fraction of particles from the first CN, which are emitted during the equilibration process. The computed pre-equilibrium fractions, as a function of incident α-particle energy, required to reproduce the experimental excitation functions for some of the target nuclei studied [12] are shown in figure 6.3.

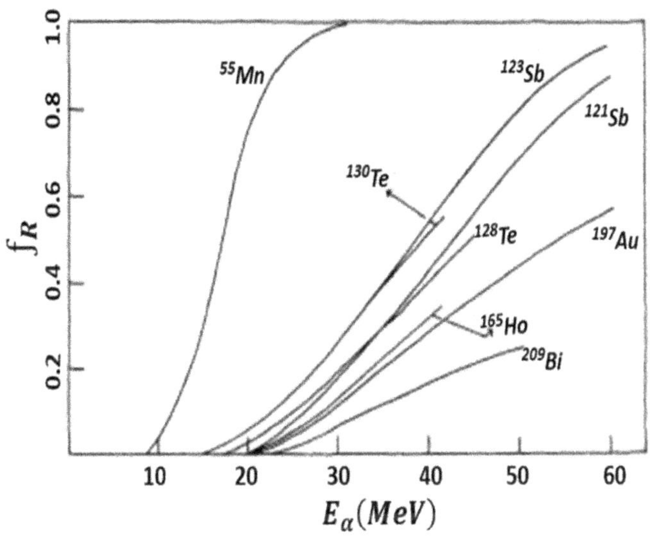

Figure 6.3. Deduced pre-equilibrium fractions as a function of alpha particle energy E_α.

It may be observed from this figure, that the fraction of pre-equilibrium emission increases rapidly with incident α-particle energy. Also, it may be inferred that the f_R tend to reach towards saturation value for the target nuclei under consideration but at different beam energies. The pre-equilibrium fraction is found to rapidly reach towards unity value at lower value of E_α in the systems involving lower atomic mass number (A). Thus, in lighter nuclei the reaction proceeds mainly through pre-equilibrium emission as the beam energy increases. In cases of antimony (Sb) and tellurium (Te) targets where two enriched isotopes of each were studied, the threshold E_α energy for pre-equilibrium emission is found to be different for different isotopes and being lower, respectively, for the isotope of higher mass number. This observation may be due to: (i) the relatively larger value of neutron excess ($A-Z$) and lower value of Coulomb barrier V_b, for the systems involving higher atomic mass number as compared to the systems of lower atomic mass number; and (ii) the single particle state density 'g' which is obtained from the level density parameter 'a' which in turn for each target was taken from the values of Dilg *et al* [13], in the calculations done with code ACT. Further, the internal transition rates for two-body interactions in the pre-equilibrium formalism have been calculated using expression $\overline{|M|}^2 = F_M A^{-3} E^{-1}$. Since, the matrix element is taken to be atomic mass number dependent, the pair creation rates depend on the mass number. As such, the pre-equilibrium emission in the heavier isotope may set in at a relatively lower energy than for the lighter isotope. From figure 6.2, it may be observed that the f_R at a given beam energy and its rate of increase are, in general, higher for target of lower atomic number. Further, the variation of f_R with incident alpha particle energy observed in figure 6.3, is markedly different for two extreme cases of manganese and bismuth isotopes.

6.3.2.1 Excitation energy per nucleon dependence

In the case of pre-equilibrium process only a few nucleons are assumed to take part. It may be assumed that in the initial stages of the de-excitation process, total available excitation energy may get distributed among all the nucleons of the compound system. As such, excitation energy per nucleon may play an important role in pre-equilibrium emission and may give a better systematics. To test the role of the average excitation energy per nucleon the calculated f_R values are plotted as a function of excitation energy per nucleon E^*/A, of the compound system in figure 6.4. It may be seen from this figure that there is substantial decrease in the spread of curves for different targets, particularly for manganese and bismuth isotopes which supports the above assumption. It is observed from this figure that pre-equilibrium emission sets in at a lower value of excitation energy per nucleon for heavier nuclei, while it sets in at relatively higher value. Further, the trend of pre-equilibrium fraction increasing monotonically with excitation energy per nucleon is observed.

6.3.2.2 Coulomb barrier and surface nucleon dependence

In the case of pre-equilibrium emission, it may be mentioned that the actual energy which is shared among the nucleons may not be the total excitation energy, because a part of the incident energy may be used up in overcoming the Coulomb barrier at the time of interaction. As such, the excitation energy in excess of the Coulomb barrier $\left(E^*-E_{V_b}\right)$ may be distributed among the nucleons of the CN.

Further, it is reasonable to assume that the available excess energy $\left(E^*-E_{V_b}\right)$ is initially shared with the nucleons present at the surface of the CN, some of which may be emitted as pre-equilibrium particles in the initial stage of the reaction. Also, the number of surface nucleons is proportional to the surface area of the nucleus which in turn is proportional $A^{2/3}$. As such, to observe the role of this parameter, the

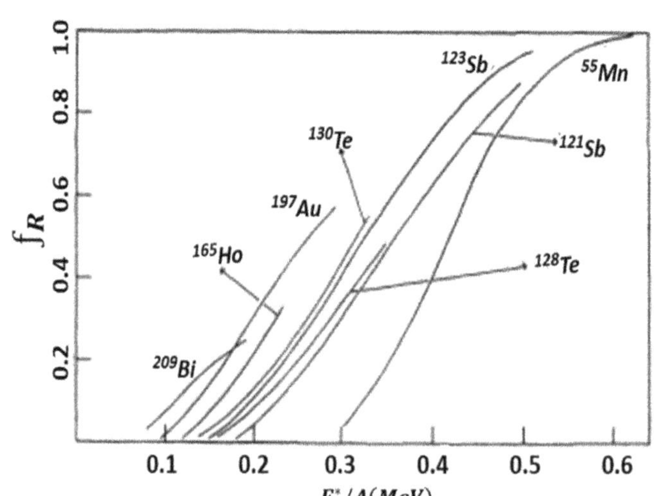

Figure 6.4. Variation of pre-equilibrium fraction as a function of excitation energy per nucleon E^*/A(MeV) of the compound system.

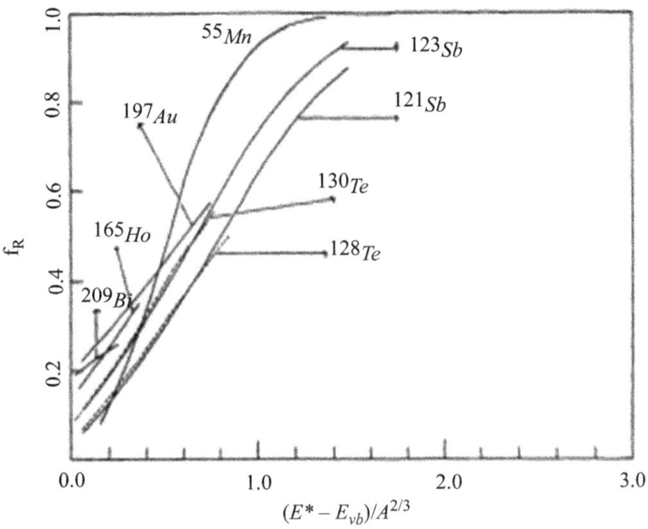

Figure 6.5. Variation of pre-equilibrium fraction as a function of excitation energy above Coulomb barrier per surface nucleon, $(E^*-E_{V_b})/A^{\frac{2}{3}}$.

pre-equilibrium fraction values are plotted as a function of excitation energy above the Coulomb barrier per surface nucleon $(E^*-E_{V_b})/A^{\frac{2}{3}}$. Considerable reduction in the spread of data in figure 6.5 compared to figure 6.4 indicates a strong dependence of pre-equilibrium fraction on factor $(E^*-E_{V_b})/A^{\frac{2}{3}}$, i.e., on excess excitation energy above Coulomb barrier and surface nucleons. It may also be derived from the above discussion that nucleons on the periphery of the CN are more likely to be emitted as pre-equilibrium particles. The above description is reasonable as the transmission coefficient of incident α-particle depends on the Coulomb barrier and the process of energy distribution starts from the surface nucleons.

6.4 Dependence of relative magnitudes of f_R and f_{CN} on energy

In order to observe the relative magnitude of pre-equilibrium emission fraction defined by f_R with respect to CN emission fraction (f_{CN}), the ratio of the two may be estimated at each energy. The CN emission fraction (f_{CN}) is defined as the relative strength or contribution of CN emission with respect to the sum of compound and pre-compound emission. The CN emission fraction f_{CN}, at a given energy may be obtained as;

$$f_{CN} = \frac{\sigma_{(CN)}}{\sigma_{(CN+PE)}} \times 100\% \tag{6.3}$$

The value of $\sigma_{(CN+PE)}$ is taken from the best fit of the experimental data over the entire energy range using the theoretical calculations, which includes both the CN as well as pre-equilibrium emission, into consideration consistently with the same set of parameters of the code. However, the $\sigma_{(CN)}$ values are the pure CN model calculations. The f_R/f_{CN} values are plotted as a function of incident α-particle energy in figure 6.6, for several targets. It may be seen from this figure that the ratio f_R/f_{CN}, in general, increases rapidly with energy for all the systems presented, with relative magnitudes which are quite different.

In the case of ^{55}Mn the ratio f_R/f_{CN} varies rapidly and approaches a value of nearly a hundred at about ≈35 MeV alpha beam energy E_α, indicating that above this energy the pre-equilibrium emission dominates the process. In the cases of targets ^{121,123}Sb, ^{128,130}Te, ^{165}Ho and ^{197}Au, the f_R/f_{CN} variations follow almost similar trends with respect to incident α-particle energy. In the case of ^{209}Bi the value of f_R/f_{CN} approaches 0.3 at ≈50 MeV alpha energy, indicating that in the heaviest system studied here the pre-equilibrium contribution with respect to CN emission is quite low. This may be because of the fact that during energy sharing in heavy system due to relatively larger number of nucleons, the average energy per nucleon of the CN is quite low and the system moves towards equilibration rather rapidly. Further, the general trend of the variation of f_R/f_{CN} with energy is the same as already indicated, i.e., pre-equilibrium emission for the lighter systems is more and increases rapidly with incident energy.

The presently derived values of pre-equilibrium fractions obtained from the semi-classical analysis of the excitation functions and the systematics obtained indicate that in the first few steps of the de-excitation process the excitation energy in excess of the Coulomb barrier $(E^*-E_{V_b})$ gets distributed among the surface nucleons of the CN. The pre-equilibrium emission, in all systems presently covered here, originates at some sort of a given threshold value of this excess excitation energy per surface nucleon $(E^*-E_{V_b})/A^{\frac{2}{3}}$. The pre-equilibrium fraction is found to increase quite rapidly with the excess excitation energy per surface nucleon and it varies nearly in the same way except that its relative values are different for different targets. It may, however, be remarked that the trends of deduced f_R as well as f_R/f_{CN} may not be conclusive, in

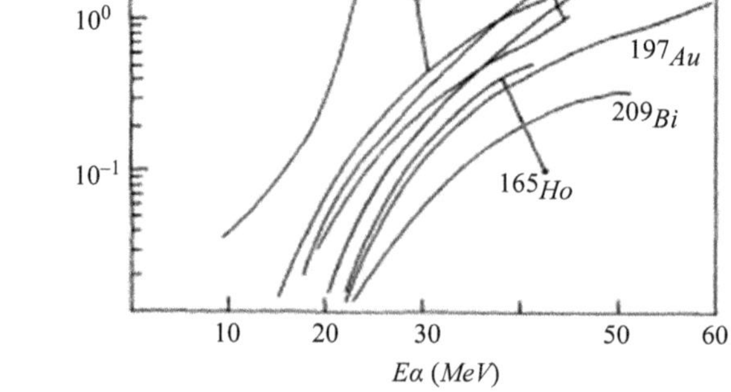

Figure 6.6. Variation of ratio of pre-equilibrium to compound fraction with beam energy.

view of a few cases presented here. However, a large set of experimental data needs to be analysed systematically within a consistent approach and framework of theoretical models covering the wide range of atomic mass number of target nuclei, to further confirm the observed systematics and the effects.

6.5 Systematics of pre-equilibrium fraction in odd *Z* and odd *A* nuclei

6.5.1 In the case of *A* = 139–203 nuclei

As mentioned above and especially depicted in figure 6.1, for an individual reaction channel the pre-equilibrium emission may be estimated and its energy dependence may be derived by subtracting the CN contribution from the measured cross-section data simulated employing pre-equilibrium models at each energy. In a recent report such an analysis is done and the pre-equilibrium fractions have been obtained as detailed above. Some of the reactions for which such an estimation is done are ^{139}La $(\alpha,n)^{142}$Pr, ^{159}Tb $(\alpha,n)^{162}$Ho, ^{181}Ta $(\alpha,n)^{184}$Re, ^{197}Au $(\alpha,n)^{200}$Tl, and ^{203}Tl$(\alpha,n)^{206}$Bi. In this analysis odd *A* and odd *Z* heavy target nuclei are studied [14] to develop systematics for pre-equilibrium emission reactions. The deduced pre-equilibrium fractions for these reactions are shown in figure 6.7, as a function of centre of mass energy. It may be seen from this figure that, the percentage of pre-equilibrium fraction for all these reactions increases with the centre of mass energy for each of these systems. It is also observed that a small variation in the value of E_{CM} produces a substantially large change in f_R values. The values of centre of mass energy at which the pre-equilibrium emission starts and attains the maximum value are different for different targets presently studied. As may be seen from figure 6.7, f_R starts increasing at a relatively lower value of centre of mass energy for lighter mass

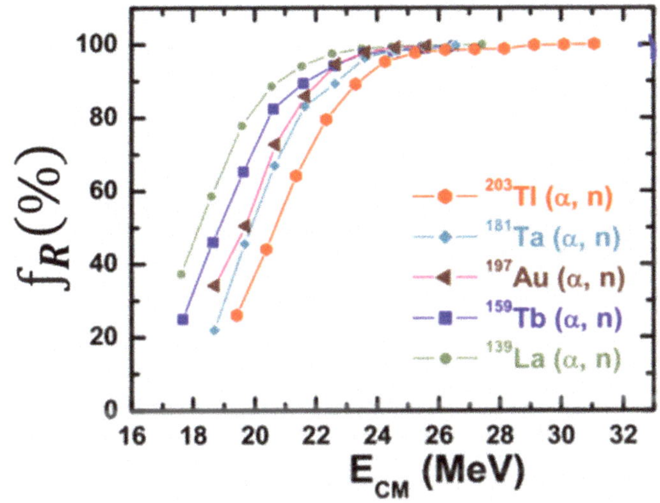

Figure 6.7. Variation of pre-equilibrium fraction as a function of the centre of mass energy E_{CM} for (α,n) reactions in different odd *Z* and odd *A* targets (adapted from [14] with permission of Springer).

target nuclei from ^{139}La to ^{203}Tl isotopes. However, the deduced values of pre-equilibrium fraction for the target nuclei ^{197}Au do not follow the trend similar to the other system. As such, no clear dependence on the centre of mass energy could be found for the studied reaction systems, and one needs some other parameter to get a proper systematics.

Further, with a view to obtain systematics for the pre-equilibrium fraction with mass number, the f_R values for the reactions ^{139}La(α,n)^{142}Pr, ^{159}Tb(α,n)^{162}Ho, ^{181}Ta (α,n)^{184}Re, ^{197}Au(α,n)^{200}Tl, and ^{203}Tl(α,n)^{206}Bi are plotted as a function of excitation energy in figure 6.8. Curves appear to merge closer for all the nuclei studied and presented in this figure. It may be observed from this figure that though the variation is small, the threshold excitation energy at which the pre-equilibrium fraction sets in is relatively lower for the target ^{203}Tl. In fact, from this figure also no systematic trend for f_R could be obtained with mass of the target nuclei and the excitation energy.

In the case of pre-equilibrium emission, it is reasonable to assume that surface nucleons of the composite system participate in higher strength as compared to the nucleons well inside. In view of this the excitation energy per nucleon available at the surface of the composite system $E^*/A^{2/3}$, may be used as another important parameter that may influence the pre-equilibrium emission. As such, the f_R, values have been plotted as a function of the above parameter in figure 6.9.

It may be observed from this figure that a systematic trend of the 'f_R' in terms of mass of the target nuclei and the available excitation energy per nucleon at the surface of the composite system is observed for the nuclei presented. The observed trends indicate that in the case of pre-equilibrium emission all the nucleons of the composite system are not involved, rather surface nucleons play an important role.

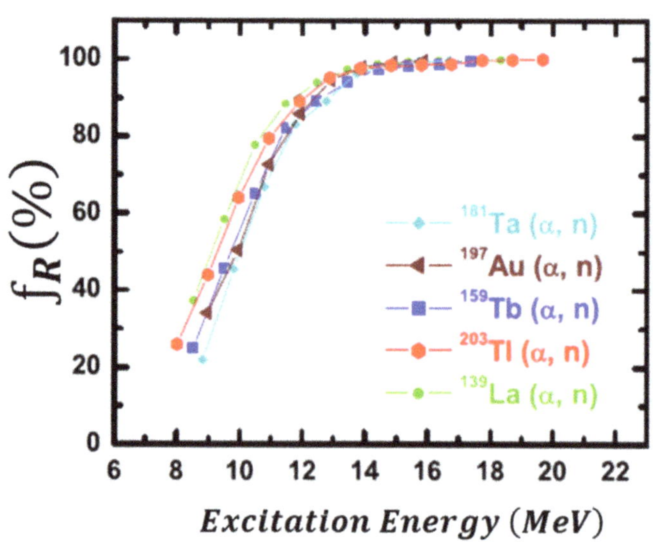

Figure 6.8. Variation of pre-equilibrium fraction as a function of the excitation energy E^* for (α,n) reactions in different odd Z and odd A targets (adapted from [14] with permission of Springer).

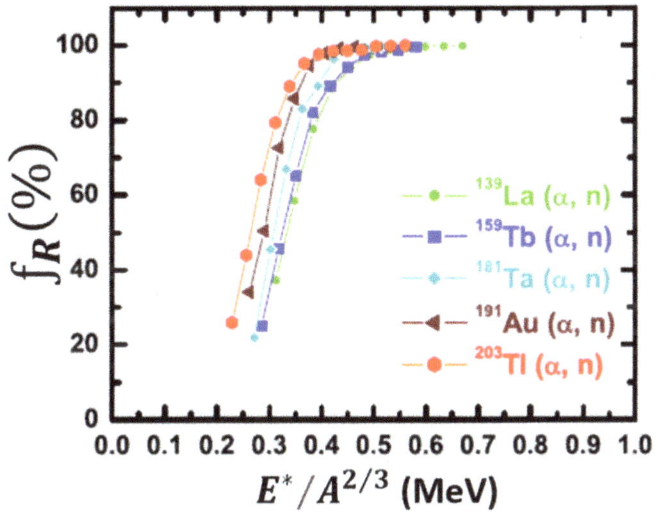

Figure 6.9. Dependence of f_R on the excitation energy per surface nucleons, $E^*/A^{2/3}$ for these reactions (adapted from [14] with permission of Springer).

As such, the pre-equilibrium emission may have a significant effect and contribution from the surface interactions. In other words, particles interacting in the nuclear periphery may have a better chance to be emitted as pre-compound particles, where an average lower density of nucleons is present, as compared to the particles passing through the entire diameter of the target nucleus and therefore underlying the effect of a higher density nuclear matter region [14].

Here, the above description indicates few points to be concluded for alpha-induced reactions, as presented in figure 6.10. In this figure, the variation of the pre-equilibrium fraction (f_R) with mass number of target nuclei (A) at five different values of the excitation energy per surface nucleons, $E^*/A^{2/3} = E_s$(= 0.300, 0.325, 0.350, 0.375 and 0.400 MeV, respectively) is shown. It may be observed from this figure that 'f_R' increases linearly with atomic mass number (A) of the target nuclei at each value of $E^*/A^{2/3}$ for the presently studied systems. As indicated in figure 6.10, the fitted lines for different values of $E^*/A^{2/3}$ with the deduced 'f_R' are not parallel to each other, rather they have different slopes. The slope of the observed lines goes on decreasing with the increase in the value of $E^*/A^{2/3}$. It means that the probability of the pre-compound emission increases as the atomic mass number of the target nuclei increases at a given value of excitation energy per surface nucleons. At relatively higher values of the $E^*/A^{2/3}(=E_s)$, these straight lines start conversing indicating that the contribution of the pre-equilibrium process gets saturated and approaches some maximum value for each target nucleus for one-neutron emitting channels. Further, with the excitation energy per surface nucleons E_s, the probability of two-neutron emission may increase. From the above it may be concluded that the systematics presented in figure 6.10 is interesting and allows one to get an additional insight into the understanding of the pre-equilibrium emission process with the surface nucleons playing an important role.

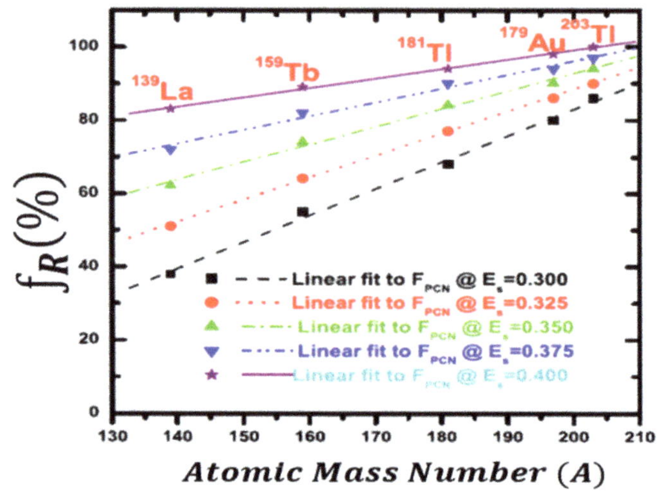

Figure 6.10. Deduced pre-equilibrium fraction 'f_R' as a function of atomic mass number A, of the target nuclei. The pre-equilibrium fraction 'f_R' increases linearly with the mass number of the target nuclei from $A = 139$ to $A = 203$. The deduced 'f_R' values for presently studied targets are shown by different symbols with different lines (adapted from [14] with permission of Springer).

6.5.2 In case of A = 63–109 nuclei

It was earlier mentioned that in the case of pre-equilibrium emission there is an involvement of few degrees of freedom that manifests the importance of lower density region at the surface of the CN. The pre-equilibrium emission of light nuclear particles is expected to be favored from this lower nuclear matter density region [14]. With a view to develop systematics for the pre-equilibrium emission for a wide range of target nuclei at low energies analysis of data has been done [15] for (α,n) reactions on odd A ($\approx$63–109) and odd Z target nuclei. In this range the nuclei chosen are ^{63}Cu, ^{65}Cu, ^{69}Ga, ^{71}Ga, ^{85}Rb, ^{89}Y, ^{93}Nb, ^{103}Rh, ^{107}Ag and ^{109}Ag, respectively. The choice of odd A and odd Z target nuclei in these reactions are taken to wash out any ambiguity arising due to odd–even effects. The deduced pre-equilibrium fractions as a function of centre of mass energy, obtained from the best fit of the experimental data taken [16–18] for the reactions ^{69}Ga$(\alpha,n)^{72}$As, ^{71}Ga$(\alpha, n)^{74}$As, ^{85}Rb$(\alpha, n)^{88}$Y, ^{89}Y$(\alpha, n)^{92}$Nb, ^{93}Nb$(\alpha, n)^{96}$Tc, ^{103}Rh$(\alpha, n)^{106}$Ag, ^{107}Ag$(\alpha, n)^{110}$In and ^{109}Rh$(\alpha, n)^{112}$In, are shown in figure 6.11.

As observed earlier, and also may be seen from figure 6.11, initial value of f_R increases very rapidly with centre of mass energy. This sharp increase in the f_R values indicate dependence of pre-equilibrium emission sensitively on the energy, as expected. Further, it may also be observed that a small variation in the value of centre of mass energy, produces a relatively large change in the f_R values. It may be also observed from figure 6.11 that the f_R attains a maximum value, for different targets at different energies. As indicated in this figure, f_R starts increasing at a lower value of centre of mass energy for heavier mass target nuclei (^{109}Ag) and vice versa, i.e., the higher the mass of nuclei, a relatively smaller energy is expected to be required for emitting a neutron as a pre-compound particle from the composite

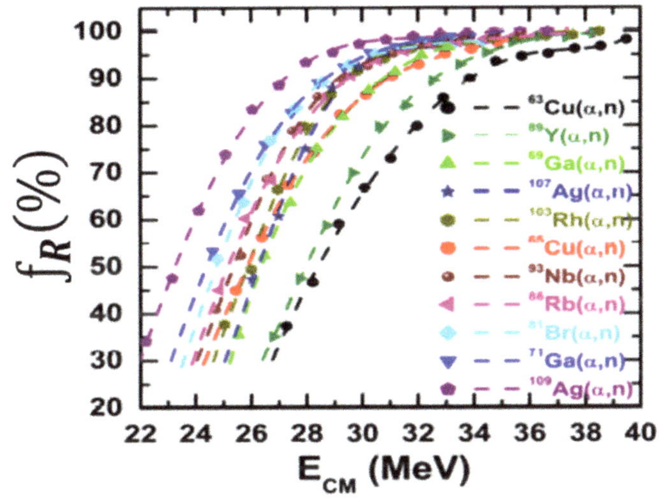

Figure 6.11. The deduced pre-equilibrium fractions for one-neutron emission channels for several odd A and odd Z nuclei, as a function of centre of mass energy (adapted from [15] with permission from American Physical Society).

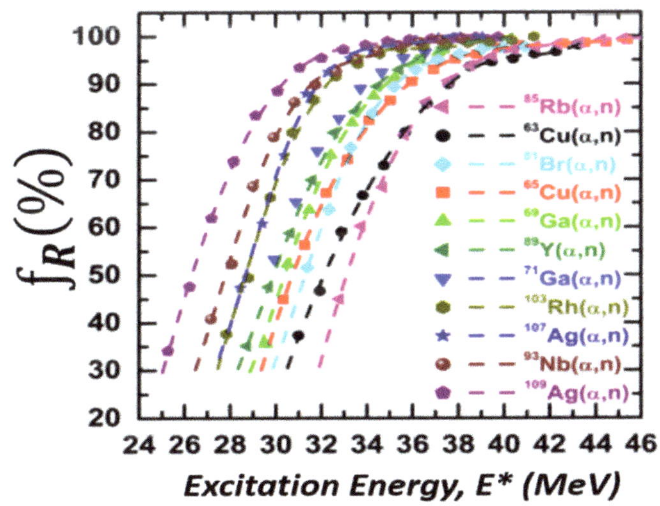

Figure 6.12. The deduced pre-equilibrium fractions for one-neutron emission channels for several odd A and odd Z nuclei as a function of excitation energy (adapted from [15] with permission from American Physical Society).

system. This is rather expected because the Q-value of neutron emission for ^{109}Ag is around −6.376 MeV and is less negative as compared to lightest studied target nucleus ^{63}Cu, for which the Q-value is about −7.502 MeV. It may be remarked that the trends of f_R for the nuclei ^{69}Ga, ^{71}Ga, ^{85}Rb, ^{89}Y, ^{93}Nb, ^{103}Rh and ^{107}Ag do not follow a systematic trend as a function of centre of mass energy. In order to obtain a mass dependence in a systematics, the f_R values are plotted as a function of excitation energy for these nuclei in figure 6.12.

It may be observed from this figure that the threshold value of the excitation energy at which pre-equilibrium fraction starts is lower for the heaviest target studied, i.e., ^{109}Ag but not in the case of lower mass target nucleus ^{63}Cu, as higher value of excitation energy corresponds to the ^{85}Rb as compared to ^{63}Cu target nucleus. Also, the nuclei ^{71}Ga and ^{93}Nb are not following the systematic trend in this figure. These results dashed the expectations of Blann [19] and Hodgson [20] in terms of some meaningful systematic trend for pre-equilibrium emission with excitation energy.

Apart from the role played by the nucleons at the surface of the nucleus the Coulomb effect (Z_1Z_2) may also be assumed to be important for the parameterization of such reactions. As such, it is expected that the available excitation energy in excess of the Coulomb barrier $\left(E^{*}-E_{V_b}\right)$ per surface nucleon of composite system, i.e., $\left(E^{*} - E_{V_b}\right)/A^{2/3}$ may influence the pre-compound emission process significantly [12, 22]. Keeping in view the above, the deduced f_R values are plotted as a function of $\left(E^{*} - E_{V_b}\right)/A^{2/3}$, in figure 6.13, for the presently studied reactions. It can be seen from this figure that a systematic trend of f_R in terms of mass number A of the target nuclei and available excitation energy above the Coulomb barrier per surface nucleon has been observed. Here, a successive decrease in the value of $\left(E^{*} - E_{V_b}\right)/A^{2/3}$ with increase in the mass number (A) of target nuclei from $^{63}_{29}$Cu to $^{109}_{47}$Ag, respectively is found, indicating the involvement of surface nucleons in pre-compound emission.

From figure 6.13, an interesting mass–energy systematics in the pre-equilibrium emission has also been deduced and is shown in figure 6.14. The variation of $\left(E^{*} - E_{V_b}\right)/A^{2/3}$ with mass numbers $63 \leqslant A \leqslant 109$ for a definite value of the pre-compound fraction (say 50%) is shown in figure 6.14. It is found that the values of

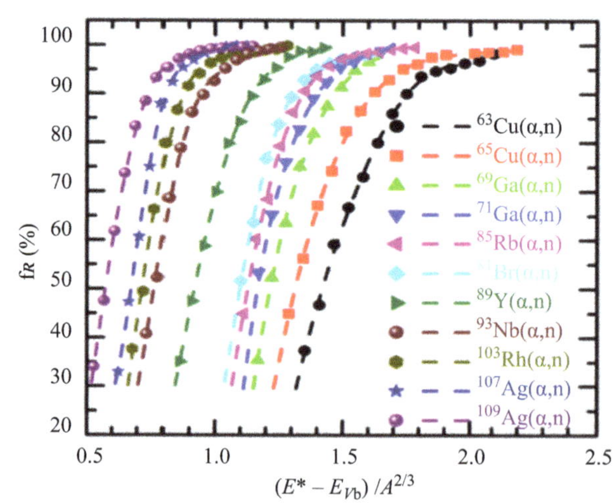

Figure 6.13. The variation of pre-equilibrium fraction for one-neutron emission channels for several odd A and odd Z nuclei as a function of excitation energy above Coulomb barrier per surface nucleon (adapted from [15] with permission from American Physical Society).

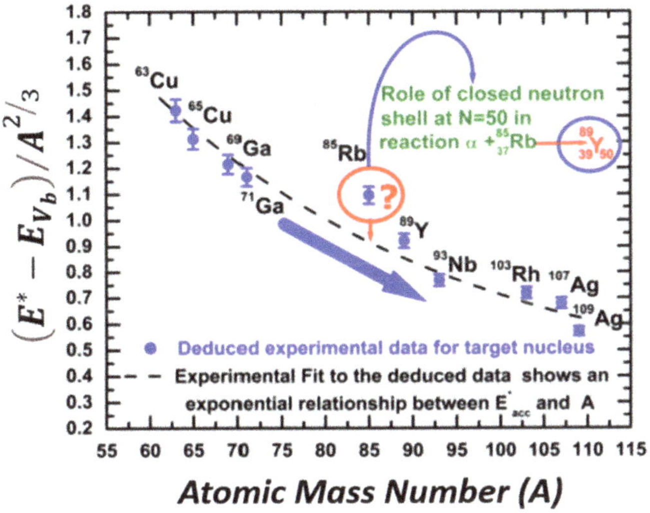

Figure 6.14. Mass–energy systematics indicating that excess excitation energy $\left(E^* - E_{V_b}\right)/A^{2/3}$ is an exponential function of atomic mass number of target nuclei for a wide mass ($A \approx 63$–107) region for a given value of pre-equilibrium fraction (adapted from [15] with permission from American Physical Society).

excess excitation energy above Coulomb barrier per surface nucleon, i.e., $\left(E^* - E_{V_b}\right)/A^{2/3}$ decrease nearly in an exponential fashion with mass number (A) of the target nuclei for a given value of pre-equilibrium fraction. As such, a lower value of available excitation energy above Coulomb barrier per surface nucleons is needed for emitting a neutron as a pre-equilibrium particle from target nucleus of higher atomic mass number and vice versa.

This exponential behaviour in pre-equilibrium emission is quite important as it gives the way to correlate the pre-equilibrium contribution for (α,n) reactions at low energies, where emission via CN process is generally considered to be important. Here, the target nucleus $^{85}_{37}$Rb needs special mention. The higher value of $\left(E^* - E_{V_b}\right)/A^{2/3}$ corresponding to $^{85}_{37}$Rb nucleus may be due to the role of nuclear structure and reaction dynamics of neutron emission in the pre-compound emission. This deviation of $\left(E^* - E_{V_b}\right)/A^{2/3}$ for $^{85}_{37}$Rb nucleus may be explained on the basis of neutron shell closer at $N = 50$ (magic number) for excited CN $^{89}_{39}$Y formed as a result of fusion of α-particle with target nuclei $^{85}_{37}$Rb (i.e., $\alpha + {}^{85}_{37}\mathrm{Rb} \Rightarrow {}^{89}_{39}\mathrm{Y}_{50}$). The CN $^{89}_{39}\mathrm{Y}_{50}$ has closed neutron shell magic structure ($N = 50$) and hence it may require relatively higher excitation energy for ejecting one neutron as compared to the neighbouring nucleus $^{89}_{39}\mathrm{Y}_{50}$, even when it has been used as a target in $\alpha + {}^{89}_{39}\mathrm{Y}_{50} \Rightarrow {}^{93}_{41}\mathrm{Nb}_{52}$, reaction. To conclude, it may be stated that particles interacting on the nuclear periphery may have a better chance to be emitted as pre-compound particles, where an average lower nucleon density is present, as compared to the particles passing through the entire diameter of the target nucleus and therefore underlying effect of a lower density nuclear matter region. As such, the systematics obtained from the present analysis is interesting and throws an additional insight into the existing reaction dynamics of the low energy pre-compound emission process

6.6 Systematics for heavy-ion induced reactions

With the availability of heavy-ion accelerators, efforts have been made to study the reaction dynamics of pre-equilibrium emission, specifically at energies where the CN emission is likely to be dominant. At moderate excitation energies in heavy-ion reactions, one expects an interplay between the CN and the pre-equilibrium emission processes. In such processes, the heavy reaction residues may be populated with atomic mass number and nuclear charge equal to or one unit less than the composite nucleus, via *xn* and *pxn* emission channels, respectively. At relatively higher beam energies, the incomplete fusion or the break-up fusion reactions are also likely to be dominant modes of reaction as compared to the CN and pre-compound emission processes [23–25]. The beams of^{12}C and ^{16}O ions, at moderate energies, having finite probability of projectile break-up, are reported [26] to be suitable candidates for study in various reaction processes, including pre-equilibrium emission, in heavy-ion interactions.

In heavy-ion interactions the processes that play an important role may be broadly classified on the basis of impact parameter of the two interacting ions [27]. At relatively higher values of the impact parameter that corresponds to the direct reactions, transfer of a few nucleons may take place. On the other hand, at relatively smaller values of impact parameters, the complete fusion, incomplete fusion [28] and pre-equilibrium emission process are likely to happen. With the availability of medium-energy light heavy-ion beams, it has become possible to study the above-mentioned reaction processes.

In order to study the pre-equilibrium emission process in a consistent and systematic way, the strength of pre-equilibrium emission defined by its fraction, f_R has been deduced from the measurement and analysis of excitation functions for three-neutron emission channels in the ^{12}C +^{128}Te, ^{12}C + ^{169}Tm, ^{16}O + ^{159}Tb, ^{16}O + ^{169}Tm, and ^{16}O +^{181}Ta systems. These experiments were performed using the stacked foil activation technique. The details of these experiments are presented in chapter 5. Here, relatively low energies have been used that vary from near the Coulomb barrier, V_B to nearly ≈7 MeV/nucleon. At these energies, generally, the CN emission processes are expected to dominate. The experimental data has been analysed within the framework of CN model using code PACE4 [29]. In order to explain the high-energy tail portion of the excitation functions, calculations for pre-equilibrium emission have been performed using code ALICE-91 [30], employing the hybrid and/or geometry dependent hybrid models. Analysis of the data for the presently studied reactions provides a significantly important confirmation of emission of pre-equilibrium neutrons prior to the attainment of equilibration of the composite system in the range of energies ≈5–7 MeV/nucleon.

The values of deduced pre-equilibrium fraction, f_R are presented as a function of excitation energy for the reactions ^{128}Te(^{12}C,3*n*)^{137}Ce, ^{169}Tm(^{12}C,3*n*)^{178}Re, ^{159}Tb (^{16}O,3*n*)^{172}Ta,^{169}Tm(^{16}O,3*n*)^{182}Ir, and ^{181}Ta(^{16}O,3*n*)^{194}Tl, respectively in figure 6.11.

It may be observed from this figure that the threshold energies for emission of pre-equilibrium particles are different for each projectile–target combination. The pre-equilibrium emission threshold is lower if the target atomic mass number is low, as expected. The pre-equilibrium fraction f_R for all these reactions rises sharply with excitation energy. The initial relatively sharp rise in f_R values may indicate that the statistical distribution of excitation energy among the nucleons participating in pre-equilibrium emission in the initial stages may be quite rapid.

It may also be observed from figure 6.15, that with the increase of the excitation energy f_R for all these reactions attains a saturation value that depends on the mass of the composite nucleus. For the ^{137}Ce nucleus formed in the fusion of ^{12}C + ^{128}Te reaction, the saturation value ≈100% of the pre-equilibrium fraction is observed at ≈70 MeV of excitation energy, while in the case of other reaction residues, the saturation is observed at relatively higher excitation energies. Further, no consistent behaviour of f_R with excitation energy of the composite nuclei of different atomic mass number is observed. It can be seen in the figure that in system (^{12}C + ^{169}Tm) the curve for pre-equilibrium fraction for three-neutron emitting channel starts at lower excitation energy per nucleon value and is well separated from the corresponding curve for system (^{16}O + ^{159}Tb), which is at relatively higher excitation energies.

In figures 6.16 and 6.17 variation of pre-equilibrium fraction is shown as a function of excitation energy per surface nucleon of the composite system and the excitation energy per nucleon at the surface of composite system, respectively. It can be observed from these figures that no systematic trends of f_R in terms of atomic mass number of composite system and excitation energy per surface nucleon is observed. This indicates that excitation energy per nucleon may not be a

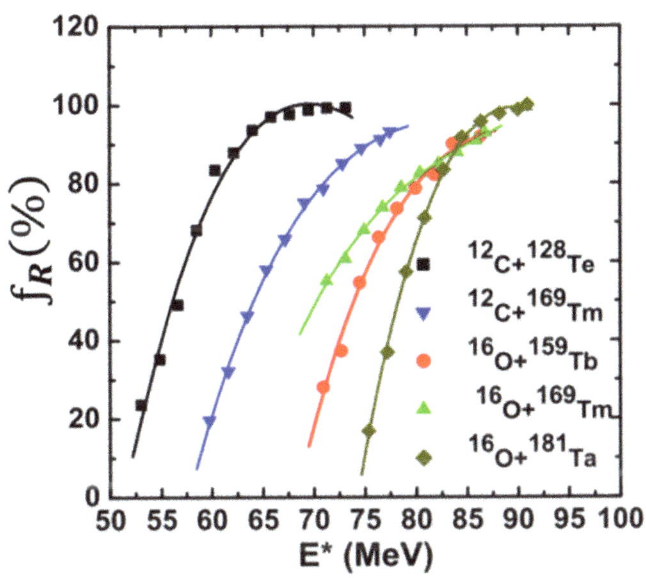

Figure 6.15. Pre-equilibrium fraction as a function of excitation energy E^* of the composite system (adapted from [25] with permission from American Physical Society).

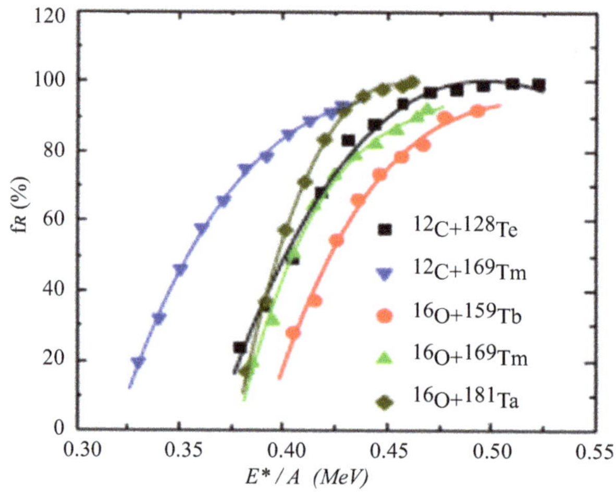

Figure 6.16. Variation of pre-equilibrium fraction with the excitation energy per surface nucleon of the composite system (adapted from [25] with permission from American Physical Society).

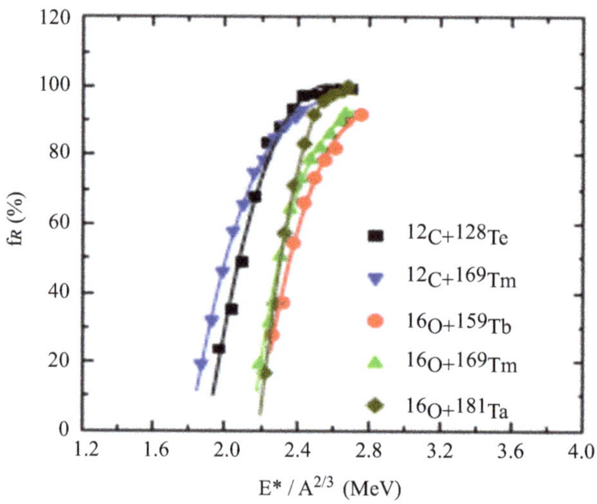

Figure 6.17. Variation of pre-equilibrium fraction as a function of excitation energy per nucleon at the surface of the composite system (adapted from [25] with permission from American Physical Society).

suitable parameter to characterise the pre-equilibrium emission. However, it may be more reasonable to assume that the particles passing through the nuclear periphery may have relatively better chances for inducing pre-compound emission rather than the particles crossing the entire diameter of the target nucleus.

With a view to study the effect of surface nucleons on the pre-equilibrium emission and to wash out the effect of the Coulomb barrier, if any, the f_R is plotted as a function of excitation energy in excess of the Coulomb barrier per surface nucleon of the composite system, i.e., $(E^* - E_{V_b})/A^{2/3}$ and is shown in figure 6.18.

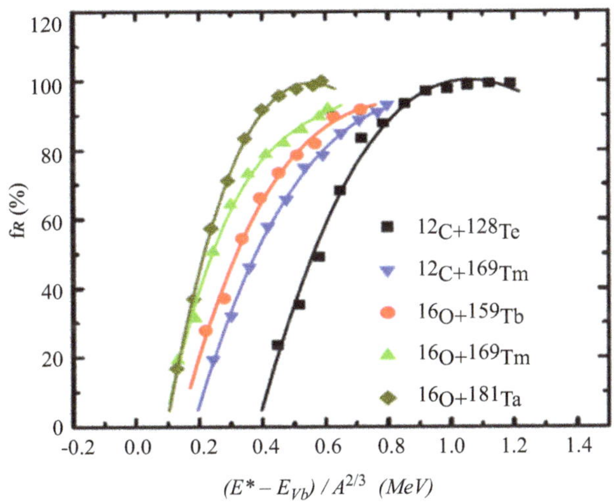

Figure 6.18. Variation of pre-equilibrium fraction as a function of excitation energy above Coulomb barrier per surface nucleons of the composite system (adapted from [25] with permission from American Physical Societ).

It may be observed from this figure that for heavier composite systems pre-equilibrium fraction begins at a smaller value of the excess excitation energy above the Coulomb barrier per surface nucleon of the composite system. However, for relatively lighter composite systems, it begins at relatively higher values of $(E^* - E_{V_b})/A^{2/3}$. This systematic trend indicates the significance of the excess excitation energy above the Coulomb barrier per surface nucleon in pre-equilibrium processes. In other words, one can say that this excess energy disseminates among the nucleons on the nuclear surface, which is likely to influence the pre-equilibrium emission strongly, even in the case of heavy-ion induced reactions.

6.7 Pre-equilibrium component from recoil range distribution (RRD)

As already mentioned, one of the ways to determine and to distinguish the relative strength of compound and pre-compound components in nuclear reactions is from the comparison of the enhancement in the intensity of emitted nuclear particles in the forward cone over the backward cone. In the previous sections, we have seen that such a pre-equilibrium component may also be obtained by analysing the measured excitation functions for deviations from the statistical predictions and to attribute this deviation as due to the PE emission. As a matter of fact, the emission of pre-equilibrium nuclear particles in the forward direction may reduce the momentum of the residual nucleus. As such, the measurement of the momentum transfer during the reaction may also be used as an important method for investigating the reaction dynamics of pre-equilibrium emission. One such attempt [31] to separate out the compound and pre-equilibrium components in nuclear reactions has been made by measuring the recoil range distribution of heavy residues populated as a result of these processes. The experimental details have already been presented in chapter 5, however, for the sake of completeness a brief description of the same and results obtained are presented here.

The distributions of ranges of recoiling heavy nuclei produced by particles emitted via CN and pre-equilibrium emissions in (^{16}O +^{159}Tb), (^{16}O +^{169}Tm) and (^{16}O +^{181}Ta) systems have been experimentally measured [31] . In the recoil range distribution experiment, a thin target followed by a packet of (nearly fifteen) thin Al catcher foils of varying thickness (≈16–45 μg cm^{-2}), prepared by the vacuum evaporation technique, was mounted in the scattering chamber normal with respect to the beam direction. Depending on the momentum carried by the product residues, the recoiling residues are expected to be trapped at different ranges in the stack of thin Al foils.

The measured yields of the residues have been plotted against the cumulative catcher foil thicknesses to obtain the experimental recoil range distributions. As a typical example, the measured recoil range distribution of residues populated in the reaction ^{169}Tm(^{16}O,2*n*)^{183}Ir is shown in figure 6.19. The solid curve shown in this figure is to guide the eye to the experimental data. The peaks observed in the distribution arise because of the overlap of ranges of the heavy reaction residues produced as a result of particles emitted due to pre-equilibrium and CN processes. It may be observed from this figure that the recoil range distribution for the reaction ^{169}Tm(^{16}O,2*n*)^{183}Ir may be deconvoluted into two peaks. One peak is observed at a relatively lower value (≈225 μg cm^{-2}) of cumulative catcher thickness and the other peak is observed at ≈330 μg cm^{-2}. The peak at higher cumulative catcher thickness (≈330 μg cm^{-2}) corresponds to the fraction of residues produced through the CN process and is consistent with that estimated due to the full momentum transfer events in the complete fusion reactions. However, the well separated peak at

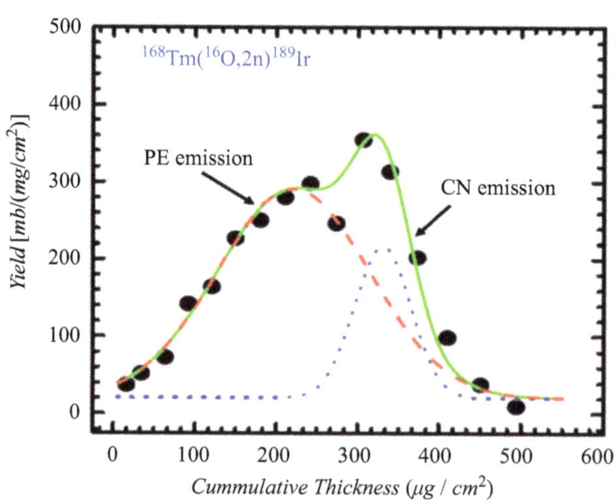

Figure 6.19. Measured recoil range distribution for the residue ^{183}Ir populated in the reaction ^{169}Tm(^{16}O,2*n*), at incident beam energy ≈88 MeV. The deconvolution of measured distribution resolves into two well separated peaks that may be assigned to two distinct processes: compound emission and pre-equilibrium emission (adapted from [31] with permission from American Physical Society).

relatively smaller cumulative catcher thickness, i.e., ≈225 μg cm^{-2} may be assigned due to the fact that the residues ^{183}Ir may also be produced prior to the establishment of thermodynamical equilibrium. Such emission of two pre-equilibrium neutrons takes away a significant part of the momentum as compared to the CN emission process. Thus, this results in the reduced momentum of the composite system and hence decrease in the range of recoiling residues in the stopping medium, as observed in figure 6.20. This gives a clear method of deciphering the compound and pre-equilibrium components in a nuclear reaction, and may be considered to be model independent.

In the same way the recoil range distribution for the residues produced in the reaction ^{159}Tb(^{16}O,3n)^{172}Ta, has also been measured and is shown in figure 6.20. Again, it may be observed that the recoil range distribution has two well-separated components, indicating the production of ^{172}Ta residues due to different reaction mechanisms. The peak at higher cumulative catcher thickness is produced as a result of CN emission due to complete momentum transfer events, however the peak at lower thickness is due to the same residues produced as a result of decreased momentum, because the pre-equilibrium emission of particles takes away a significant amount of momentum. In order to obtain the relative contributions of the pre-equilibrium emission and the CN processes in this reaction, the areas under the Gaussian peaks corresponding to these processes have been obtained using the ORIGIN software. In the case of reaction ^{159}Tb(^{16}O,3n)^{172}Ta at ≈90 MeV, the relative contributions of PE emission and CN emissions are found to be ≈35% and ≈65%, respectively, within the experimental uncertainties. It may be seen that the measurement of recoil range distributions may be used as a sensitive tool for deciphering the CN and pre-equilibrium emissions.

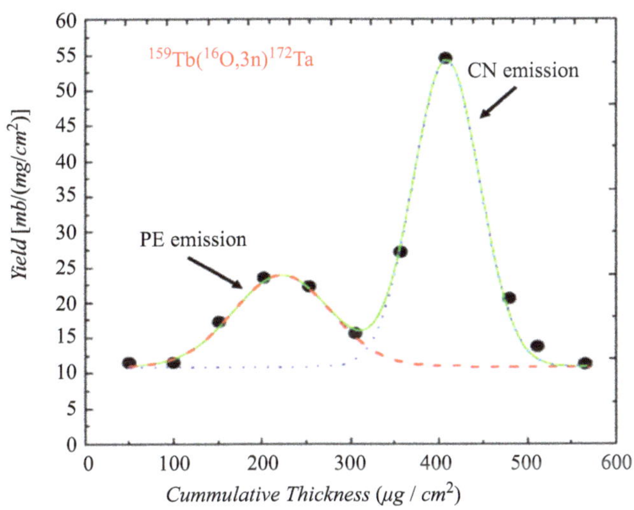

Figure 6.20. The measured recoil range distribution for the reaction ^{159}Tb(^{16}O,3n)^{172}Ta at incident beam energy ≈90 MeV. The deconvolution of measured distribution indicates two well separated peaks as a result of two distinct processes of CN and pre-equilibrium emissions (adapted from [31] with permission from American Physical Society).

6.8 Pre-equilibrium component from spin distribution measurement

In a typical nuclear reaction, particles emitted prior to the establishment of equilibrium take away a significant part of the angular momentum. As such, the angular momentum associated with the residues as a result of pre-equilibrium emission is relatively lower than that associated with the residues populated in the CN process. Therefore, the residues populated via pre-equilibrium emission of particles have relatively less strength of high spin states as compared to the spin states of the residues populated via CN process. Therefore, if one measures the spin population in such reaction residues, it may reflect the effect of the competition between compound and pre-equilibrium emissions. In order to obtain the spin population of residues during the de-excitation process, a particle-γ coincidence experiment was performed at the Inter University accelerator Centre (IUAC), New Delhi, India. The measured spin distributions [32] of the reaction ^{169}Tm(^{16}O,2n)^{183}Ir at ≈88 MeV and ^{159}Tb(^{16}O,2n)^{173}Ta at ≈93 MeV have been measured using the Gamma Detector Array (GDA) along with a Charged Particle Detector Array (CPDA) setup. The experimental conditions are detailed in a paper reported by Singh *et al* [32]. However, a brief account of the experimental details is given here for ready reference.

The gamma detector array is an assembly of 12 Compton suppressed, high-resolution HPGe *γ-ray* spectrometers arranged at angles 45°, 99°, and 153° with respect to the beam axis direction and there are in all four such detectors at each of these indicated angles. The CPDA is a group of 14-phoswich detectors which are contained in a 14-cm-diameter chamber, and covers about 90% of the entire solid angle. The populated residues in a given reaction have been identified from their characteristic prompt γ-radiations. The proton and the α-emitting channels are presented in reference [32], and have been demarcated from the particle (Z =1,2)-gated γ spectra. However, neutron-emission channels have been obtained from the singles spectra of prompt γ transitions observed at forward 45° and backward 153° angles with respect to the beam direction.

The relative production yield values of the reaction residues, i.e., the observed area under the peak of the experimentally measured prompt γ lines, have been represented as a function of observed spin J_{Observed} corresponding to prompt γ transitions [32]. In order to get the idea of the involved input angular momenta and to obtain the entry state spin population of the reaction ^{169}Tm(^{16}O,2n)^{183}Ir, the measured relative yields have been normalized with respect to the minimum observed spin ($J_{\text{Observed}}^{\text{minimum}}$) at the highest yield ($Y_{\text{Observed}}^{\text{maximum}}$). The experimentally measured spin distributions derived from the prompt γ rays recorded in forward and backward directions at 88 MeV for the reaction ^{169}Tm(^{16}O,2n)^{183}Ir are shown in figures 6.21 and 6.22, respectively [23]. It may be observed from these figures that the measured spin distribution and hence its decay pattern for the reaction ^{169}Tm (^{16}O,2n)^{183}Ir obtained in the forward and backward directions are widely different from each other, indicating the involvement of distinctly dissimilar reaction mechanisms. These distinctly different configurations in forward and backward directions indicate that the two processes may be quite different in nature. To explain it, it may be remarked that in the case of CN emission it is likely that higher

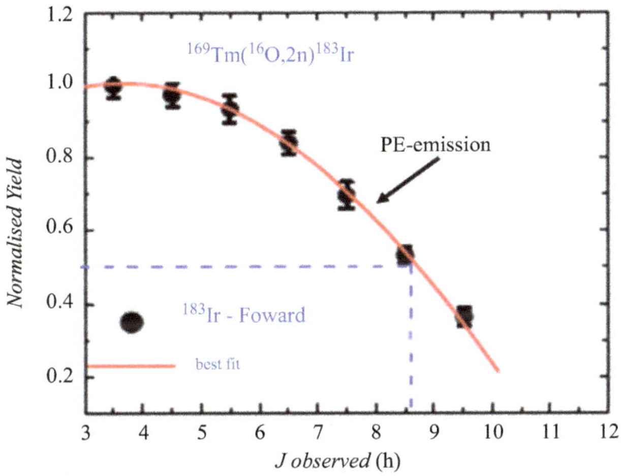

Figure 6.21. Experimentally measured spin distribution for reaction ^{169}Tm(^{16}O,2*n*)^{183}Ir in forward cone. The solid curve through data points represents the best fit (adapted from [31] with permission from American Physical Society).

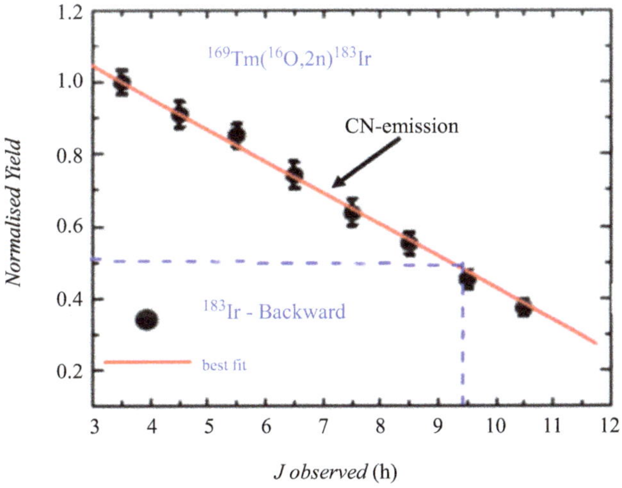

Figure 6.22. Experimentally measured spin distribution for reaction ^{169}Tm(^{16}O,2*n*)^{183}Ir in the backward cone. Solid line through data points represents the best fit (adapted from [31] with permission from American Physical Society).

spin states in the residues are populated, giving rise to linearly increasing populations of lower spins due to their feeding from the higher spin states. While in the case of pre-equilibrium emission, the emitted particles carry substantial amount of angular momentum and energy, the residues are left with relatively lower excitation energy and hence with lower spin entry state; therefore, the feeding of lower spin sates is not linear. The measured spin distributions for the CN and the pre-equilibrium contributions may be designated by the mean input angular momentum associated with the process, and obtained as the spin where the normalized yield decreases to half of its maximum value at lowest observed spin. The observed lower value of the mean input angular momentum in the forward direction may be attributed to the fact that emission of pre-equilibrium neutrons takes away a significant part of the angular momenta. As such, it may be concluded that distinctly different spin distributions in the forward and the backward cones give a direct confirmation of the pre-equilibrium emission process.

6.9 Pre-equilibrium emission and synthesis of SHE

The phenomenon of pre-equilibrium emission may be exploited to synthesise super heavy nuclei (SHE). As we know, the heaviest naturally occurring and stable nucleus is uranium ($_{92}U$). The heavier nuclei with atomic number higher than uranium up to fermium ($Z = 100$) have been successfully synthesised using nuclear reactions in which the lower atomic number nucleus captures a neutron and the product nucleus undergo successive beta decays to yield the desired higher atomic number nuclei, as shown by the following example,

$$ {}_0^1n + {}_{92}^{238}\mathrm{U} \rightarrow {}_{92}^{239}\mathrm{U} \xrightarrow[23\,\mathrm{min}]{\beta^-} {}_{93}^{239}\mathrm{Np} \xrightarrow[2.355\,\mathrm{d}]{\beta^-} {}_{94}^{239}\mathrm{Pu} $$

However, for nuclei with $Z > 100$, this process cannot be continued because of the short half-lives of the product nuclei that may undergo fission and may also emit alpha particles immediately after formation. On the other hand, elements with $Z >$ 100 may be synthesised by the fusion of heavy nuclear projectiles and targets. In general, there are two fundamental challenges in synthesising of super heavy elements: (i) the Coulomb barrier between the projectile and the target nucleus has to be overcome by suitable bombarding energy; and (ii) the excitation energy of the CN formed by the fusion of heavy ions has to be reduced to prevent their spontaneous fission. With the availability of heavy-ion accelerators which may accelerate any ion from helium to uranium to sufficiently high energies, it is possible to overcome Coulomb barrier between two heavy ions and produce a CN lying in super heavy range. However, the super heavy nucleus so formed generally has high excitation energy, but pre-equilibrium emission of nucleons and or clusters in the initial stage of the reaction, rapidly dissipates excitation energy, cooling it down below the fission barrier. This line of action (fusion of two heavy ions and pre-equilibrium emission) has led to the discovery of elements like nobelium ($_{102}No$) with $Z = 102$, to seaborgium ($_{106}Sg$) with $Z = 106$ at the Lawrence Berkeley National Laboratory (BNL), California. For the synthesis of super heavy nuclei, the incident heavy projectile may be bombarded on actinides targets having $Z = 89$–90 with sufficient energy to overcome their mutual Coulomb barrier. The excitation energy of the CN formed by the fusion of two heavy ions may be of the order of 40–50 MeV. Since the heavy nucleus formed has sufficiently high excitation energy, the emission of a few high energy neutrons from the CN as pre-equilibrium particles rapidly reduces the available excitation energy of the residual nucleus below the fission barrier and thus may help in the formation of a super heavy nucleus.

6.10 Future perspective

In recent years, a variety of experimental techniques have been developed to identify and detect neutrons as well charged particles that may be used to determine cross-sections for reactions responsible for the emission of the identified particles. The data of neutron- and proton-induced reactions are very important for a number of technical applications. The data may also be used to develop theoretical models to explain the reaction mechanism involved and to give accurate prediction of cross-sections as well. The cross-section data of not only neutron and proton induced reactions but also of heavy-ion induced reactions is important for the development of nuclear reactor technology while estimating the nuclear transmutation rates, nuclear heating, radiation damage, surface wear effects etc. The proton-and alpha-induced reaction cross-sections over a large range of energy are significantly important for the production of radioisotopes using cyclotrons that are used in medical treatments. For this purpose dedicated cyclotrons are built to produce isotopes of interest. Recently, a Medical Cyclotron Facility [33] Cyclone-30 has become operational at the Variable Energy Cyclotron Centre (VECC), Kolkata, India. The evaluation of nuclear data is eventually carried out based on experimental data and the theoretical model predictions. As a matter of fact, it is practically impossible to experimentally determine the cross-section for all possible reactions at wide energy range for all the targets available in the periodic table. In many cases there are experimental limitations to measure the cross-section for a given reaction and hence accurate predictions from nuclear models are needed. Therefore, nuclear reaction model calculations play an important role in the nuclear data evaluation. The details given in the book may be very helpful for the above purpose and may also be useful for those who wish to work in this promising area of nuclear reaction dynamics. Such studies may also be extended for radioactive ion beam projectiles which may give a better and additional insight of the reaction dynamics.

Bibliography

[1] Blann M 1975 Pre-equilibrium decay *Annu. Rev. Nucl. Sci.* **25** 123–66

[2] Bohr N 1936 Neutron capture and nuclear constitution *Nature* **137** 344

[3] Blann M 1972 Importance of the nuclear density distribution on pre-equilibrium *Phys. Rev. Lett.* **28** 757

[4] Feshbach H, Kerman A and Koonin S 1980 The statistical theory of multi-step compound and direct reactions *Ann. Phys.* **125** 429

[5] Gruppelaar H, Nagel P and Hodgson P E 1986 Pre-equilibrium processes in nuclear reaction theory: the state of the art and beyond *La Riv. del Nuovo Cim.* **9** 1

[6] Gupta J P, Bhardwaj H D and Prasad R 1985 Pre-equilibrium emission effect in (*n*, p) reaction cross-sections at 14.8 MeV *Pramana – J. Phys.* **24** 637–42

[7] Griffin, J J 1966 Statistical model of intermediate structure *Phys. Rev. Lett.* **17** 478

[8] Blann M Hybrid model for pre-equilibrium decay in nuclear reactions *Phys. Rev. Lett.* **27** 337

[9] Bhardwaj H D 1985 A study of pre-equilibrium emissions in some nuclear reactions *PhD Thesis* Aligarh Muslim University

[10] Uhl M and Strohmaier B 1981 *Report IRK 76/01* France: NEA Data Bank, Cedex
[11] Musthafa M M 1996 Equilibrium and pre-equilibrium emission in some reactions induced by light and heavy ions *PhD Thesis* Aligarh Muslim University
[12] Singh B P 1991 Equilibrium and pre-equilibrium emission in alpha induced reactions at moderate excitation enrgies *PhD Thesis* Aligarh Muslim University
[13] Dilg W, Schantl W and Vonach. H 1973 Level density parameters for the back-shifted Fermi gas model in the mass range $40 < A < 250$ *Nucl. Phys.* A **217** 269
[14] Sharma M K, Kumar M, Shuaib M, Sharma V R, Yadav A, Singh P P, Singh D P, Singh B P and Prasad R 2018 Analysis of experimental cross-section for (α, n) reactions in odd A and odd Z heavy nuclei: a systematics on pre-compound emission *Eur. Phys. J.* A **54** 205
[15] Sharma M K *et al* 2021 A new experimental approach for developing mass–energy systematics for pre-compound emission *Phys. Rev. Lett.* C **104** L031601
[16] Navin A *et al* 2004 Direct and compound reactions induced by unstable helium beams near the Coulomb barrier *Phys. Rev.* C **70** 044601
[17] Singh N L, Patel B J, Somayajulu D R S and Chintalapudi S N 1994 Analysis of the excitation functions of (α, xnyp) reactions on natural copper *Pramana – J. Phys.* **42** 349
[18] Levkovskij V N 1991 Act. Cs. By Protons and Alphas, Moscow, USSR
[19] Blann M 1971 Hybrid model for pre-equilibrium decay in nuclear reactions *Phys. Rev. Lett.* **27** 337
[20] Hodgson P E 1981 Pre-equilibrium processes in nuclear reactions *Nature* **292** 671
Liu Z-H and Bao J-D 2014 Neutron emission on the fusion process process and its effect on the formation of superheavy nuclei *Phys. Rev.* C **89** 024604
[21] Cavinato M, Fabrici E, Gadioli E, GadioliErba E, Vergani P, Crippa M, Colombo G, Redaelli I and Ripamonti M 1995 Study of the reactions occurring in the fusion of ^{12}C and ^{16}O with heavy nuclei at incident energies below 10 MeV/nucleon *Phys. Rev.* C **52** 2577
[22] Vergani P, Gadioli E, Vaciago E, Fabrici E, GadioliErba E, Galmarini M, Ciavola G and Marchetta C 1993 Complete and incomplete fusion and emission of preequilibrium nucleons in the interaction of ^{12}C with ^{197}Au below 10 MeV/nucleon *Phys. Rev.* C **48** 1815
[23] Sharma M K, Unnati B K, Sharma B P, Singh H D, Bhardwaj R, Kumar K S, Golda and Prasad R 2004 Complete and incomplete fusion reactions in the ^{16}O + ^{169}Tm system: excitation functions and recoil range distributions *Phys. Rev.* C **70** 044606
[24] Cavinato M, Fabrici E, Gadioli E, GadioliErba E, Vergani P, Crippa M, Colombo G, Redaelli I and Ripamonti M 1995 Study of the reactions occurring in the fusion of ^{12}C and ^{16}O with heavy nuclei at incident energies below 10 MeV/nucleon *Phys. Rev.* C **52** 2577
[25] Sharma M K, Singh P P, Singh D P, Yadav A, Sharma V R, Bala I, Kumar R, Unnati, Singh B P and Prasad R 2015 Systematic study of pre-equilibrium emission at low energies in ^{12}C and ^{16}O induced reactions *Phys. Rev.* C **91** 014603
[26] Hodgson P E 1976 *Nuclear Heavy-ion Reactions* (Oxford: Oxford University Press)
[27] Prasad R and Singh B P 2018 *Fundamentals and Applications of Heavy Ion Collisions: Below 10 MeV/Nucleon Energies* (Cambridge: Cambridge University Press)
[28] Nörenberg W and Weidemüller H A 1976 *Introduction to the Theory of Heavy Ion Collisions* (Berlin: Springer)
[29] Gavron A 1980 Statistical model calculations in heavy ion reactions *Phys. Rev.* C **21** 230
[30] Blann M 1991 *Report PSR-146, NEA Data Bank* (France: Gif-sur-Yvette)
[31] Sharma M K, Singh P P, Sharma V R, Shuaib M, Singh D P, Yadav A, Unnati, Kumar R, Singh B P and Prasad R 2016 Pre-compound emission in low-energy heavy-ion interactions

from recoil range and spin distributions of heavy residues: a new experimental method *Phys. Rev.* C **94** 044617

[32] Singh P P, Singh B P and Sharma M K *et al* 2009 Probing of incomplete fusion dynamics by spin distribution measurement *Phys. Lett.* B **671** 20

[33] REDWARE level scheme (http://radware.phy.ornl.gov/agsdir1.html) Medical cyclotron cyclone 30 (http://dae.gov.in/node/1120)

www.ingramcontent.com/pod-product-compliance
Ingram Content Group UK Ltd.
Pitfield, Milton Keynes, MK11 3LW, UK
UKHW051857100926
8556IPUK00005B/671

9 780750 350754